Integrative Ecology

Marc Salomon

Integrative Ecology

A Federation of Sciences as an Explanatory Key for the Sustainability of our Life-Bearing Planet

Springer

Marc Salomon
Sorbonne Université, Faculté des
Sciences et Ingénierie Computational,
Quantitative and Synthetic Biology, UMR
7238, Institut de Biologie Paris Seine
Paris, France

ISBN 978-3-031-90655-8 ISBN 978-3-031-90656-5 (eBook)
https://doi.org/10.1007/978-3-031-90656-5

This Springer imprint is published by the registered company Springer Nature Switzerland AG
The registered company address is: Gewerbestrasse 11, 6330 Cham, Switzerland

To my dear wife Françoise, for her patience, care and encouragement.

In memory of François Terrasson (1939-2006) who was my inspirer and mentor.

Preface

Why write an *n*th academic book on ecology after so many remarkable essays on the pedagogy of this science? Some readers would sensibly ask this question. I will try to answer it. I believe several fundamental definitions of concepts should gain more accuracy and recent findings need to be integrated. Moreover, I think that reminding the readers of some points borrowed from other sciences can help them understand many non-trivial aspects of ecology. This book is intended to be a helpful contribution to future practical academic textbooks on ecology. This work has been primarily inspired by my lecture notes as a university teacher at Rennes 1 and Paris Sorbonne universities, respectively, to continuous vocation trainees, and second and third year university students, much enlarged by personal reflections and added bibliographic and personal research. To guide and support the reader, this book comprises (i) detailed tables of contents, figures and tables, (ii) a list of acronyms and abbreviations, (iii) the main text illustrated with 73 figures and 21 tables, (iv) a series of quizzes and problems to solve at the end of several chapters, (v) a glossary and (vi) author, taxonomy, and subject indexes. A site containing supplementary materials containing appendices and solutions of the quizzes and problems is associated to this book.

Ecology is often presented as a diverse and complex array of apparently disconnected scientific studies concerning living beings. What is in common, for example, between the biogeochemical cycle of oxygen, a food web, the social structure of a beehive, the description of biomes, the structure-functions-dynamics of a pond community, the composition and evolution of bacterial microfilm, the age structure of a population, the mathematical modelling of an ecosystem? These examples, though, were rightfully published as ecological phenomena. Thus, ecology is a science that explains a whole set of phenomena related to living beings.

We also incorporate in the concept of ecology citizens' concerns about the different types of pollution or damage inflicted by human activities, either directly or indirectly, upon such natural systems and their endeavour to slow down or even stop these destructive trends. Hence, ecology has a fundamental part of a scientific nature, called fundamental ecology, that deals with phenomena like those described above, and an applied part, termed applied ecology, inferred from fundamental ecology and preoccupying society and politics.

This book will mainly treat general aspects of fundamental ecology, with the viewpoint that ecosystems constitute the main object of the discipline. An integrative approach to ecology means that the subject of ecology is presented as a system whose parts are interlinked and which is open to other sciences (biology, genetics, geology, physics, chemistry, mathematics and statistics, informatics, the humanities) from which many concepts and reasoning can be imported but also to which the findings of ecology might foster concepts and valuable reasoning. After a brief introduction outlining the book's structure, it is divided into three parts: (i) the definition and scope of ecology, (ii) the history of ecology, (iii) the structure, functions, and dynamics of ecosystems (with the energetical, material, and biological approaches). This book ends with a short epilogue and some prospective.

Paris, France
August 25, 2025

Marc Salomon

Acknowledgements I am deeply indebted to my colleagues and naturalist photographers who graciously allowed me to include their photographs in this book. I would like to express my sincere gratitude to Octavio Aburto (UCSD), Charlotte Duval (MNHN), Angela Falciatore (IBPC), Will Fennie (AFSC), Gwenaëlle Gain (IBPC), Marlin Harms, Stig Bjarte Haugen (IMR), Linda Ianiello, Chad King (NOAA), Sandra Lindstrom (UBC), Steve Lonhart (NOAA), Olivier Maire (EPOC), Lars-Johan Naustvoll (IMR), Peter Pearsall (USGS), the Roscoff Culture Collection (SU), Robert Schwemmer (NOAA), Kevin Stierhoff (NOAA), and Luke Thompson (NOAA). Their works have significantly enhanced the content of this publication, and I am honoured to have their support.

Moreover, I would like to express my thanks to Joël Bried (CRBPO), Bruno Dassy (SU), John A. Endler (JCU), Gwenaëlle Gain (IBPC), Michel Leconte (CRBPO), and Grégoire Sergeant-Perthuis (SU). for having agreed to read parts of the manuscript in their field of expertise with care and interest. Their approbation, encouragement, and, when needed, corrections and invaluable comments were crucial to the development of this book.

Competing Interests The author has no competing interests to declare that are relevant to the content of this manuscript.

Contents

List of Figures

List of Tables

List of Acronyms, Abbreviations, and Symbols

AFS	Abiotic factor system
AFSC	Alaska Fisheries Service Center, NOAA, USA
AMS	Arcachon Marine Station, Arcachon, France
CEM	Commission on Ecosystem Management (of the IUCN) and between DRC and EOL
CRBPO	Centre de Recherche sur la Biologie des Populations d'Oiseaux (Research Centre on the Biology of Birds Populations), Paris, France
d	day
DRC	Democratic Republic of the Congo
EFG	Ecosystem Functional Group
EOL	Encyclopedia of Life
EPOC	Environnements et Paléoenvironnements Océaniques et Continentaux (Oceanic and Continental Environments and Paleoenvironments), UBo, Bordeaux, France
GOE	Great Oxidation Event
HCS	Hierarchical Classification System
IBPC	Institut de Biologie Physico-Chimique (Institute of Physico-Chemical Biology), Paris, France
IMR	Institute of Marine Research, Bergen, Norway
IUCN	International Union for Conservation of Nature, Gland, Switzerland
IR	Infrared
JCU	James Cook University, Townsville, Queensland, Australia
kJ	kilojoule
m	metre
MNHN	Muséum National d'Histoire Naturelle (National Museum of Natural History), Paris, France
NOAA	National Oceanic and Atmospheric Administration (US Department of Commerce), USA
RCC	Roscoff Culture Collection, Biological Station, Roscoff, France
SU	Sorbonne University, Paris, France

UBC	University of British Columbia, Canada
UBo	University of Bordeaux, France
UCSD	University of California San Diego, USA
USGS	US Geological Survey, USA
UV	Ultraviolet
WWF	World Wide Fund for Nature

Chapter 1
Introduction: The Structure of the Book

This book is mainly about fundamental ecology and is divided into three parts.

- Part I deals with the scope of ecology and is divided into five chapters.

  The aim of Chap. 2 is to propose that (i) the primary objects of ecology are the ecosystems, and (ii) the scope of the discipline is the study of the ecosystems' structure, function and dynamics. Not all ecologists share this formal definition. I give it after a historical overview of the conceptions of ecology since the second half of the nineteenth century when the term "ecology" was coined. The historical overview also includes the evolution of the concept of ecosystem.

  The effort to define this discipline seemed necessary to me. Its scope has been ill-circumscribed for historical reasons because ecology stems from different disciplines with heterogeneous approaches. For example, microbial ecology has been studied within the realm of microbiology; mathematical ecology has been considered a branch of applied mathematics; conventional ecology has its origins in natural history (geology, botany, zoology and ethnology) and has consequently mainly been applied to macroscopic organisms. Thus, from these examples, we can see that the different branches of ecology have been disconnected. In addition, ecology needs more consensus about the interpretation of many ecological phenomena due to the existence of different schools of thought and philosophical backgrounds. Thus, I propose contributing to a unifying methodology and integrative, hence the book's title.

  Chapter 3 points to the fact that ecological systems other than ecosystems and their components exist, e.g. the biosphere on the one hand and biomes and biome-like categories (or biome-likes) on the other hand. We will see in this chapter that contrary to ecosystems and their components, which are functional concepts related to the cycling of matter and the transfer of energy flowing from definite sources, biome-likes are defined by their physiognomy. A need was strongly

Supplementary Information (SI) The online version contains supplementary material available at https://doi.org/10.1007/978-3-031-90656-5_1.

M. Salomon, *Integrative Ecology*, https://doi.org/10.1007/978-3-031-90656-5_1

felt to devise a hierarchical classification system of the biomes and biome-likes for conservation purposes: the IUCN endeavours to conserve entire ecosystems based on this hierarchical classification system. Summary treelike descriptions of these biome-likes will be performed.

Chapter 4 deals with the interactions between ecosystem components. I propose a typology of these interactions that is different from the usual one. Namely, I divide the various interactions within biocoenoses into four categories: trophic relationships, coevolution, competition and decomposition/mineralisation.

Chapter 5 illustrates the theoretical points developed in Chaps. 2–4 using a thoroughly and extensively studied example: the ecology of a subbiome: the kelp forests off Central California.

In Chap. 6, I describe ecosystems using compartmental models, which I associate with a specific relation, subsumption. I address the issue of complexity by organising the metastructure of the compartments that model ecological systems in a broad sense. These systems are mutually exclusive, nested, and encompassed within the broadest compartment of the metastructure, the biosphere. I propose a definition of the biosphere that is different from the current acceptance, seeing it as the resultant of the existing ecosystems. In addition, I set up a comprehensive landscape that encompasses all the main physical and biological sciences, highlighting the interdisciplinary nature of this research and its relevance to various fields, including ecology

- Part II is over the history of ecology.
  Ecology has been elaborated after a long process of philosophical and scientific maturation that depended on humanity's historical evolution. Its emergence as a scientific field is a response (i) to the degradation of our environment as a result of our scientific and technological development and (ii) to a set of behaviours influenced by a philosophical background that reflects a number of prejudices, some originating in one's specific culture and others from fears within a society which has increasingly become more urban and less directly related to the natural world [1–3].

  The development of ecology has also been tributary to the state of scientific knowledge. Ecology could not tell the same story before and after discovering the role of DNA or before and after the awareness of the great role of microorganisms in ecosystem functioning. Part II is divided into two chapters.

  Chapter 7 is about a very long period, from the prehistorical times through the ancient Greeks and the Middle Ages until the beginning of the Modern Ages when great scientific explorers discovered the wealth of nature across the Planet. This was a time when much empirical knowledge of the relationship between animals, sometimes plants or fungi, and their environment was collected. This period ended with the Darwinian revolution of thought. Chapter 8 deals with the emergence of modern ecology with the advent of evolutionary thinking, genetics, microbiology, modern statistics, populational reasoning, and systemics.[1] It was during that period that ecology became a scientific discipline.

- Part III treats the structure, functions and dynamics of the ecosystems with three entangled approaches—the energetical, the material and the biosociological[2] ones.
  - Chapter 9 treats the energetical approach. (i) Some thermodynamic principles are reminded, useful to describe the thermodynamic specificity of the living beings: the biotic part of the ecological entities are in perpetual thermodynamic disequilibrium, whereas the abiotic physical objects not related to biology generally tend to thermodynamic equilibrium. (ii) This approach helped us to explain the specificity of the thermodynamical system of living beings and, thus, the ecological systems.
  - Chapters 10 and 11 treat the material approach, respectively (i) the distal origin of the chemical species entering the ecosystems and the biogeochemical cycles, and (ii) the physical parameters (e.g. temperature, humidity, light, pH)
  - Chapters 12 and 13 address the biosociological approach.[2] Fundamental properties of the biotic agents (individuals, populations, taxa, communities), the relays of circulating matter and energy, are mainly twofold:

    (1) they are multiscale and organise themselves spatially and temporally,
    (2) they emerge from similar entities with a potential to evolve, develop, stabilise, senesce, and eventually disappear.

    - Chapter 12 will first address point (1), which is the motor of point (2), and then treat the nature of the "bricks" of the communities (or of the biocoenoses), the species. An evolutionary panorama, extremely summarised, will fill half of the chapter.
    - Chapter 13 will tackle the complex problem of how living beings fundamentally group themselves, i.e. in populations and communities of heterospecific populations. I warn the readers that some parts involve some mathematical skills. Nevertheless, this chapter contains by-passes to allow a reader who does not wish to delve into quantitative reasoning to be able to follow the essentials.

We will end the book with an epilogue, which is not a conclusion but a summary of what this essay has tried to achieve and proposals to address evoked issues in the future.

---

[1] Not to be confused with systematics. Systemics is the science dealing with systems, whereas systematics deals with the scientific classification of the living beings.

[2] The biosociological concept that I propose is not to be confused with the sociobiological concept, a theory set up by several neodarwinian theorists (e.g. Hamilton or Wilson), see Chap. 6, footnote 8.

Quizzes and Problems

1.1 Give at least four examples of studies exemplifying ecological phenomena.

1.2 What are the two most current acceptations of the term *ecology*, and what is this book about?

1.3 What is the rationale and organisation of this book?

1.4 Why was it that ecology was, in the opinion of the author, relatively ill-circumscribed?

1.5 What are the properties opposing the concepts of (i) ecosystems and their components on one hand, and (ii) biomes and biome-likes on the other hand?

1.6 What are the four types of interactions involved in ecosystems according to the author?

1.7 What is the basis of models describing ecological categories? How is called the relationship between the categories A and B, where A encompasses B with the constraint that B has an integration level lower than A (for example, B can be a cell and A a tissue containing it; or B can be a tree and A a forest?)

1.8 Why is it important for an ecologist to have some knowledge of the history of ecology since the earliest periods of humanity?

1.9 What are the three approaches of the functional structural study of ecosystems?

References

1. Terrasson F (1988) La peur de la nature (In French). Sang de la terre, Paris
2. Terrasson F (1994) La civilisation anti-nature (In French). Edition du rocher, Paris
3. Terrasson F (2002) En finir avec la nature (In French). Edition du rocher, Paris

Part I
The Scope of Ecology

This part is composed of six chapters.

- Chapter 2 proposes to define ecology as a discipline whose primary object is the ecosystem, with the difficulties inherent to this concept. My definitions are given after a historical overview showing the heterogeneity of the definitions proposed by our colleagues since the term "ecology" was coined. I assert that ecology is the science that primarily studies the ecosystems, their structure, functions and dynamics.
- Chapter 3 treats the difference between ecosystem and biome-like categories (or biome-likes) after a historical overview of the evolution of the biome concept. On the one hand, ecosystems are categories explaining how matter and energy circulate to make the life of living beings, organised in different heterospecific populations, possible. On the other hand, biome-like categories have another meaning; they have a physiognomy, conveying the wealth, diversity and specificity of the biological structural system linked to drivers. A typology is proposed, followed by the different biome-likes, as they are acknowledged.
- Chapter 4 deals with the interactions between the components of the ecosystems.
- Chapter 5 presents an example of a well-studied marine ecosystem and illustrates its interactions. This ecosystem occurs in a biodiverse subbiome and shows the great role played, in particular, by microorganisms.
- Chapter 6 discusses the recursive compartmental models describing ecosystems. I define the biosphere as the resultant of the existing ecosystems, and I insert the ecological systems, particularly ecosystems, in a network of compartments describing bio-physico-chemical categories. (I thus locate ecology amidst the biological and physicochemical disciplines.)

Chapter 2
The Definition and Scope of Ecology

There has been little consensus on a clear definition and scope of ecology. In this chapter, I will present a short historical overview of the subject since the name of this science was coined. Following this, I will share my personal definition of ecology and my view about its scope.

2.1 Historical Overview of the Conception of Ecology

2.1.1 Early Views

By the conception of ecology, I mean its definition and scope. The German zoologist, philosopher and free thinker Ernst Haeckel (1834–1919) was the first known author to have ever published the term "ecology"[1] in his masterwork on the general morphology of organisms published in 1866 [22, p. 286], and defined it as follows:

> By ecology, we mean the science of the relations of organisms to their outer world where we recognise the conditions of existence . . . Some are of an inorganic nature; first of all, the physical and chemical conditions of the habitat,[2] climate, soil characteristics . . . In the

Supplementary Information (SI) The online version contains supplementary material available at https://doi.org/10.1007/978-3-031-90656-5_2.

[1] The word "ecology" stems from two Greek roots οιχος meaning *home*, and λογος, *discourse*; thus, discourse on the homes (=habitats) of the living beings. For the decipherment of the Greek letters, see Appendix A (SI).

[2] Generally speaking, *habitat* is a term defining the place where a community of organisms live.

M. Salomon, *Integrative Ecology*, https://doi.org/10.1007/978-3-031-90656-5_2

> conditions of existence, we also integrate the mutual relationships between organisms, some favourable, others unfavourable ... Parasite organisms must be included in the conditions of existence.[3]

This concept of ecology—the idea that organisms are forged and constrained by their relationships with an environment, in a certain way their home and economics—was directly derived from the Darwinian concept of natural selection, the force driving evolution, of which Haeckel was a staunch and early partisan. This idea of ecology became very popular already at the end of the 1800s. It developed during the first half of the twentieth century, e.g. through associations, scientific societies, clubs, and journals. We will mention three scientists of that epoch who laid down early scientific theories treating ecology: the American botanist Frederic E. Clements (1874–1945), the Russian chemist Vladimir I. Vernadsky (1863–1945), and the British zoologist Charles S. Elton (1900–1991).

- Clements forged basic concepts like *vegetal successions*, *climax*, *biotic communities* and *biomes* [6, 7]. Importantly, he considered a *formation* (or *biotic community*) a complex organism, arguing that the latter passes "through a cycle similar to that of a plant": it arises, grows, matures, reproduces, possesses functions and structures, adapts to changes of the environment, and eventually dies [7, pp. 3, 30 and 124].
- Vernadsky's work introduced the biosphere concept, a novel idea that revolutionised ecological theory. He established a comprehensive theory on the *pathways of matter and energy* through biological species that act as relays [51].
- Elton's definition was laconic, I quote [17, p. 1]:

  > Ecology is a new name for a very old subject. It simply means scientific natural history.

  Despite this broad definition of ecology, Elton was the first to introduce key ecological concepts, e.g. a modern conception of the distribution of animal communities, food web, ecological succession, ecological niche, or ecological gradient.

2.1.2 Tansley's Concept of Ecosystem

The author who initiated the concept of *ecosystem* and coined the word in 1935 was the British botanist Arthur G. Tansley (1871–1955) [46, p. 299]. The invention of this concept resulted from a reaction against the notion of complex organism proposed by Clements [7] (mentioned above) in 1916 and his disciple, the South African botanist John F. V. Phillips (1899–1987) [35] in 1931. As adepts of holism, these authors considered biotic organisms as complex organisms, thus actual organisms. Tansley entirely disagreed with this notion, considering that the

[3] Translated and summarised by me from the German original text.

members of such a 'biotic community' should instead be processed as interacting parts of a system in the physical sense. In sum, the association of animals and plants in a given region constitutes, for Tansley, a system that he called biome (a term borrowed from Clements [7]). A Tansleyan biome associated with the system formed by the physical factors on which it depends forms a supersystem Tansley called *ecosystem*—a concept gradually adopted, increasingly successful and improved by the scientific community [46, p. 289].

2.1.3 Discussion on the Centrality of the Ecosystem Concept in Ecology

Seven years later, a clearer picture of the structure of an ecosystem was proposed by a young American ecologist Raymond L. Lindeman (1915–1942), strongly influenced by the theory of Vernadsky already mentioned, in an article he published in the Journal Ecology, the year of his death at the age of 26, I quote [29, p. 400]:

> The ecosystem may be formally defined as the system composed of physical-chemical-biological processes active within a space-time unit of any magnitude, i.e., the biotic community plus its abiotic environment. The concept of the ecosystem is believed by the writer to be of fundamental importance in interpreting the data of dynamic ecology."

Further on, the cited author describes the ecosystem as the seat of the *trophic relationships*, leading the flow of energy and cycling of matter through a "biotic community" in the following steps (phrases in quotes are passages directly cited from Lindeman's article):

(1) the intake of energy from solar energy by producers ("autotrophic plants") and, via a portion of it, the biosynthesis of "complex organic substances" from "inorganic substances",
(2) the relaying of a portion of this energy and transfer of organic matter within the producers to consumers ("animals and heterotrophic plants"),
(3) the decomposition after death accompanying the decrease of potential energy with a hypothesis he set forth that some bacteria might transform organic compounds to inorganic compounds,[4] and
(4) the disposal of these released inorganic compounds to plants while all the energy necessary for the cycle of matter is dissipated.

Lindeman introduced the concept of trophic level, which relates to the second principle of thermodynamics. At each step, the author noticed that the higher the trophic level, the more the energy dissipates as heat, and the stronger the dissipation during the decomposition stages after death. This author was ahead of his time, unheeded by his contemporaries and rediscovered only in 1977 [8]. His seminal

[4] This hypothesis posed by a preceding publication [52] was verified later and subsequently called mineralisation.

article [29] setting up the basis of the modern ecosystem concept was almost rejected then, considered too theoretical, but eventually and unexpectedly accepted thanks to its strong support by George E. Hutchinson (1903–1991), the inventor of ecological niche in its modern acceptation [42, p. 178].

Meanwhile, the concept of ecosystem, as seen by Tansley, was accepted and popularised in the 1950s thanks to the masterwork of Eugen P. Odum (1913–2002) and colleagues, *Fundamentals of Ecology*, published in five editions [32–34]. In the 2nd edition of the Fundamentals [33, p. 3], the authors wrote an introduction where they stated that, in sum, ecology has been practised by humanity since the earliest periods without knowing the term, as the need of people, I quote,

> to have an intelligent knowledge of the environment if our complex civilization is to survive, since the basic "laws of nature" have not been repealed.

A little further, the authors wrote:

> Because ecology is concerned especially with the biology of groups of organisms and with functional processes on the lands, in the oceans and in fresh waters, it is more in keeping with the modern emphasis, to define ecology as the study of the structure and function of nature (it being understood that mankind is a part of nature).

The concept of ecosystem, redefined by Odum & Odum as a functional system containing a biotic community (all populations in a given area) and the 'non-living environment', is of the utmost importance for the authors; it is the basis of the phenomena described in the book, e.g. *food web*, *homeostasis*, *ecological niche*, *biogeochemical cycles*, *energy flux* and *matter cycling*, *limiting factors*, *population* and *community ecology*. However, though the concept of ecosystem was so essential to the authors and fostered by them, the latter did not use it as a fundamental object in their definition of ecology. The *Fundamentals* substantially impacted environmental policy and activism since the early 1950s; it gave them their scientific basis and justification. (E. P. Odum was himself a staunch activist of environmental causes and one of the promoters of conservation biology.) The *Fundamentals* was a successful textbook and had many followers.

From the 1960s and onwards, the theory proposed by the *Fundamentals* raised criticism among several scientists who rejected the central role of the ecosystem in ecology. The Australian entomologist Herbert Andrewartha (1907–1992) [1, p. 3] focused ecology on the importance of the abundance of animal populations in relation to simple observable factors like weather conditions, vegetation type or soil characteristics. He gave Elton (mentioned above) the paternity of the idea that he made his own of

> the importance of studying the distribution and abundance of animals in nature and to demarcate this as a proper field for ecology.

For this author, the study of ecosystems, though interesting for explaining the "ontogeny of communities" and of "the limitations imposed on the development of communities" does not help the naturalist understand "the distribution and abundance of animals in nature which Elton saw as the essence of animal ecology"

[1, p. 5]. Interestingly, Andrewartha refers to the works of an author prior to the *Fundamentals*.

Charles J. Krebs (1936–), following the approach of Andrewartha, defined ecology [27, p. 4] as:

> the scientific study of the interactions that determine the distribution and abundance of organisms.

David Tilman (1949–) overlooked the concept of ecosystem and willingly adopted a mathematical stand to describe all the possible trophic or competitive relations between species within different types of organismal communities [47].

The three authors mentioned above proposed a practical and functional but non-comprehensive definition of ecology, criticising their forerunners whom I have mentioned above for having proposed a comprehensive but non-practical and non-functional definition.

From the 1980s and onwards, the need of a practical and functional but also *comprehensive* definition of ecology was felt; an increasing number of major ecologists realised the crucial role played by ecosystems in the definition of ecology.

- The American evolutionary ecologist Eric R. Pianka (1939–2022) defined ecology in his book *Evolutionary Ecology* in 2011 as follows [36, p. 1]:

  > The biological discipline of ecology deals with a myriad of ways in which organisms (plants, animals, and other heterotrophs such as bacteria and fungi) interact with, influence, and are in turns influenced by natural surrounding.

  Interestingly, this author mentions that ecology not only applies to animals and plants, but also to microorganisms.[5] He defines ecosystems, six pages, later as follows.

  > The climate, soils, bacteria, fungi, plants, and animals at any particular place together constitute an ecosystem. Ecosystems have both abiotic (nonliving) and biotic (living) components. All of the biotic components of an ecosystem, or all the organisms living in it, taken together, comprise an ecological community.

  Pianka evokes an environment roughly containing a biological and a physicochemical part. But the conception of Pianka, in my opinions, has two drawbacks: (i) The ecosystems are still not fundamental objects entering in the definition

[5] Most textbooks treating general ecology prior to this publication, mainly those mentioned here, overlooked the ecology of microorganisms. This tendency persists until the present. François Ramade (1934–), for example, as late as 2009 [40], excluded from the biosphere the localities on Earth devoid of plants and animals (e.g. the polar icecaps and highest mountainous strata) and did not include bacterial or archaeal ecosystems (like those existing on bare rocks or deep in sediments) in the *biosphere*. The publications that described ecosystems of microorganisms were mainly carried out by microbiologists, without a clear connection with the ecology of multicellular living beings, or by plant ecologists interested in the symbiotic relationships between plants and nitrifying bacteria.

of ecology. And, (ii) a typology of the different interactions, thus the functional aspects of ecosystems, is lacking.
- The American ornithologist Robert E. Ricklefs (1943–) did not give a formal definition of ecology in his book *Ecology* [42]. But interestingly, this author stressed the importance of ecosystems in ecological studies and devoted an entire chapter (Chap. 10) to the concept of ecosystem, making an interesting historical overview on the evolution of this concept, mentioning F. E. Clements, C. S. Elton, A. G. Tansley, A. J. Lotka, R. L. Lindeman and E. P. Odum. This overview gave me directions to write my own in the present section.

To establish ecology as a science once and for all, since it is not apparent to everyone—even in scientific circles—that ecology is a science, one needs to give a short, practical and at the same time comprehensive definition in a way that no doubt will be left about the scientific nature of the discipline. A way to achieve this is to accept that ecosystem is the central principle in ecology, a conception viewed by many authors, e.g. A. G. Tansley, R. L. Lindeman, E. P. Odum and M. Begon [4]. A very general definition in its formulation but of a philosophical nature is to consider that ecology is the physics of the conditions of life on Earth. However, one has to translate this idea in a more practical formulation, inspired by some of the abovementioned definitions but extended and simplified. A hint of what could nearly be an appropriate definition of ecology is found in the book of the Belgian botanist Paul Duvigneaud (1913–1991) [16]:

> Ecology is theoretically the study of habitat, the home of living beings (οιxoς). One considers ecology the science of relationships between living beings in their environment; since living beings are narrowly integrated into their environment, ecology is about the complex functional biological systems called ecosystems. Ecology also contains the study of the relationships between living beings.[6]

2.2 My Definition Proposal of Ecology

All the definitions of ecology mentioned in the last section contain several relevant points characterising ecology. However, I am not entirely convinced by them, and I think they are pretty composite and lack a simple object to grasp, as is the case for well-established sciences like zoology, physics, and geology. I will now propose a definition based on the feeling that ecology has to do with

(1) the mutual relationships between groups of living beings sharing the same area,
(2) the interactions between these groups and their environment,
(3) the physicochemical conditions of life, and
(4) a dynamic.

[6] Translated by the author from the French original text.

Now, I will introduce my definition proposal of ecology.

> **Definition 2.1** Definition proposal of ecology
> Ecology is the science that studies *ecosystems*, namely (i) their structure and functions and (ii) their dynamics. Properties (i) and (ii) are intimately interlinked.

Property (i) is the conception of a given ecosystem at time t; and property (ii), since an ecosystem spontaneously and inherently changes across time, the apprehension of its evolution.

Now, the problem is, with that definition, to comprehend the concept of ecosystem, which is moving and changing according to the problem studied. Since Tansley's definition, the concept of ecosystem has remained relatively unchanged. Interestingly, a formal and philosophical approach of the definition of ecosystem, was proposed by Stanley T. A. Pickett and Mary L. Cadenasso, in 2002 [37]. These authors wished to complete the early definitions of ecosystem by introducing an inner complexity. The concept of complexity would be divided, by definition, into three separate dimensions: meaning, model and metaphor.

- The meaning of ecosystem is given by the definition of Tansley, which we already mentioned in Sect. 2.1.2, and to which two properties are added: an ecosystem is scale-independent on the one hand, and it is not necessarily in equilibrium on the other hand.
- Inspired by thermodynamics and economics studies, models involving energy and nutrients implicitly define ecosystems.
- Ecosystems are metaphors of places on the Earth's surface.

Though these discussions are interesting and valuable, I'm afraid I have to disagree with the authors' introduction of them into a definition. Their view of a definition is too complex and incorporates practices and philosophical reflections that should be carried out after having posed a simple definition.

I will propose two definitions of ecosystem according to the study context, but beforehand, briefly describe the basic structure of an ecosystem, whatever its definition.

2.3 The Basic Components of an Ecosystem

Definition 2.2 (Biocoenosis) Within a given spatiotemporal unit, a biocoenosis is a system comprising several populations of distinct interacting species.[7] Importantly, two different concepts of biocoenosis will be respectively processed in our two definitions of ecosystem:

- *Limited biocoenosis* (LB): This concept is involved in case studies. An LB comprises a small number of interacting populations of different species whose relationships allow the community they build up to persist.
- *Total biocoenosis* (TB): This concept is used in ecological *modelling* or conservation biology. A TB consists of all the interacting species of the spatio-temporal unit studied[8]

The literature contains several synonyms for the term 'biocoenosis'. For example, what I call 'total biocoenosis' has been called 'biome',[9] 'biotic community' or simply 'community' [7, 29, 33, 36, 46]. However, I did not find any definition in textbooks or articles equivalent to my concept proposal of *limited biocoenosis* versus *total biocoenosis*.

Definition 2.3 (Abiotic Factors (AF)) Within the spatiotemporal unit delineating a given biocoenosis, *abiotic factors* are the non-biological elements that sustain and—so to say—fuel the functioning of the biocoenoses. They are, namely, the chemical and physical agents of the environment (chemical constituents in given proportions allowing to sustain life, thus any given biocoenosis; physical parameters like, e.g., temperature, pressure, humidity, light, pH, whose values must lie in intervals corresponding to the life conditions of the organisms forming the biocoenoses).

In return, the abiotic factors essential for the existence of biocoenoses in a given ecosystem are actually generated by the organisms themselves. This process forms a co-adaptive supersystem, where the abiotic factors constitute a subsystem of the environment (see Definition 2.4). They interact with each other and with the components of the biocoenoses.

---

[7] The term *biocoenosis* was coined in 1877 by the Prussian zoologist Karl Möbius in a book on the ecology of the Oyster in northern Germany [31]. For Möbius, a biocoenosis was the community of the species of plants and animals among which the oysters lived. For François Ramade, a biocoenosis is an association of "species showing various types of population fluctuations and modes of interactions" [40, p. 290]. I propose here a definition inspired by these two seminal definitions.

[8] This definition should mention a difficulty: it is impossible to ascertain the totality of a biocoenosis and ensure that every species was considered in a biocoenosis survey. Besides, TBs are generally very complex, making describing the relations between the constitutive species immensely intricate. Hence, the interest of studying LBs. The opposition between TB and LB is analogical to that existing, for example, between fundamental niche and realised niche sensu Hutchinson (cf. Chap. 4).

[9] In recent publications, the term 'biome' has been applied with other meanings (see Sect. 2.6 of the present chapter and Chap. 3).

Table 2.1 Vital chemical elements (symbols in brackets) included in organic matter or as factors, cofactors or redox molecules, and the broad category of their containing molecules: AT = atmospheric element, BR = biologic role, CH = carbohydrates, HO = hormones, LI = lipids, LI = lithospheric element, NA = nucleid acids, NF = non-organic factors or cofactors necessary for life (e.g. vitamins, growth factors, homeostatic factors), OC = organic compounds, PR = proteins, RE = molecules involved in redox, * = small quantities (<5%)

Elements	Type	OC, BR	Elements	Type	OC, BR
Arsenic (As)	LI	RE*	Molybdenum (Mo)	LI	PR*
Boron (B)	LI	NF	Nickel (Ni)	LI	PR*
Calcium (Ca)	LI	NF	Nitrogen (N)	AT	PR, NA, LI*
Carbon (C)	AT	all OC	Oxygen (O)	AT	PR, NA, CA, LI
Chlorine (Cl)	LI	PR*, NF	Phosphorus (P)	LI	NA, ATP/ADP
Cobalt (Co)	LI	NF	Potassium (K)	LI	ATP/ADP, RE
Copper (Cu)	LI	PR*, RE	Selenium (Se)	LI	PR*
Hydrogen (H)	AT	PR, NA, CA, LI	Silicon (Si)	LI	NF
Iodine (I)	LI	HO	Sodium (Na)	LI	PR*, NF, ATP/ADP
Iron (Fe)	LI	PR*	Sulphur (S)	LI	PR*, NF
Magnesium (Mg)	LI	ATP/ADP	Vanadium (V)	LI	PR*
Manganese (Mn)	LI	NF	Zinc (Zn)	LI	PR*

Abiotic factors (AFs) can be categorised into two groups.

- The first-order AFs are the AFs directly assimilated by the members of a biocoenosis; they consist of the chemical elements presented in Table 2.1 and physical parameters like e.g. temperature, pressure, pH, and hygrometry. The available chemical elements are generally directly or indirectly assimilated by the primary producers (e.g. plants, cyanobacteria) in the form of ions or mineral salts. In an example (a representative sample of a freshwater plant community), it was shown that the atmospheric elements (oxygen, carbon, hydrogen, nitrogen) account for the overwhelming majority of the elements in percentage, making up about 97.4% of the biomass on average, the remaining elements of the biomass being lithospheric. Some latter ones are as vital as the atmospheric elements and form a group accounting for 2.33% of the biomass (silicon, calcium, potassium, phosphorus, magnesium, sulphur, chlorine, sodium, iron). Other elements (boron, manganese, zinc, copper, molybdenum, cobalt), though also very important for the organisms thus for the biocoenoses, are merely traces (the rest, i.e. 0.27% of the biomass) [24, Table 7.4].
- The second-order AFs are not available to living beings; they leave the ecosystems as layers of calcic sediments or dead organic matter that, over time, mineralise and transform into deposits of complex hydrocarbons (oil, coal, asphalts).

The abiotic factors form a system called the abiotic factor system (ABS), which several authors, mainly from continental Europe, termed as *biotope* [9, 20, 39, 40].

Nevertheless, I will not choose this terminology because of its polysemy [50]; for example, several authors used it as a synonym for habitat [48].

Definition 2.4 Within the spatiotemporal unit of a biocoenosis, the *environment* (EN) of the latter is defined differently according to the concept of the biocoenosis studied.

- In the case of an LB, the EN comprises the other limited biocoenoses (OB) and the abiotic factors supporting all the limited biocoenoses.
- In the case of a TB, the EN is exclusively restricted to the abiotic factors.

Definition 2.5 (Interactions) These are the mutual relationships existing between elements of an ecosystem, i.e.,

(1) the interactions between the components of biocoenoses, e.g. trophic relationships, parasitism, mutualism, symbiosis, and decomposition/mineralisation;
(2) regulatory dynamics between abiotic factors;
(3) all types of relationships and feedback between abiotic factors and biocoenoses during life history (for example, the biogenesis of gaseous dioxygen by photosynthetic microorganisms and, in return, the availability of this gas to the respiration of the aerobic organisms).

Axiom 2.1 The trophic components of a biocoenosis
Any limited or total biocoenosis comprises its trophic components [36]—the primary producers, consumers, and detritivores/mineralisers.

- The *primary producers* are *autotrophic.*
  They extract the carbon atoms needed for the construction of their organic matter (the bulk of organic compounds[10] that constitute them) from the carbon dioxide molecules (CO_2) present in the atmosphere or dissolved in the hydrosphere. In parallel, the other elements of the organic compounds [e.g. hydrogen, oxygen, see footnote 10] are inserted in their organic material exclusively from simple inorganic compounds (mineral salts and ions) present in the environment through a process called *osmotrophy* [2], i.e. their absorption through the membranes of their cells.[11]

[10] As a reminder, an organic compound is a chemical species containing one or more carbon atoms covalently linked, either directly or indirectly, to atoms of vital elements whose distal origins are either atmospheric (hydrogen, oxygen, nitrogen) or lithospheric (mainly sulphur, phosphorus, calcium, potassium and iron; but also oligo-elements like chrome, cobalt, copper, fluorine, manganese, molybdenum, selenium, silicon, tin, vanadium and zinc). The main organic compounds in every living being are nucleic acids, proteins, fats and carbohydrates. Such molecules extracted from any living being on our planet are similar; they have many structures in common with varying homologous parts reflecting life's evolution from a common ancestry. A hypothetical last universal common ancestor (LUCA) of all living beings on Earth has been proposed [26].

[11] The carbon atoms and the molecules absorbed via osmotrophy that eventually release the elements listed in footnote 10 are called *nutrients*.

- An autotroph is said to be a *photoautotroph* if it uses, as primary energy, solar energy (via photosynthesis). Among the primary producers that are photoautotrophs, we can count the organisms of the green lineage (green algae and plants), the brown algae, and phytoplankton[12] (phototrophic unicellular eukaryotes and cyanobacteria).
- *Chemoautotrophs* are organisms whose energy source is a redox chemical reaction (e.g. , bacteria living in the deep parts of the ocean not reached by the sun's rays, close to deep-sea hydrothermal vents). These prokaryotes absorb the energy they need by oxidising hydrogen, sulphur, iron, nitrogen or carbon monoxide [13].

• The *consumers* are *heterotrophic*.
They extract the carbon atoms needed to construct their organic matter from complex organic compounds.

- Heterotrophic prokaryotes absorb the organic compounds dissolved in water via *osmotrophy*.These organic compounds may be glucose, asparagine[13] or cellulose [44].
- Heterotrophic eukaryotes ingest the organic material from the body or production (e.g. eggs, fruits) of other living beings, generally not belonging to the same species. Heterotrophic protists do that via *phagotrophy*, heterotrophic multicellular eukaryotes via consumption: fungi via *absorptive nutrition* and animals via eating and drinking.

The organic material ingested by a heterotroph is digested, i.e. broken down into simple chemical compounds similar to those found by the primary producers in their environment for their biosynthesis. From the final products of the digestion, which are similar to the chemicals available to the primary producers during the biosynthesis process, the consumers will edify their proper organic material (see footnote 10, p. 10).

- A few heterotrophic microorganisms are photoheterotrophs: they take in their energy from sunlight, like some purple non-sulphur bacteria [43].
- However, apart from these rare cases, the remaining unicellular prokaryotic heterotrophs (bacteria, archaea), as well as the heterotrophic eukaryotes (protists, fungi and animals), are chemoheterotrophs: their primary energy

---

[12] In fact, the only completely photoautotrophic organisms are the most evolved taxa of the green lineage (vascular plants save the carnivorous plants), which developed a sophisticated organ system, relatively late during evolution, to perform photosynthesis under all circumstances. Simpler organisms, like the phytoplankton species, normally photoautotrophic, can occasionally switch to heterotrophy in environmental stress situations and were, therefore, called mixotrophs by several authors [14, 19]. Nevertheless, they belong to the primary producers as they are the most abundant agents in the marine habitat, which incorporate carbon from environmental carbon dioxide into the biocoenoses.

[13] Asparagine is an amino acid. There exist twenty types of amino acids. As a reminder, the amino acids are the monomers of the *proteins* (see Appendix B (SI, link p. 1)).

stems from the redox reactions of energetic nutrients, e.g. glucose at the basis of Krebs cycle, the main pathway for the distribution of energy in the aerobic organisms.

Consumers are distributed into a finite but not fixed number n of trophic levels.

- The *primary consumers* or *consumers 1* divert their nutrients from primary producers to construct their organic compounds. They distribute themselves into different subcategories: (i) the grazers; (ii) in terrestrial habitats, the frugivores; and (iii) in the aquatic ecosystems, the detritivores and planktivores[14] [5]. They are composed of herbivores or planktivores. They constitute prey species, acting as relays in the pathway of matter and energy from the producers to organisms of a higher trophic level, and are thus *secondary producers.*
- The secondary consumers or consumers 2 are predators consuming primary consumers. Thus, they are composed of carnivores or omnivores, eating herbivores or planktivores, and also primary producers if the consumer 1 is omnivorous.
- The tertiary consumers or consumers 3 prey on secondary consumers. They are composed of carnivores eating other carnivores or herbivores. In terrestrial habitats, tertiary consumers are frequently at the top of a food chain,[15] which means that they are not preyed by another organism. Such predators are called *apex predators*. But this is not a general rule. Often in aquatic biocoenoses, tertiary consumers are not apex consumers, and apex consumers can be consumers 4, consumers 5, etc. [54].

- The detritivores and mineralisers constitute a chain of organisms, generally heterotrophs whose main function is to break down the complex organic molecules to its simplest components.
  - *Detritivores* are heterotrophs feeding on decaying carcasses or excreta (dead biomass), decomposing the organic matter. They are organised in a chain of species, leaving as by-products an increasingly decomposed biomass. These organisms are scavenging or burrowing animals (vertebrates or invertebrates), or *saprophagous* fungi.
  - At the end of this chain, we find the *mineralisers*, i.e. organisms that finish the total decomposition of the organic compounds, eventually breaking them down to the simple compounds and ions available to the autotrophs via osmotrophy. The agents performing this are prokaryotes (bacteria or archaea) [3].

[14] This means that they can be partially secondary consumers since some producers are themselves, on occasion, partial consumers (thus mixotrophs) [14].

[15] Let a_1 be a primary producer, a_2 a consumer 1 preying on $a_1, \ldots,$ and a_n an apex predator preying on a_{n-1}, the food chain is $(a_1, a_2, \ldots, a_n)$, a n-uple; the food chain length will be $n - 1$.

I will now propose two definitions of ecosystem according to the study object (see Definitions 2.6 and 2.7).

2.4 Two Ways of Defining Ecosystem According to Study Context

One of the difficulties of a proper definition of ecosystem is its scalability. For example, the ecosystems in a given forest address the content of 1 g of soil, the entire forest, and all the habitats intermediate in size between these two extremes. Each of these ecosystems comprises a number of agents (biotic and abiotic, they can be few or numerous) underpinning their functioning—thus ensuring their sustainability. Lately and until present, an ecosystem has been defined as an ecological unit that comprises a biotic complex, an abiotic complex, the interactions between and within them, and occupies a finite physical space [25]. I will slightly modify this definition and present a twofold definition of ecosystem.

Definition 2.6 (First Definition of Ecosystem) An ecosystem is a system delineated in a spatiotemporal unit and composed of two interacting subsystems—the *limited biocoenosis* (LB) under study and its *environment* (EN). Let us recall that the environment, here, is composed of two sub-subsystems, the *other biocoenoses* (OB) sharing the same spatiotemporal unit and the *abiotic factor system* (AFS) supporting the ecosystem.

Definition 2.6 has been used since the first ecological studies and is still the most frequently applied. The notions of ecosystem and its components mentioned above arose from this type of research, cf. an example of a very simple ecosystem (EC) [28, 45] illustrated in Fig. 2.1. It is composed of a limited biocoenosis (LB) formed by (i) a Canadian lynx (*Lynx canadensis*) population, (ii) a Snowshoe hare (*Lepus americanus*) population, and (iii) the plant resources Pl_1 (spruce needles, and woody parts of alder, aspen, birch and willow in winter; grass and leaves of *deciduous trees in summer* [53]*) that feed the hares*. Within the LB, each component interacts with each other (here, via trophic interactions[16]), involving biotic and abiotic factors of the environment (EN) [respectively members of the other biocoenoses of the ecosystem (OB) and abiotic factors (AF)]. All the interactions between the members of the LB, or between the latter and those of the EN, constitute an intricate web of relationships. The members of the LB can be related to those of the OB by other types of interactions than the trophic ones (competition, mutualism)[16] and,

[16] The different types of interactions are explained in Chap. 4 in detail.

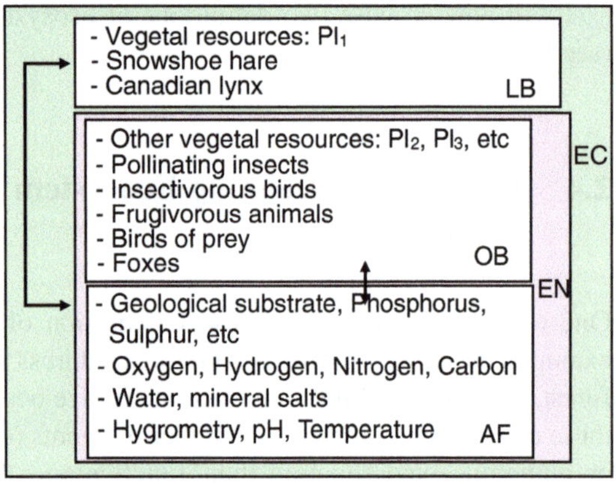

Fig. 2.1 An arctic ecosystem: Canadian lynx and Snowshoe hare. EC = Ecosystem, LB = Limited biocoenosis under study, EN = Environment, OB = Other biocoenoses, AF = Abiotic factors. The double-headed arrows represent some of the interactions

of course, to the AFs since all organisms depend on specific chemical constituents and physical parameters of the environment for their existence.

In Fig. 2.1, one can notice trophic chains parallel to the one under study and sharing the same habitat, e.g. "Plant resources Pl_i $(i \neq 1)$—a frugivorous animal species—a raptor" or "Plant resources Pl_j $(i \neq 1$ and $i)$—a pollinating insect species—an insectivorous avian species—artic fox". As for the abiotic factors, one can easily understand that an ecosystem locally varies with, a.o., the atmospheric elements, the geological substrate, the water content of the environment, and the pH of the soil. These elements constitute interacting parameters that secure the life conditions structuring the biocoenoses.

Now, I shall define my second definition of ecosystem.

Definition 2.7 An ecosystem is a system composed of two interacting subsystems, the total biocoenosis (TB) and the environment that is, as a reminder, exclusively restricted to the AFs supporting the TB.

Definition 2.7 is the original definition of ecosystem. It is rarely encountered in case studies but frequently found in the literature for modelling entire *communities* [11]. One of its drawbacks is that such a definition lends itself to a description that is too complex to explain the functional aspects of an ecosystem. And one should keep in mind the remark of footnote 8 (p. 8). Nevertheless, ecology based on the study of total biocoenoses has several merits. It opposes two subsystems equally important in the description of ecosystems. It is obvious that total biocoenoses are essential parts since they gather all the interacting species sharing the same habitat. The study of their structures and functions place ecology within biology, and we will situate ecology within the realm of life sciences in Sect. 6.4 (p. 6.3). However, this consideration may convey the spurious idea that the physicochemical environment of an ecosystem is a secondary domain of ecology. In fact, the study of the abiotic factor system (ABS) is equally important and will be the subject of Sect. 2.5.

Another advantage of Definition 2.7 is that it can be applied as a convenient tool in conservation biology and policy (see Sect. 3.4).

2.5 Importance of the Abiotic Factor System

Biocoenoses have the capacity of capturing physicochemical factors that are useful or necessary to them, thus for their benefit; but in return, they ensure the existence of abiotic parameters that could not exist in a world without life. It suffices to compare Earth with the other telluric planets[17] of the Solar System, deprived of life, to fully appreciate the specificity of our planet's physical and chemical characteristics. Indeed, the interactions on Earth between the biological and physicochemical worlds entail a relative content of chemical elements and compounds that are very characteristic of our planet. Without life, the relative proportion of the different chemical bodies would be very different, as is observed in dead planets like Venus or Mars. The proportion of different gases in a volume of gas mixture composing their atmosphere were compared between the Earth, Venus and Mars [15, 18, 30, 38, 49], and one can easily notice that the Earth is singular (cf. Table 2.2).

- Compared to the atmospheres of Venus and Mars, the Earth is characterised by a much higher VMR of gaseous nitrogen and dioxygen.
- The atmospheres of Venus and Mars are nearly exclusively composed of carbon dioxide (more than 95%), whereas, in comparison, the proportion of that gas around the Earth is negligible, despite its crucial importance for life and, thus for ecosystems.
- The contrast is very striking if we look at the trace gases:
  - Argon, the most abundant rare gas in our atmosphere in proportion, is comparable to that of Mars and 50 times more abundant than in Venus.
  - Water vapour is about 100 times denser on Earth than on the other two planets.

[17] The planets of the Solar System fall into two categories, i.e. the telluric (or terrestrial or rocky) planets and the gaseous (or Jovian) planets.

- Telluric planets are roughly characterised by their rocky surface which forms a very thin surface, the crust, above a massive bulk of silicate rocks, the mantle. The latter embeds a metallic, generally ferrous core. In the Solar System, this category of planets comprises the inner ones, i.e. Mercury, Venus, the Earth, and Mars.
- On the contrary, gaseous planets are mainly composed of gaseous hydrogen and helium, organised in an outer layer surrounding a liquid metallic hydrogen layer covering a very ponderous molten rocky core. These characteristics give these planets their gigantic mass [10]. In the Solar System, gaseous planets are the outer planets, i.e. Jupiter, Saturn, Uranus, and Neptune.

Relatively to gaseous planets, telluric plants are characterised by far smaller volumes, a much higher density and a much smaller number of satellites [18].

Table 2.2 Volume mixing ratio (VMR), ppmv and ppbv of several gases within a volume of gas mixture composing the atmosphere of three terrestrial planets (in %)

	Volume mixing ratio[a]		
Gas	Earth	Venus	Mars
Gaseous nitrogen N_2	78.08%	3.50%	2.70%
Dioxygen: O_2	20.95%	0	0.14%
Water vapour: H_2O	1–5%	<0.01%	0.03%
Argon: Ar	0.934%	70 ppmv[b]	1.60%
Carbon dioxide: CO_2	0.04%	96.5%	95.3%
Methane: CH_4	1.8 ppmv	28 ppmv	1 ppbv
Dihydrogen: H_2	0.5 ppmv	$\simeq 10^{-3}$	? [c]
Carbon monoxide: CO	0.2 ppmv	26 ppmv	0.8%
Ozone: O_3	0.01–0.1 ppmv	10–20 ppbv	10–350 ppbv[d]

[a] Volume mixing ratio of a gas $\mathcal{G} = \mathcal{V}(\mathcal{G}) = \dfrac{\text{Number of moles of } \mathcal{G} \times 100}{\text{100 moles of the atmosphere containing } \mathcal{G}}$

[b] $1\,\text{ppmv} = 10^{-4}\,\mathcal{V}(\mathcal{G})$. For example, $\mathcal{V}(\text{Ar}) = 9340\,\text{ppmv} = 0.934\%$

[c] Unavailable data for Mars; but indirect evidence gives credit to the presence of vast amount of this gas in the proto-atmosphere of this planet [41]

[d] $1\,\text{ppbv} = 10^{-3}\,\text{ppmv} = 10^{-7}\mathcal{V}(\mathcal{G})$. For example, in the atmosphere of Mars, $\mathcal{V}(O_3) < 350\,\text{ppbv} = 3.50 \times 10^{-5}\%$

– Carbon monoxide is negligible on Earth, whereas this gas is not negligible on Venus or Mars.
– Gaseous dihydrogen exists in far less negligible proportion in our atmosphere than in Venus. Data on this chemical were lacking for Mars in the abovementioned cited articles.

To conclude, suppose an exoplanet were discovered characterised by relative amounts of chemical bodies similar to those found on Earth. That might be a serious indication of the existence of life on that planet. Thus, we can reasonably assert that biocoenoses are agents that control the composition of the abiotic factor system.

2.6 Are the Ecosystems the Only Objects of Ecology?

The answer is no. Any ecological object is not necessarily an ecosystem in the sense of Definitions 2.6 and 2.7. We will now focus on ecological systems, a vast class of objects that includes ecosystems.

Definition 2.8 An ecological system is a spatiotemporal unit comprising heterospecific populations and, possibly, related abiotic factors, whose existence is due to ecological drivers as defined below (cf. Definition 2.9 and Sect. 3.3.2).

Definition 2.9 Drivers of a biological or ecological phenomenon are dynamic parameters explaining its transformation; they can be natural or anthropogenic. Drivers can be biotic or abiotic factors.

For instance, the ecological drivers of ecosystems are the interactions described in Definition 2.5. We can mention, among ecological systems other than ecosystems, *biomes* and the *biosphere*. These ecological systems are defined by drivers other than those defining an ecosystem. Biomes and the biosphere are ecological systems intimately dependent on ecosystems. Moreover, they *subsume*[18] ecosystems, are tightly controlled by them, and, in return, control them. That is why I maintain that ecosystems are the main basic objects of ecology. Hereunder, I will formally propose provisory definitions of the concepts of biosphere and biome—to give the reader a feel for why, in my opinion, the three concepts of ecosystem, biosphere and biome are distinct. Formal definitions will be given later in the book.

- The biosphere, the most extensive ecological system on Earth, results from the structure, functions, and dynamics of the ecosystems it subsumes. In return, a biosphere exerts feedback on its constitutive ecosystems, ensuring their sustainability and reinforcing its significance in the ecological balance.

  At present, the biosphere is a layer including the quasi-entire surface of the planet (lithosphere + hydrosphere) and an extremely thin sheath of the subterranean and atmospheric regions on both sides of the surface (extremely thin relative to the volume of the Planet), where life is present. The biosphere was more limited in the Archaean aeon (−4.03 to −2.5 Ga) than in the last geological periods. According to different opinions based on paleontological evidence, the earliest evidence of life (ca −3.8 Ga) indicates that the biosphere of that epoch was restricted to a system either of land hot springs [12], or of hydrothermal vents at ocean bottoms [21]. During the geological periods, the biosphere has expanded to reach its present position.
- A biome is one of the ecological systems (or units) described by a given physiognomy. In the modern sense, a biome conceived by David A. Keith and colleagues is a unit at a given level of a hierarchical classification system (HCS) [23, 25]. I will call these units biome-likes throughout the book (for details, see Chap. 3). The biome-likes are nested and interrelated by a relationship called subsumption, which I will define more precisely in Sect. 6.4. In that sense, ecosystems are also biome-likes. In a certain way, this classification is a typology of the large ecological systems that we can find in the lithosphere, hydrosphere and atmosphere. This typology does not include all microbial ecosystems, such as the microbiota and their environment, within multicellular organisms. We will resume our ecosystem analysis and develop in detail their motor interactions in Chap. 4.

[18] The meaning of the verb *subsume* is given on Chap. 6.

Quizzes and Problems

2.1 What are the two definitions of ecosystem expounded here? In which situation are each of these definitions appropriate?

2.2 Questions about the concept of abiotic factors and the abiotic factor system in an ecosystem:

(a) Give ten examples of abiotic factors in an ecosystem.
(b) Is the abiotic factor system a minor subsystem of an ecosystem? If not, explain why as concisely as possible?

References

1. Andrewartha HG (1970) Introduction to the study of animal populations, 2nd edn. Chapman and Hall Ltd, London.
2. Balch WM et al (2023) Osmotrophy of dissolved organic compounds by coccolithophore populations: Fixation into particulate organic and inorganic carbon. Sci Adv 9:eadf6973, https://doi.org/10.1126/sciadv.adf6973
3. Bazylinski DA, Frankel RB (2003) Biologically controlled mineralization in prokaryotes. Rev Mineral Geochem 54:217–247, https://doi.org/10.2113/0540217
4. Begon M et al (2006) Ecology: from individuals to ecosystems, 4th edn. Blackwell Publishing, Oxford.
5. Carr MH, Reed DC (2016) Shallow rocky reefs and kelp forests (ch. 17). In: Mooney H, Zavaleta E (eds) Ecosystems of California, University of California Press, pp 311–336.
6. Clements FE (1904) The development and structure of vegetation. In: Pound R, Woods AF, Marsland H (eds) Botanical survey of Nebraska, University of Nebraka, Lincoln.
7. Clements FE (1916) Plant succession: an analysis of the development of vegetation. Carnegie Institution of Washington.
8. Cook RE (1977) Raymond Lindeman and the trophic-dynamic concept in ecology. Science 198 (4312):22–26.
9. Dahl F (1908) Grundsätze und Grundbegriffe der biozönologischen Forschung (In German). Zool Anz 33:349–353.
10. D'Angelo G, Lissauer JJ (2018) Formation of giant planets. arXiv preprint, arXiv:1806.05649v1.
11. Dézerald O et al (2013) Food-web structure in relation to environmental gradients and predator-prey ratios in tank-bromeliad ecosystems. Plos One 8:e71735, https://doi.org/10.1371/journal.pone.0071735.
12. Djokic T et al (2017) Earliest signs of life on land preserved in ca. 3.5 Ga hot spring deposits. Nat Commun 8:15263, https://doi.org/10.1038/ncomms15263.
13. Dobrinski KP et al (2005) The carbon-concentrating mechanism of the hydrothermal vent chemolithoautotroph *Thiomicrospira crunogena*. J Bacteriol 187:5761–5766, https://doi.org/10.1128/JB.187.16.5761-5766.2005
14. Dolan J (2012) Microzooplankton: the microscopic (micro) animals (zoo) of the plankton. Institut Océanographique, Monaco. https://www.oceano.org/wp-content/uploads/2020/02/20.Microzooplankton_Dolan.pdf.
15. Donahue TM, Hodges RR Jr (1993) Venus methane and water. Geophys Res Lett 20:591–594.
16. Duvignaud P (1974) La synthèse écologique (In French). Doin, Paris.
17. Elton CS (1927) Animal ecology. McMillan Co, New York.

18. Encrenaz T, Coustenis A (2017) Composition and chemistry of the atmospheres of terrestrial planets: Venus, the Earth, Mars and Titan. In: Deeg HG, Belmonte JA (eds) Handbook of exoplanets, Springer International Publishing.
19. Flynn KJ et al. (2013) Misuse of the phytoplankton-zooplankton dichotomy: the need to assign organisms as mixotrophs within plankton functional types. J Plankton Res 35:3–11, https://doi.org/10.1093/plankt/fbs062
20. Gams H (1918) Prinzipien der Vegetationsforschung (In German). Vierteljahrssch Naturforsch Ges Zürich 63:293–493.
21. Georgieva MN et al (2021) The history of life at hydrothermal vents. Earth Sci Rev 217:103602, https://doi.org/10.1016/j.earscirev.2021.103602.
22. Haeckel E (1866) Generelle Morphologie der Organismen. G. Reimer, Berlin.
23. IUCN (2020) IUCN global ecosystem typology 2.0. Descriptive profiles for biomes and ecosystem functional groups. IUCN, Gland (CH).
24. Jørgensen SE, Fath BD (2011) Fundamentals of ecological modelling. Application in environment management and research. 4th edn, Elsevier.
25. Keith DA et al (2013) Scientific foundations for an IUCN red list of ecosystems. PLos One 8:e62111.
26. Koonin EV (2003) Comparative genomics, minimal gene-sets and the last universal common ancestor. Nature Rev 1:127–136, https://doi.org/10.1038/nrmicro751.
27. Krebs CJ (1985) Ecology. Harper & Collins, New York.
28. Krebs CJ et al (1995) Impact of food and predation on the snowshoe hare cycle. Science 269:1112–1115.
29. Lindeman RL (1942) The trophic-dynamic aspect of ecology. Ecology 23:399–417.
30. Marcq E et al (2019) Discovery of cloud topozone on Venus. Icarus 319:491–498.
31. Möbius KA (1877) Die Auster und die Austernwirtschaft (In German). Wiegandt, Hempel & Parey, Berlin
32. Odum EP, Odum HT (1953) Fundamentals of ecology, 1st edn. Saunders, Philadelphia and London.
33. Odum EP, Odum HT (1959) Fundamentals of ecology, 2nd edn. Saunders, Philadelphia and London.
34. Odum EP, Barrett GW (2005) Fundamentals of ecology, 5th edn. Thomson Brooks, Cole (CA).
35. Phillips J (1931) The biotic community. J Ecol 19:1–24.
36. Pianka ER (2011) Evolutionary ecology. Author account (ebook).
37. Pickett STA, Cadenasso M. L. (2002) The ecosystem as a multidimensional concept: meaning, model, and metaphor. Ecosyst 5:1–10, https://doi.org/10.1007/s10021-001-0051-y
38. Prinn RG, Fegley B (1987) The atmospheres of Venus, Earth and Mars: a critical comparison. Ann Rev Earth Planet Sci 15:171–212.
39. Ramade F (1982) Elément d'écologie, écologie appliquée: action de l'homme sur la biosphère (In French). Edition Mc Graw-Hill, Paris.
40. Ramade F (2009) Eléments d'écologie: Ecologie fondamentale (In French) 4th edn, Dunod, Paris.
41. Ramirez RM et al (2014) Warming early Mars with CO_2 and H_2. Nature Geosci 7:59–63, https://doi.org/10.1038/ngeo2000
42. Ricklefs RE (1990) Ecology. Freeman & Co, New York.
43. Sakarika M et al (2020) Purple non-sulphur bacteria and plant production: benefits for fertilization, stress resistance and the environment. Microb Biotechnol 13:1336–1365, https://doi.org/10.1111/1751-7915.13474
44. Southam G, Westall F (2007) Geology, life and habitability. In: Spohn T (ed) Treatise on Geophysics, vol. 10: Planets and Moons, Elsevier, pp 421–437.
45. Stenseth NC et al (1997) Population regulation in snowshoe hare and Canadian lynx: Asymmetric food web configurations between hare and lynx. Proc Natl Acad Sci USA 94:5147–5152, https://doi.org/10.1073/pnas.94.10.5147.
46. Tansley AG (1935) The use and abuse of vegetational concepts and terms. Ecology 16:284–307, https://doi.org/10.2307/1930070.

47. Tilman D (1982) Resource competition and community structure. Princeton University Press, Princeton, NJ.
48. Tischler W (1949) Grundzüge der terrestrischen Tierökologie (In German). Friedr. Vieweg & Sohn, Brunswick (Germany).
49. Trokhimovskiy A et al (2020) First observation of the magnetic dipole CO_2 absorption band at 3 μm in the atmosphere of Mars by the ExoMars trace gas orbiter ACS instrument. A & A 639:A142, https://doi.org/org/10.1051/0004-6361/202038134
50. Udvardy MFD (1959) Notes on the concepts of habitat, biotope and niche. Ecology 40:725–728.
51. Vernadsky VI (1997) The biosphere. English translation from Russian, original published in 1926. Springer Verlag, New York.
52. Waksman SA (1941) Aquatic bacteria in relation to the cycle of organic matter in lakes. In: Symposium on Hydrobiology, University of Wisconsin Press, Madison (WI), pp 86–105.
53. Wolff JO (1978) Food habits of snowshoe hares in interior Alaska. J Wildl Manage 42:148–153.
54. Zainordin AF, Hamid SA (2017) Determination of trophic structure in selected freshwater ecosystems by using stable isotope analysis. Trop Life Sci Res 28:9–29, https://doi.org/10.21315/tlsr2017.28.2.2.

Chapter 3
Biomes

3.1 Ecosystem vs Biome

This chapter will be dedicated to the biome-like categories. In the last section of Chap. 2, we saw that an ecosystem, as we defined it, can also be considered a biome-like category. Ecosystems fit the lowest integration levels of the Hierarchical Classification System (HCS) in the definition by Keith and colleagues [36, 79]. This system was mainly aimed at fulfilling conservation policies, i.e. protecting complete ecosystems rather than individual species, as were the traditional policies of the IUCN in the past (before the last two decades). We will now (i) make an overview of the evolution of the concept of biome, (ii) proceed to a formal definition of the biome-like categories, and (iii) present a typology of the three highest ranked biome-likes, the realm, the biome and the subbiome [or ecosystem functional group (EFG)] sensu Keath and coll. When the term *biome* was coined in 1916, a long lineage of naturalists had proposed a couple of terms for more than a century to describe the biotic parts[1] of the habitats [58], e.g. *association*, *flora*, *formation*, *biocoenosis* and *life zone*.[2]

Association, formation, and life zone concepts were closely linked to a type of physiognomy (cf. Definition 3.1) and could be considered as forerunners of the modern biome concept. In contrast, the concepts of flora and biocoenosis were not associated with any physiognomy and thus historically led to the ecosystem concept.

Supplementary Information The online version contains supplementary material available at https://doi.org/10.1007/978-3-031-90656-5_3.

[1] In these early times, the ecological systems were depicted in terms of plant societies.

[2] The idea of constructing a typology of plant societies dates back to the ancient Greeks [Theophrastus (372–288 BC)] [39]; but we will focus on the development of the concepts in the modern era.

M. Salomon, *Integrative Ecology*, https://doi.org/10.1007/978-3-031-90656-5_3

Definition 3.1 Physiognomy, a key concept in our discussion, can be defined by three different, non-mutually exclusive criteria:

(1) Physiognomy can be defined in terms of key (or characteristic) taxa. In this case, we will use the phrase *partially taxonomic physiognomy* for this type of physiognomy.
(2) Physiognomy can be defined in terms of ecogeographical parameters, e.g. climatic variables (temperature, precipitation height, evapotranspiration) or edaphic attributes (parameters describing soil[3] structure and functions). We will then speak of *ecogeographical physiognomy*;
(3) Physiognomy can be defined in terms of vegetation's internal structure, for example

- shrubby, woody, grassy, barren, or
- category of forests, marshes, and deserts.

In this case, one may refer to this type of physiognomy as *architectonic physiognomy*.

3.2 The Forerunners of the Biome Concept

3.2.1 The Concept of Association

The *association* concept was proposed in 1805 by Humboldt and Bonpland [34]—the first botanists in history to have carried out a thorough geographical and sociological analysis of the plants.[4] The cited authors remarked that a certain number of plants were asocial (individuals growing isolated from one another), whereas others on the contrary, were social (constituting societies). Fifteen species were exemplified as asocial, and 20 as social (see Sect. 7.4.1.1). These would constitute more or less loose associations characterised by key genera—thus, a partially taxonomic physiognomy. Fifteen associations (or forms) within a set of associations forming a universal distribution scheme on Earth are presented in Sect. 7.4.1.2. From a field study along a *transect* across the Andes, comparatively to their knowledge of the vegetation in Europe and North America, Humboldt and Bonpland hypothesised that associations in the tropics would be strictly associated

[3] We will see later that soil is a complex system involving soil granulometry, specific biocoenoses involved in the mineralisation of organic compounds and chemicals whose distal origin is the bedrock.

[4] Indeed, past and contemporaneous colleagues, e.g. the British botanist Richard Bradley (1688–1732) [8] and Carl Linnaeus (1707–1778) [48], were aware that different plant species always clustered together and pertained to a given geographical area. However, they used this fact only as an empirical means to locate the plants they wished to collect and not to study these groups per se [62].

with altitudinal strata showing specific, nonoverlapping florae. Conversely, in cold or temperate areas, the same plant species can be present at different altitudinal levels, the florae being thus less specific and more overlapping [35]. The climatic factors, determining floral provinces, were indirect and never proceeded as direct explanatory factors of the florae.

The discipline of phytogeography was born. Section 7.4.1 provides a thorough description of Humboldt and Bonpland's phytosociological theory.

Humboldt had a number of disciples who applied his concept of association [62, p. 97]: (i) The Danish Botanist Joakim Frederik Schouw (1789–1852) is known for having invented "a nomenclature for associations by adding '-etum' to the generic name of the dominant plant" [e.g. *Fagetum* refers to an association where beech (genus *Fagus*) is the dominant tree] [78]. (ii) The German naturalist Franz Julius Meyen (1804–1840) was the first author, in 1836, who attempted to correlate the distribution of plants with physical variables (temperature, humidity) [54], followed by two Austrian botanists, Josef R. Lorenz (1825–1911) [49] and Arthur Kerner von Marilaun (1831–1898)) [40]. (iii) We can mention the Swiss botanists Oswald Heer (1809–1883) [28] and Jules Thurmann (1804–1853) [86], and the French botanist and geologist Henri Lecocq (1802–1871) [46]. (iv) The Scandinavian botanists Humpus von Post (1822–1911) [91] and Ragnar Hult (1857–1899) [33] are noted for having applied the Humboldtian method to the Scandinavian vegetation. This approach has influenced the Scandinavian approach of phytosociology until present.

3.2.2 *Gradualism sensu Philippi vs Formation sensu Grisebach*

About 20 years later, a debate arose among two early phytogeographers who followed the tracks of Humboldt and Bonpland, Rudolf Amandus Philippi (1808–1904) [69] and August Heinrich Rudolf Grisebach (1814–1879) [26].

Philippi weakened the Humboldtian association concept by considering that the association of plant species was conjunctural, thus equating the associations to florae at a given time t (the analysis is made on the geographical variation of the families of plants measured in percentage). On the contrary, Grisebach strengthened the concept of Humboldtian association, i.e. clusters containing characteristic species always present in them, thus solidly fixing diagnostic key species in them. These two points of view were radically opposed. Indeed, these two scientists were the ancestors of two schools of thought, which appeared at the end of the nineteenth century and had fiercely debated until the modern concepts of ecosystem and biome.

- Rudolf Amandus Philippi was a gradualist: the change of taxonomic composition between the florae is gradual, with a smooth transition between the neighbouring florae. The stress is put on the individual species, their groups are not individualised with much emphasis. One century later, this view would evolve into the theory of individualism, with H. A. Gleason (1882–1945) as leader.[5]

- August H. R. Grisebach proposed to study the geographical variation of entities more closed than the Humboldtian associations, the *formations*. In his view, neighbouring formations are isolated as far as their taxonomic composition is concerned and separated by narrow and abrupt transition areas or even no transition zones. Grisebach initiated a school of thought which eventually led—one century later—to holism, an interdisciplinary theory set up by the South African military leader and statesman Jan Christian Smuts (1870–1950) and defended in ecology at the same epoch, as we saw in Sect. 2.1, by Clements and followers.[5] In more recent days, the term *formation* has still remained popular with other meanings [31].
- From these early days until the present, a number of authors remained neutral concerning the question invoked above.

3.2.3 Life Zones

This concept has been proposed by the American zoologist Clinton H. Merriam (1855–1942) [53]. *Life zones* are areas delineated by isotherms and containing a specific type of flora and fauna. Animals and plants are restricted in their possibility to multiply and spread indefinitely: one efficient constraint against unlimited dispersion would be the quantity of heat necessary for the physiological activity of the plants of a given life zone, thus also of the animals depending on them, during the season of growth and reproduction [53, p. 233]. An isotherm would be associated to the value of a parameter, i.e. the sum of the normal mean daily temperature[6] above 6 °C.

In a study of the main life zones of North America, Merriam determined five of such areas, i.e. from north to south,

- the boreal life zone below the isotherm 5500 °C;
- the transition zone between the isotherms 5500 and 6400 °C;

[5] More details on these pioneering ecologists and biogeographers are found in Chap. 7.

[6] The normal mean daily temperature is the average temperature of a given day as defined in the calendar, e.g. 15th March, calculated from a 30-year period. As an example, let, in a certain life zone, the minimum mean daily temperature (i) be below 6 °C before a = 15th March (codified 3), (ii) reach 6 °C or above from a until b = 15th October (codified 10) included, and (iii) be below 6 °C after b. Let, for a given day d, the temperatures during the year of the study be θ_0 and the ones of the previous 29 years be $\theta_0, \theta_1, \cdots \theta_{29}$. Then the normal mean daily temperature of day d is $\bar{\theta}_d = \sum_{i=0}^{29} \frac{\theta_i}{30}$; and the estimation of the total quantity of heat during the season of growth and reproduction is $\sum_{d=a}^{b} \bar{\theta}_d$.

- the upper austral life zone between the isotherms 6400 and 10,000 °C;
- the lower austral life zone between the isotherms 1000 and 14,500 °C;
- the tropical life zone above the isotherm 14,500 °C.

These main life zones are subdivided into smaller life zones. The ecological system's physiognomy is mixed. It is simultaneously partially taxonomic and ecogeographical (with a criterion derived from temperature). Merriam, considering humidity a secondary issue, did not carry out an analysis of humidity.

3.3 The Evolution of the Concept of Biome

3.3.1 From Clements to Whittaker

The term *biome* has evolved in its meaning [58] since it was invented in 1916 by Clements [17, p. 319] as a synonym of *biotic community*—in the mind of the cited author, a plant or animal community conceived as an organic entity located in a definite space (see Sect. 2.1).

The term *biome* proposed by Clements was extremely inconspicuous since it occurred only once in his bulky book of 658 pages, and not in a central position in his analysis but just in a small subsection about cultural relics of prehistorical humankind within a section devoted to the animal fossils of the Pleistocene. I quote the passage:

> Human evidence of past climates and biotic communities, or *biomes*, must come to be of very great value when our methods of interpretation have improved.

For Clements, the formation of plants and animals on the one hand, and habitats (e.g. geological factors, climate) on the other hand, are [17, preface] :

> two inseparable phases of a development which terminates in a climax controlled by climate.

Vegetation would constantly develop an ecological succession ending up with an equilibrium called *climax*, the final stage. An ecological succession is defined as the spontaneous and gradual development of vegetation across time and space from the simplest form (nearly bare rock or barren landscape) to the more complex one (forest) via a continuity of increasingly complex intermediate stages as far as species richness and vegetation internal structure are concerned. As a partisan of holism, Clements was convinced that an ecological succession was [17, p. 349]

> a complex organism, possessing functions and structures, and passing through a cycle of development similar to that of the plant.

All stages before the forest stage were hypothesised to be juvenile, and the forest stage, the climax, to be mature and correspond to the biotic community or biome. Thus, Clementsian biomes are considered organic entities, like plants or animals. In

Fig. 3.1 Clements' model of biome. Each curve of different color represents different heterospecific populations of the biocoenosis composing the biome *sensu Clements*. The adjacent biocoenoses are separated by a transition biocoenosis, the ecotone

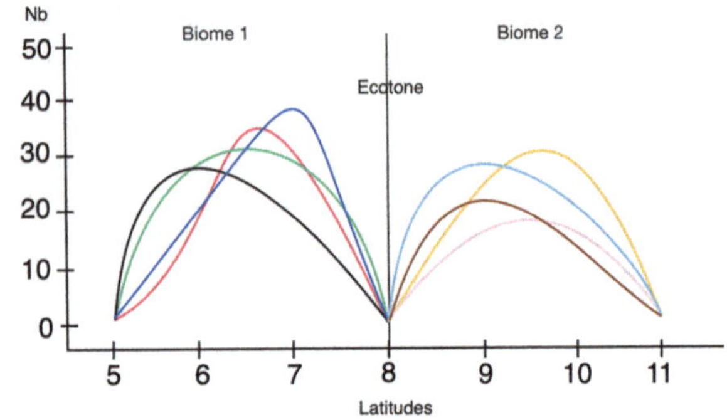

the holistic view of Clements, neighbouring biomes are separated and individualised by a narrow transitional ecosystem called *ecotone* analogous to a membrane.

Figure 3.1, in sum, illustrates the concept of the *Clementsian biotic community* or *biome*:

- A biome is represented by a set of finite curves separated from a neighbouring biome (similar set of curves) by a vertical line representing a narrow passage—a sharp transition corresponding to an ephemeral and labile ecosystem called *ecotone*. Each biome looks like a skein.
- Each curve of different colour represents the distribution of a population of a definite and distinct species (e.g. density as a function of latitude) that interact via trophic, coevolutionary, competitive relationships, depending on the same abiotic elements. They not only coexist but are necessary components for their existence. The perennity of the group is the essential condition for the survival of the species forming the group. Hence, the idea that functionally, each curve would be the analogue of cells in a live body.
- Each curve of a definite colour has a specific shape at a given time t, but this shape is not static. It varies across time, reflecting the dynamic nature of biomes. The only constant elements of each curve are its beginning and end. Hence, its skein-like aspect.

The concept of biome gradually evolved. A thorough overview [58] allowed me to sketch briefly several steps that led to the modern biome concept of *biome* as defined by the IUCN.

- In the paper where he introduced the ecosystem concept in 1935 and coined the term [82], Tansley still equated *biotic community* and *biome*, albeit with a different meaning than that of Clements and his followers. Tansley's divergence from the Clementsian school of thought, particularly his disagreement that biomes were climaxes, is a significant point of academic debate. He believed that successions were not necessarily progressive but could be regressive, and they were not only intrinsic but could also be caused by external factors.
- In 1945, the American ecologist Victor E. Shelford (1877–1968) still equated biome and biotic community with a Clementsian conception [80]. However,

the novelty of this paper was that (i) it applied Clementsian biomes to large-scale climactic communities (in the biogeographical sense)[7] with a more explicit inclusion of the faunae, and (ii) it compared the merits of such a concept with that of the life zones of Merriam (see above). Though still relatively conservative, this paper fostered reflections on the biome concept.

- The view that the aspect and structure of a plant community would depend on climatic conditions is an old idea, dating back to writings of ancient Greeks, paradoxically overlooked until the mid-twentieth century. Biomes, in their modern acceptation, appeared with the publications of Leslie R. Holdridge [31] and Robert H. Whittaker [95]. Holdridge devised a typology of 100 *biomes* he called *formations* covering most of the terrestrial part of the Earth and based on climatic (temperature, annual precipitation height, humidity classes) and topographical (altitude, latitude) variables. This classification was expounded in a figure (redrawn below in Fig. 3.2) within a short article published in 1947 [31, Fig. 1].

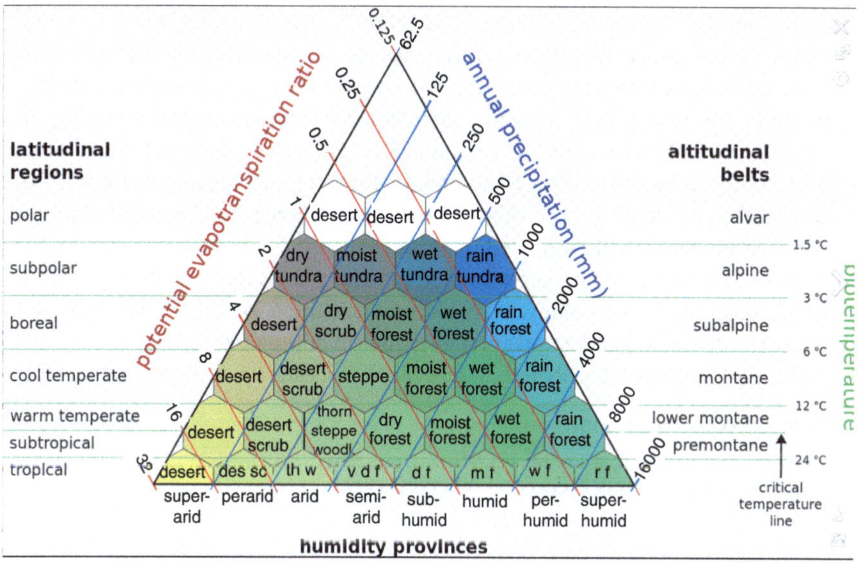

Fig. 3.2 Biome typology by Holdridge (1947). Redrawn figure of Fig. 1 in [31] from English wikipedia, creative Commons BY SA (any version). Abbreviations: des sc = desert scrub, th w = thorn steppe wood, v d f = very dry forest, d f = dry forest, m f = moist forest, w f = wet forest, f f = rainforest

[7] Large-scale in the biogeographical sense corresponds to small-scale in the cartographic sense—measured as the ratio $\rho = \frac{d(\mathrm{A,B})}{d(\mathrm{M,N})}$ where A and B are respectively the representative points on a map of the points M and N in the real world, and $d(\mathrm{X, Y})$ the distance between two points X and Y.

The American ecologist Robert Harding Whittaker (1920–1980) introduced the term *biome* by nearly equating it to *formation* sensu Holdridge [95, p. 51], with a slight difference between the two concepts:

- A formation is restricted to plant societies.
- A biome is a formation plus the animals, which the former shelters.

For clarity, I will formally present his definition [95].

Definition 3.2 (Biome sensu Whittaker) A biome is a community described by a given *physiognomy*[8] (see Definition 3.1).

We saw that physiognomy can be defined by three different, not mutually exclusive criteria, i.e. *partially taxonomic physiognomy*, *ecogeographical physiognomy* and *architectonic physiognomy* (cf. Definition 3.1). These criteria historically depended on the technological advancement of the time. Until the end of WWI, the naturalists practising an ecological field study could only do it on foot, rarely from balloons, and overwhelmingly relied on partially taxonomic physiognomy. Some kinds of ecological physiognomy could be practised in earlier times, like the description of vegetation assemblage as woody, shrubby, and so forth. Since the 1920s, aeroplanes associated with photography could be used to make general aerial views of entire ecological systems, making it possible for researchers to apply architectonic ecological physiognomy, like e.g. (i) tropical rainforests, or (ii) humid or dry savannahs [39]. Two biomes were assessed identical if they had identical physiognomies and distinct if these were different; the degree of this difference determined the degree of similitude between the biomes incriminated.

Like Holdridge, Whittaker restricted his analysis to the terrestrial part of the Earth, resulting in a typology of 25 biomes in relation to climatic variables, e.g., temperature (in °C) and annual precipitation (in cm) (see Fig. 3.3).

These 25 biomes distributed into 14 biome types are as follows:

Type 1: (i) tropical rainforests
Type 2: (ii) tropical seasonal forests
Type 3: temperate rainforests:
(iii) plain –; (iv) montane –
Type 4: (v) temperate deciduous forests
Type 5: (vi) temperate evergreen forests
Type 6: (vii) taiga; (viii) elfin woodlands;
Type 7: (ix) thorn woodlands (x) thorn scrubs
Type 8: (xi) temperate woodlands
Type 9: (xii) temperate shrublands

[8] For R. H. Whittaker, the physiognomy of a community is the equivalent of morphology (thus structure) and function used to classify organisms in systematics. The mentioned author illustrates rather than defines this concept.

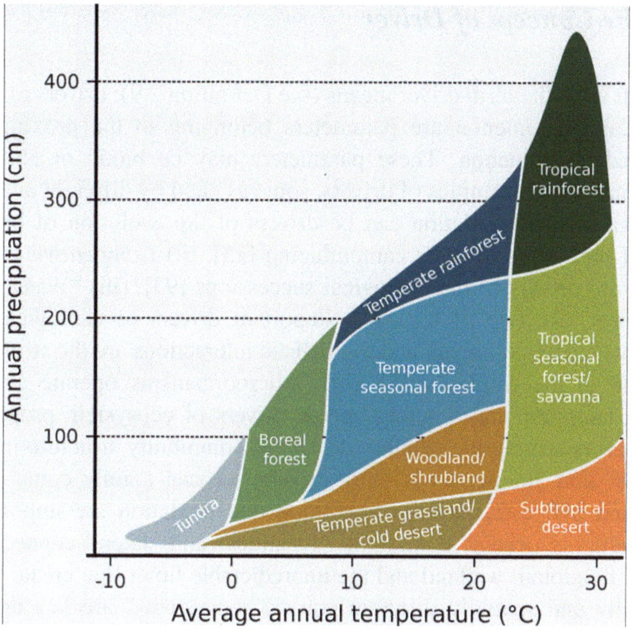

Fig. 3.3 Biome typology by Whittaker (1970). Redrawn figure of Fig. 3.8 [95, p. 55] from English wikipedia, creative Commons BY SA (any version). Nine biome types, function of temperature and rain precipitation, are represented

Type 10: (xiii) savannahs
Type 11: (xiv) temperate grasslands
Type 12: (xv) alpine grasslands; (xvi) tundras grasslands
Type 13: deserts:
(xvii) tropical and subtropical –; (xviii) warm-temperate –; (xix) cool-temperate –; (xx) arctic-alpine –
Type 14: Hydric communities:
(xxi) cool-temperate sphagnum bogs; (xxii) tropical fresh-water swamps; (xxiii) temperate fresh-water swamps; (xxiv) mangrove swamps; (xxv) saltmarshes of temperate coasts.

For a detailed description of this biome typology, see Appendix C (SI, p. 1).

Whittaker's concept of biome and his classification scheme of this ecological system approach the notion's modern acceptation. However, the latter had to be improved. Indeed, only two climatic variables, temperature and rain precipitation, were involved in Whittaker's view. A large number of subsequent authors [58] added to the climatic variables other ecogeographical physiognomies, such as, for example, edaphic parameters, which would give a more synthetic picture of the biomes. Two essential concepts would strengthen the principle of biome in the modern sense: the driver concept and hierarchy theory.

3.3.2 The Concept of Driver

Let us recall what the term driver means (see Definition 2.9): drivers of a biological or ecological phenomenon are parameters belonging to the proximal cause of the observed phenomenon. These parameters may be biotic or abiotic factors. Hereunder are a few examples of drivers, some as cited by different authors.

(i) Parasitism and predation can be drivers of the evolution of bright or dull plumage of prey birds towards camouflaging [25]. (ii) *Competitive inhibition*[9] or *facilitation* are drivers of the ecological successions [93]. (iii) "Water and energy, including heat, are two of the most important drivers of ecological function"; "predator-prey, host-pathogen, and mutualistic interactions are the strongest drivers of evolution in the tropics" [75]. (iv) "Microorganisms operate at a range of spatial and temporal scales acting as key drivers of ecosystem properties" [90]. (v) "Site factors remain the primary drivers of community structure in high arctic tundra soils" and among the primary drivers, we can mainly count "soil parent materials and pH" [92]. (vi) "Competition and predation are important drivers of change altering oceanic community dynamics" and "lateral connectivity to the river of the floodplain wetland and the unpredictable flows that create wide ranges of temporally and spatially different aquatic ecosystems" are key drivers to the very rich biological diversity of the Australian wetlands [37]. (vii) The concept of ecological driver has been used as a factor that generated important ecological events during early geological periods [15, 27, 50]. (viii) The need to protect entire threatened 'ecosystem' types (ecosystem taken in the sense of Tansley) and not only species was a novelty proposed by an international team of 34 researchers [38]. This article was a tool enabling the protection of these 'ecosystem' types. Drivers of these units were enumerated throughout the paper, though the word driver only appeared three times. (ix) The physicochemical properties of minerals within a spatiotemporal unit constitute key drivers of microbial communities within microbial ecosystems (and the biome subunits defined by these drivers) [88] (x) Drivers structuring the distribution of planktonic communities have been thoroughly studied [76], a paper very useful to assess the impact of climate change on the oceans.

3.3.3 Hierarchy Theory and Formal Definitions of Biome and Biome-Like Categories

A novel axiomatic theory that appeared in the 1980s, *hierarchy theory*, defines ecological systems as nesting ecological systems [2, 42, 64], each corresponding to

[9] In this type of competition, "one species has a negative effect on the other. The inhibitor slows or arrests successional change by preventing the establishment of species in the next stage" [93, p. 209].

an integration level, an overall description and a set of physical parameters (abiotic and biotic) organising them, the drivers as defined in Sect. 3.3.2.

Mucina nicely defines the biome at its proper integration level [58, p. 98]

> Macroclimate, soil, water and, finally, disturbance are the major ingredients from which biomes—in all their appearance, complexity and functioning—are 'cooked'.

The four global factors defined above by Mucina form sets of ecological drivers.

The overview in Sect. 3.3.2 showed that the term *driver* had a different meaning among the ecologists mentioned above. Definition 2.9 (p. 22) and hierarchy theory will contribute to my definition of biome and biome-like category, respectively, a biome or a category subsuming or subsumed by a biome. The IUCN constructed a hierarchical classification system from these categories in 2020 [36].

Definition 3.3 (Biome-Like Category) A biome-like category or biome-like is an ecological system defined within a treelike structure called a hierarchical classification system (HCS) at a given *integration level*. The construction of the biome-likes are based on a set of traits and drivers whose number is fixed empirically in the context of assembly theory [36, p. 18], namely by a parameter called the assembly index [52].

The higher an integration level is, the broader its spatial scale and the less numerous the number of its attributes will be.

Definition 3.4 (Biome) A biome is a biome-like, empirically defined, with a set of traits and drivers that would correspond to a spatial scale such that there would exist, on Earth, 25 biomes, a number arbitrarily fixed, corresponding to the number of biomes attributed by Whittaker.

This criterion was empirical and dictated by two arguments: (i) to conform to Whittaker's typology, and (ii) because of the close relation between "functional traits and plant dominant growth forms" [36, p. 9]. This biome concept though is broader than Whittaker's system and encompasses all the ecosystems of the planet and not only the terrestrial one.[10] For the time being, though this criterion seems somewhat arbitrary, it has received a broad consensus, and therefore, we will also adopt it.

[10] The improvements brought by the IUCN concept of biome have been inspired notably by the work of Spalding et al. [81] on marine biomes. These authors used the term biomes for entities called *ecoregions* by other authors, and proposed the concept of *realm* to designate the highest-level biome-like.

3.4 Towards a Scientific and Universal Typology of Biome-Likes

The Commission on Ecosystem Management (CEM) of the IUCN proposed such a typology in 2020 to objectively determine ecological spatiotemporal entities as appropriate objects of protection [36]. The commission is an international panel of 103 scientific advisers of the IUCN, all distinguished and highly recognised scholars in ecology or environmental sciences sensu lato. A hierarchical classification system (HCS) was adopted after a consensus between 48 researchers who met in a forum at Kings College London in 2017 [36]. The CEM took a philosophical stand based on the Freudian concept of reality principle and thus accepted the *anthropogenic*, i.e. human-induced biome-likes that gradually and dramatically tend to replace the *natural* ones. Indeed, these artificial biomes nowadays occupy an extensive portion of the Earth's surface.

Humans have exerted an impact on the habitats since prehistorical times. The anthropogenic or natural character of a biome-like is more or less a matter of dose, a principle established by the French ecologist François Terrasson (1939–2006) [83–85]: a biome-like would be recognisable as clearly natural if it results from a spontaneous development independent of human intervention; otherwise, it is anthropogenic. Nowadays, almost all ecosystems, except in Antarctica, have a certain dose of anthropogenic influence; the frontier between anthropogenic and natural is not always clear-cut. The feeling that a biome-like is anthropogenic or natural is a matter of consensus based on empiricism and a culturally and aesthetically accepted feeling that is subjective. Though, in the most conspicuous cases, the distinction is so strongly perceived by most sensible human beings, whatever their culture, via common sense and intelligence, that it becomes objective. Hereunder, we adopted the nomenclature of the CEM [36].

The HCS is a nested structure of six levels, corresponding to categories in decreasing order of magnitude (Fig. 3.4). The categories we use are those defined and described in the report of the IUCN [36]. We summarise this report; the reader is invited to read it to obtain more details. Each level has a given set of ecological traits derived from ecological drivers. All the categories given below follow the terminology of the aforementioned report.

We will focus on the three upper levels. (i) Level 1 corresponds to the category *realm*, with two subcategories: the *global realm*, and the *transitional realm*.[11] (ii) Level 2 corresponds to the category *biome*. (iii) Level 3 corresponds to the category Ecosystem Functional Group (EFG) that we will call *subbiomes* throughout this book.

[11] The phrase *transitional realm* almost looks inappropriate since it refers to intersections of global realms, which are the fundamental realms. However, the transitional realms subsume very typical and specific biomes, which justifies this category.

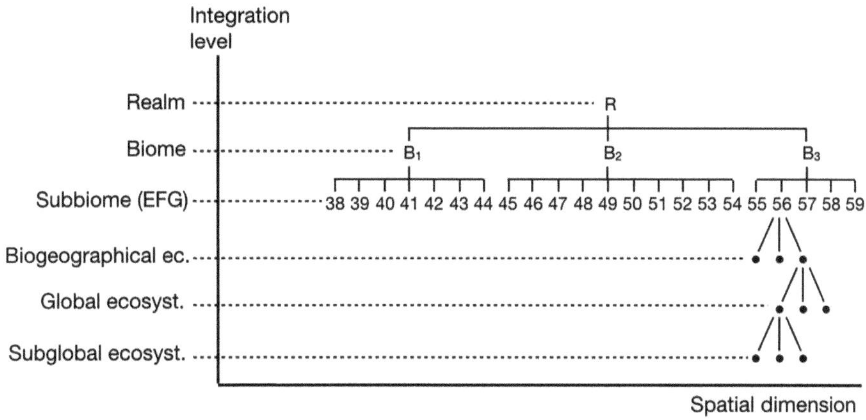

Fig. 3.4 Representation of a hierarchical classification system. Biome-like categories: realm: R; biomes: B_1, B_2, B_3; subbiomes: 38, 39, 40 …; bullets: lower biome-like categories; EFG = Ecosystem Functional Groups; ec. = ecoregions; ecosyst. = ecosystems

The subcategory *global realm* has five modalities:

R_1: the *terrestrial realm,*
R_2: the *subterranean realm,*
R_3: the *freshwater realm* (including saline water bodies on land that are disconnected from the marine realm),
R_4: the *marine realm,*
R_5: the *atmospheric realm.*

- The terrestrial realm (R_1) subsumes seven biomes:

  B_1: *tropical-subtropical forests biome,*
  B_2: *temperate-boreal forests and woodlands biome,*
  B_3: *shrublands and shrubby woodlands biome,*
  B_4: *savannahs and grasslands biome,*
  B_5: *deserts and semi-deserts biome,*
  B_6: *polar-alpine (cryogenic) biome,*
  B_7: *intensive land-use biome.*

- The subterranean realm (R_2) subsumes two biomes:

  B_8: *subterranean lithic biome,*
  B_9: *anthropognic subterranean voids biome.*

- The freshwaters realm (R_3) subsumes three biomes:

  B_{10}: *rivers and streams biome,*
  B_{11}: *lakes biome,*
  B_{12}: *artificial wetlands biome.*

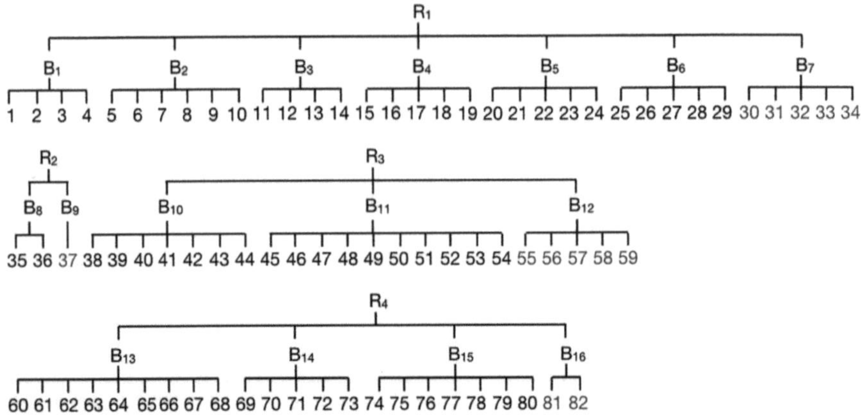

Fig. 3.5 Diagram of the HCS restricted to the global realms. The realm R_5 is not represented since no biome, hence no EFG, is known to be specific to this realm. The numbers from 1 to 82, corresponding to the leaves of the tree, are the EFGs or subbiomes (in black, the natural and in grey, the anthropogenic ones). For the significance of the numbers, see the text as follows

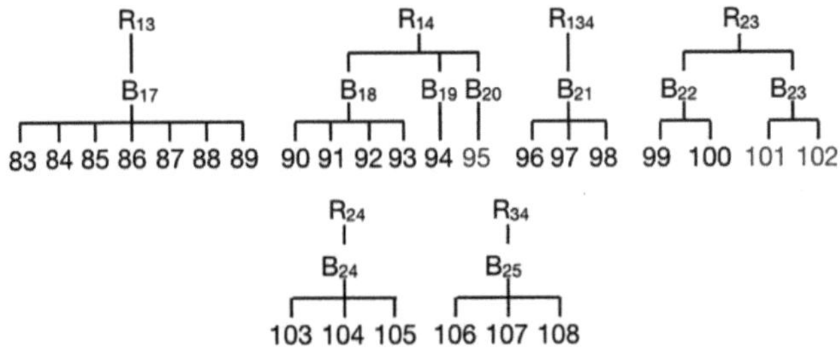

Fig. 3.6 Diagram of the HCS restricted to the transitional realms. The numbers from 83 to 108, corresponding to the leaves of the tree, are the EFGs or subbiomes (in black the natural and in grey the anthropogenic ones). For the significance of the numbers, see text as follows

- The marine realm (R_4) subsumes four biomes:

  B_{13}: marine shelf biome,
  B_{14}: pelagic ocean waters biome,
  B_{15}: deep sea floors biome,
  B_{16}: anthropogenic marine biome.

- The atmospheric realm (R_5) subsumes no specific biome known.

Each biome subsumes a variable number of subbiomes (or ecosystem function group) (Three highest levels of the HCS presented in Fig. 3.5).

We hereunder present a global list of all the subbiomes. The bold numbers identifying them are reported in Figs. 3.5 and 3.6.

1: Tropical-subtropical lowland rainforest; **2:** Tropical-subtropical dry forests and thickets; **3:** Tropical-subtropical montane rainforests; **4:** Tropical-subtropical heath forests; **5:** Boreal and temperate high montane forests and woodlands; **6:** Deciduous temperate forests; **7:** Oceanic cool temperate rainforests; **8:** Warm temperate laurophyll forests; **9:** Temperate pyric humid forests; **10:** Temperate pyric sclerophyll forests and woodlands; **11:** Seasonally dry tropical shrublands; **12:** Seasonally dry temperate heaths and shrublands; **13:** Cool temperate heathlands; **14:** Rocky pavements, lava flows and screes; **15:** Trophic savannahs; **16:** Pyric tussock savannahs; **17:** Hummock savannahs; **18:** Temperate woodlands; **19:** Temperate subhumid grassslands; **20:** Semi-desert steppes; **21:** Succulent or thorny deserts and semi-deserts; **22:** Scerophyll hot deserts and semi-deserts; **23:** Cool deserts and semi-deserts; **24:** Hyper-arid deserts **25:** Ice sheets, glaciers and perennial snowfields; **26:** Polar alpine rocky outcrops; **27:** Polar tundra and deserts; **28:** Temperate alpine grasslands and shrublands; **29:** Tropical alpine grasslands and herbfields; **30:** Annual croplands; **31:** Sown pastures and fields; **32:** Plantations; **33:** Urban and industrial ecosystems; **34:** Derived semi-natural pastures and old fields; **35:** Aerobic caves; **36:** Endolithic systems; **37:** Anthropognic subterranean voids; **38:** Permanent upland streams; **39:** Permanent lowland rivers; **40:** Freeze-thaw rivers and streams; **41:** Seasonal upland stream; **42:** Seasonal lowland rivers; **43:** Episodic arid rivers; **44:** Large lowland rivers; **45:** Large permanent freshwater lakes; **46:** Small permanent freshwater lakes; **47:** Seasonal freshwater lakes; **48:** Freeze-thaw freshwater lakes; **49:** Ephemeral freshwater lakes; **50:** Permanent salt and soda lakes; **51:** Ephemeral salt lakes; **52:** Artesian springs and oases; **53:** Geothermal pools and wetlands; **54:** Subglacial lakes; **55:** Large reservoirs; **56:** Constructed lacustrine wetlands; **57:** Rice paddies; **58:** Freshwater aquafarms; **59:** Canals, ditches and drains; **60:** Seagrass meadows; **61:** Kelp forests; **62:** Photic coral reefs; **63:** Shellfish beds and reefs; **64:** Photo-limited marine animal forests; **65:** Subtidal rocky reefs; **66:** Subtidal sand beds; **67:** Subtidal mud plains; **68:** Upwelling zones; **69:** Epipelagic ocean waters; **70:** Mesopelagic ocean waters; **71:** Bathypelagic ocean waters; **72:** Abyssopelagic ocean waters; **73:** Sea ice; **74:** Continental and island slopes; **75:** Submarine canyons; **76:** Abyssal plains; **77:** Seamounts, ridges and plateaus; **78:** Deepwater biogenic beds; **79:** Hadal trenches and troughs; **80:** Chemosynthetic-based ecosystems (CBEs); **81:** Submerged artificial structures; **82:** Marine aquafarms.

Six interfaces between the global realms have been observed, the *transitional realms*.

- The transitional terrestrial-freshwater realm (R_{13}) subsumes one biome:

  B_{17}: *palustrine wetlands biome.*

- The transitional terrestrial-marine realm (R_{14}) subsumes three biomes:

  B_{18}: *shorelines biome,*
  B_{19}: *supralittoral coastal biome,*
  B_{20}: *anthropogenic shorelines biome.*

- The transitional terrestrial-freshwater-marine realm (R_{134}) subsumes one biome:

B_{21}: *the brackish tidal biome.*

- The transitional subterranean-freshwater realm (R_{23}) subsumes two biomes:

  B_{22}: *subterranean freshwaters biome,*
  B_{23}: *anthropogenic subterranean freshwaters biome.*

- The transitional subterranean-marine realm (R_{24}) subsumes one biome:

  B_{24}: *subterranean tidal biome.*

- The transitional freshwater-marine realm (R_{34}) subsumes one biome:

  B_{25}: *semi-confined transitional waters biome.*

Each biome subsumes a variable number of subbiomes (Fig. 3.6).

83: Tropical flooded forests and peat forests; **84**: Subtropical-temperate forested wetlands; **85**: Permanent marshes; **86**: Seasonal floodplain marshes; **87**: Episodic arid floodplains; **88**: Boreal, temperate and montane peat bogs; **89**: Boreal and temperate fens; **90**: Rocky shorelines; **91**: Muddy shorelines; **92**: Sandy shorelines; **93**: Boulder and cobble shores; **94**: Coastal shrublands and grasslands; **95**: Artificial shorelines; **96**: Coastal river deltas; **97**: Intertidal forests and shrublands; **98**: Coastal saltmarshes and reedbeds; **99**: Underground streams and pools; **100**: Groundwater ecosystems; **101**: Water pipes and subterranean; **102**: Flooded mines and other voids; **103**: Anchialine caves; **104**: Anchialine pools; **105**: Sea caves; **106**: Deepwater coastal inlets; **107**: Permanently open riverine estuaries and bays; **108**: Intermittently closed and open lakes and lagoons.

3.5 Traits and Drivers Taken From Plant Ecology

I will now individually describe the realms, biomes and some subbiomes (SB_1 to SB_{14}),[12] within the hierarchical classification system (HCS) seen in the last section.

Definition 3.5 (Leaf Area Index (LAI)) We will better grasp the concept by a little sketch (Fig. 3.7).

The formal definition as presented in the glossary ("projected area of leaves as a proportion of the area of land compared to which it is measured" as illustrated in Fig. 3.7) is an extrapolation on an entire forest of this definition due to John Monteith and Mike Unsword [55, 56] when the leaves and the ground surface are flat. For these authors,

> LAI is the total one-side leaf area per unit ground surface area

[12] This will give a feel to how analyse subbiomes. The detailed analysis of the 108 subbimes is beyond the scope of this book.

Fig. 3.7 Illustration of the leaf area index (LAI) of a portion of a forest of surface A. Green surface = vertical projection of the canopy surface on the floor (supposed horizontal) = L; grey surface = land surface not under the leaves (supposed horizontal) = $A - L$. The LAI of this forest sample is equal to $\frac{L}{A}$

But in the general case, a probabilistic computation is needed to ameliorate the definition [14].

Definition 3.6 Absolute and relative growth rate.

In plant biology, the size of a plant is often evaluated by its dry mass M, a variable function of t. Let $I = [0, t]$ be a given time interval during which a plant grows, and $M(t)$ the size (dry mass) of the latter at time t.

Absolute growth rate (AGR):

$$\mathrm{AGR} = \frac{dM}{dt}(t) \tag{3.1}$$

Relative growth rate (RGR) [5]: Let $I = [0, t]$ be a given time interval during which a plant grows, and $M(t)$ the size (dry mass) of the latter at time t. Then

$$\begin{cases} \frac{dM}{dt}(t) = r\,M(t), \; r\ (= \mathrm{RGR}) \text{ is the relative growth rate.} \\ M_0 \text{ is the initial mass at } t = 0. \end{cases} \tag{3.2}$$

Theorem 3.1 *A fundamental property of the* RGR *of a plant*
Let $I = [0, t]$ be a given time interval during which a plant grows, $M(t)$ the size (dry mass) of the latter at time t, and $M_0 = M(0)$. Then

$$\mathrm{RGR} = \frac{\ln M(t) - \ln M_0}{t} = \frac{1}{t} \ln\left[\frac{M(t)}{M_0}\right] \tag{3.3}$$

Proof The proof is the object of Problem 3.1. □

Definition 3.7 Specific leaf area (SLA) [68]
Let $\mathcal{L}$ be a fresh leaf, $A(t)$ be its one-sided area, and $M(t)$ its oven-dry mass, then its specific leaf area at time t is

$$\mathrm{SLA}(t) = \frac{A(t)}{M(t)} \tag{3.4}$$

LAI, RGR and SLA, averaged on a sample of plants representative of a given subbiome, can constitute specific descriptors of the latter. The procedure to calculate the average of parameters like RGR and SLA, two strongly related parameters in the field, has been described in the paper of Natalia Pérez-Harguindeguy and coll [68, p. 181]. We will describe its general principle hereunder.

Let $I = [0, t]$ be the interval time between two harvests of leaves, which yield two samples of leaves respectively of (i) mean dry mass $\bar{M}_0 = \bar{M}(0)$ and $\bar{M}(t)$, and (ii) mean one-sided area $\bar{A}_0 = \bar{A}(0)$ and $\bar{A}(t)$. τ being the time variable taking on the time values between 0 and t,

- $\tau = 0$ corresponds to the harvest where, on average, the leaf size is minimal and
- $\tau = t$, to the harvest where the leaf size is maximal on average.

Axiom 3.1 The average SLA of the biome-like will be simply

$$\overline{\mathrm{SLA}} = \frac{1}{2}\left[\frac{\bar{A}(t)}{\bar{M}(t)} - \frac{\bar{A}_0}{\bar{M}_0}\right] \tag{3.5}$$

Now, it is interesting to relate SLA to RGR. It has been shown that these two quantities are positively related. The SLAs of the plants of a given subbiome are very specific to the latter, thus providing invaluable information on its RGR.

Theorem 3.2 *The specific leaf area (SLA) increases as the relative growth rate (RGR) increases.*

Proof This correlation has been shown empirically in the field. A mathematical proof, quite complex, has been carried out in an article [66]. □

Definition 3.8 Raunkiær's universal leaf size typology [72, p. 371]

Raunkiær's leaf classification is based on the empirical construction of six classes that distribute a large number of leaves relatively evenly (cf. Table 3.1). Even though this method is old, dating back from the early 1930s at a time when technology was far less advanced than nowadays, it has proven to be sufficiently robust to be still currently adopted, contributing to the modern classification of biome-likes now adopted by the IUCN [36]. We will now proceed with the description of the biome-likes of the highest ranks (realms, biomes, some subbiomes).

Table 3.1 Typology of size leaves (Raunkiær's scale). Recent authors added the size *notophyll* (2025; 4500 mm^2] [36]

Size classes	Size intervals (in mm^2)
Leptophyll	≤ 25
Nanophyll	$(25;\ 9 \times 25] = (25;\ 225]$
Microphyll	$(9 \times 25;\ 9^2 \times 25] = (225;\ 2025]$
Mesophyll	$(9^2 \times 25;\ 9^3 \times 25] = (2025;\ 18{,}225]$
Macrophyll	$(9^3 \times 25;\ 9^4 \times 25] = (18{,}225;\ 164{,}025]$
Megaphyll	$>164{,}025$

3.6 Description of the Realms [36, Table 3]

3.6.1 *The Terrestrial Realm*

The terrestrial realm (R_1) corresponds to dry land with a vegetation cover. It includes the proximate atmosphere and substrate to the rooting depth of plants, with associated animals and microorganisms. The realm shows the most significant variability between the constitutive ecosystem, for its aspect, taxonomic content and species richness. The terrestrial realm is specifically characterised by a great variability of autotroph types, from very complex and large vascular plants to the most minute photoautotrophic microorganisms. Autotrophs form habitats ranging from dense and extremely stratified to very sparse with no visible vegetation architectonic structure, depending on climate, soil content and resources. The trophic structure is complex: autotrophs supporting multiple heterotrophic levels and long trophic chains (with length and complexity depending on resources).

(i) Energy is non-limiting because its source is mainly solar, save at high latitudes in winter (across the polar circles) and among tree species under the shade of other trees growing faster during their intense competition for light. (ii) *Carbon* is non-limiting, since it is extracted from the air's carbon dioxide. (iii) Oxygen is non-limiting, as a perpetual byproduct of photosynthesis universally carried out in this realm. (iv) Water shows a strong disparity due to climate and topography, with a gradient from large quantities to almost none. This compound is contained in interstices of soil and rocks or obtained through evapotranspiration. (v) Nutrients are dissolved in interstitial water. A gradient of this driver is observed, from saturated to none, due, for example, to leaching or differential evaporation. (vi) Productivity is highly variable along a gradient of nutrients and periodicity. (vii) Temperature displays an extended temperature gradient, ranging from very hot (above 50 °C) to very cold (below −70 °C) along a latitudinal and continental to oceanic gradient, locally altered by topography. (viii) Seasonality is a highly variable driver, between hardly marked around the equator to very pronounced in the continental temperate regions. (ix) Fire is a main ecosystem driver whose probability increases when the climate is hotter and drier and with the flammability of the local habitat and plant species. It strongly influences life histories and habitat evolution.

3.6.2 The Subterranean Realm

The subterranean realm (R_2) contains natural and anthropogenic biomes spatially below the terrestrial realm. It lies within the earth's crust and includes rocks and sediments with interstices sheltering prokaryotic life and subsurface voids. Hardly any light penetrates this realm; life's energy source is thus necessarily chemical (from redox systems). It is sometimes in contact with the transitional subterranean freshwater realm (R_9) when standing water bodies form at their contact.[13] Within realm R_2, the producers are all prokaryotes practising chemolithoautotrophy (thus chemotrophs, lithotrophs and autotrophs, lithotrophs meaning that the source of their electrons is not an organic compound, either solar energy, or redox). The trophic structure is very simple, with most heterotrophic organisms being prokaryotes practising osmotrophic heterotrophy, phagocytic protists and symbiotic tiny invertebrates. The aspect of this realm is quasi-lunar and mineral.

3.6.3 The Freshwater Realm

The freshwater realm (R_3) senso latu corresponds to the (i) freshwater water bodies and (ii) saline water bodies not directly connected to the oceans, carving their beds on dry land. The hydrological regimes are characterised by flow velocity, frequency and duration (permanent or temporary) and the extent of inundation. The realm regulates allochthonous inputs of water and nutrients from catchments. It includes its benthos. The diversity of autotrophic types and structure complexity are variable and strictly depend on the size and connectivity of the water body. That has a direct impact on the trophic structure. In rivers, for example, the general trophic pathway is generally composed of four levels; "Plants (as primary producers, level 1) → Aquatic insects or non-predatory fish (as primary consumers, level 2) → Invertebrate predators (as secondary consumers, level 3) → Vertebrate predators (as tertiary consumers, level 4)" [97]. In many cases, level 2 can proceed to level 4.

(i) Energy sources are diversified. Upper levels are euphotic, lower levels mesophotic when the water body is deep,[14] and, in sporadic cases, aphotic.[15] Energy availability decreased with increasing turbidity gradient. (ii) Carbon is non-limiting. The CO_2 molecules dissolved in water give the carbon atoms to the immersed autotrophs; the atmospheric carbon dioxide is the carbon source of the autotrophs on and above the surface. The carbon source for the heterotrophs (detritus, plankton, prey) is carried by the inflow or mixing of the current. (iii) Oxygen is non-limiting, diffused in water available to plankton, nekton and several

[13] Thus, these underground stretches of water and subterranean rivers do not belong to realm R_2.

[14] Mean depth of Earth's lakes: 41.8 ± 0.6 m [11].

[15] ...notably at the lowest layers of the deepest lakes in the world (depth > 1000 m), e.g. Lake Baikal [10]. For the definition of the italicised terms, see glossary.

aquatic plants, and in the air, at the disposal of most air-breathing plants and animals adapted to the realm. (iv) Water is sometimes a limiting driver, as its action depends on the efficiency of catchments as regulators (from precipitation or groundwater). (v) Nutrients sometimes constitute a limiting driver. Indeed, they are dissolved in waters—autochthonous, extracted form from sediments, or allochthonous, drained from catchments. (vi) Productivity is variable and depends on the gradients of nutrients and energy, flow regime and turbidity level. (vii) Temperature varies weakly locally; the mean temperatures vary along climatic and depth gradients. (viii) Seasonality influences flow and filling/drying regimes with a strong impact on water and nutrients. (ix) Fire can occur at the level of catchments, impacting water and dissolved nutrient fluxes and turbidity and sedimentation rates. (x) The geomorphology of the terrestrial realm determines the extent of the catchments that form water flow direction and velocity, which directly influences water and nutrient supplies and flood regimes.

3.6.4 The Marine Realm

The marine realm (R_4) encompasses all interconnected saline ocean waters. It is characterised by the dynamic interplay of waves, tides, and currents, which play a crucial role in the transportation of resources and biota over short temporal and local spatial scales. Deep currents, in contrast, have facilitated global circulation for centuries. The realm also includes its benthos. The diversity of autotrophs depends on the distance from the coast and depth. The trophic structure is complex near the shore and on the continental shelf; we will see it with our example of a subbiome, the kelp forest (cf. Chapter 5.) In colder waters, at deeper ocean layers or close to the pole, the trophic structure is simplified (food chain length reduced to 2 or 3), which has a direct impact on the *trophic structure*.

(i) The energy source is either solar or of chemical redox origin. Light diminishes with depth. The upper layers are *euphotic* above ca 50 m deep on average, though the variance around this value is important from one ocean to another [47]. The medium layers, between 50 and 150 m on average, are *mesophotic*, witnessing a decreasing gradient of solar energy availability and an increasing role of chemical energy, depending on the sea [13]. Below 150 m, the ocean zone is *aphotic*, with an energy source for the producers, which is only chemical. **(ii)** Carbon is derived from the CO_2 dissolved in the sea and ocean water at the disposal of the photoautotrophs at the upper and chemoautotrophs at the lower layers. The geomorphology of the sea bottoms influences the lateral and vertical flux of organic carbon available to the heterotrophs. **(iii)** Oxygen is diffused in water in decreasing concentration as depth increases. It is at the disposal of plankton and nekton above a minimum concentration level; below a certain depth, it is rarefied, and the organism needs specific adaptations. In hydrothermal vents at very great depths where the concentration of diffused oxygen is negligible, this chemical might follow two pathways to be consumed by heterotrophs, in particular, animals: (a) thanks to anatomic and biochemical adaptations to very little oxygen concentration as is the case, for example, to some deep-sea hydrothermal scale worms (Annelia, Polynoidae) [45], or (b) via symbioses with chemoautotrophs as is the case for several species of vent crustaceans, clams or polychaetes [16]. **(iv)** Water is not a limiting driver; aquatic organisms actively absorb salted water and regulate the salt in their inner bodies by excreting the excess salt. **(v)** The concentration of nutrients diminishes with depth; thus, nutrients is sometimes limiting. The ratio C :N: P within ocean phytoplankton, called the Redfield ratio, is approximately constant in all oceans and at various depths, i.e. 106: 16: 1 [74]. **(vi)** Productivity decreases with increasing depth save in the aphotic zone next to the hydrothermal vent, where it is high. Its level also varies according to ocean circulation. **(vii)** Temperature decreases with increasing depth or increasing latitude, though the variance of this variable is much less than on dry land. This parameter influences the productivity and growth of organisms. **(viii)** Seasonality impacts productivity at surface layers, influencing the vertical flux of nutrients and carbon through the water column to the benthos. **(ix)** Fire is rarely influential except on some shorelines. **(x)** Bathymetry[16]: the morphology of the sea bottom influences currents and habitat structure, hence nutrients, carbon, oxygen and biotic processes, impacting productivity.

3.6.5 The Atmospheric Realm

The atmospheric realm (R_5) has been the least studied. No specific biomes or subbiomes have been circumscribed yet. However, it is a critical transitional habitat, namely the seat of an intense circulation of matter, energy and organisms between the different realms. Importantly, it is one of the main seats of long biogeochemical cycles. The atmospheric realm consists of a gas medium with aerosols, i.e. suspended liquid or solid particles, extending above the other realms up to the altitudinal limit of life. All the biological species living for an extended period in the air are attached to the terrestrial, freshwater or marine realm, and this is due to behavioural needs or simply to gravitation. They are distributed between (i) relatively large and actively flying animals (birds, bats, insects) and (ii) the aeroplankton, particulate organisms passively carried away in the aerosols by winds or aerial currents. In the latter species category, we can find spores, pollens or seeds of algae, mosses or liverworts—or airborne microbes (viruses, bacteria, archaea, protists and fungi [12]). Because plants circulate in this realm as seeds or spores, they are dormant and thus do not practise autotrophy. Consequently, there is no trophic interaction; the heterotrophs circulate in the air depending on their food in other realms. They are either in transit in this habitat or passively transported as particles circulating without precise interactions between them.

[16] Bathymetry is the geomorphology of the sea bottoms.

3.7 Description of the Biomes

As a reminder, biomes are biome-likes that are defined empirically to match the classification level fixed by Whittaker (see Definition 3.4, p. 37). However, contrary to the typology of the latter, which restricted the biomes to the terrestrial realm, the modern classification also includes the biomes within (i) the subterranean, freshwater and marine global realms, and (ii) the transitional realms, i.e., interfaces between global realms. The *Assembly Theory* gathers the description traits of the biomes in such a way that they constitute extensive complementary surfaces on the Globe, corresponding to 25 of such units. This classification takes into account the anthropogenic ecosystems that nowadays occupy large portions of the Earth and that most textbooks tend to disregard, probably due to unconscious aesthetic and ideological considerations. Hereunder, I present a schematic description of the biomes and their subbiomes. Precise locations on several of the latter ones are put in Appendix D (SI, p. 1). The biomes will be presented hereunder per realm; some subbiomes will be presented within their biome (in the cases of biomes B1–B3).

3.7.1 *Within the Terrestrial Realm* R_1

3.7.1.1 The Tropical-Subtropical Forest (B_1)

The *tropical-subtropical forest biome* is determined by hot weather throughout and wet condition during a large part of the year. It approximately spans the intertropical belt (with several projections outside, e.g. Southern Madagascar). Though diverse, their subbiome components share several prominent traits. They all have (i) high species richness represented by relatively few individuals in relatively limited distribution areas and (ii) great biomass. Plants under the cover of larger trees are shade tolerant. Gap-phase dynamics due to the natural uprooting of the trees by natural causes (storm, lightning, trampling by animals) allow the opening of the forests and the possibility for saplings to grow (the forest thus displaying a typical multistorey landscape). The biome comprises four subbiomes detailed hereunder.

SB_1: The tropical-subtropical lowland rainforests subbiome:

- Distribution:
  This subbiome is a large discontinuous belt around the equator, between the Tropics of Cancer and Capricorn[17] and several extensions slightly outside the intertropical zone, in America, Africa, Asia and Oceania. It covers less than 6% of Earth's land surface. SB_1 is one of the most endangered subbiomes of the planet. In normal situations, fire is a very exceptional driver. Still, in the

[17] As a reminder, the tropics are the two parallels of latitude 23°26' (N for the Tropic of Cancer and S for the Tropic of Capricorn).

last decade, anthropogenic fires largely consumed a portion of the Amazon forest (research is now carried out to evaluate the impacted surface). For the locations of representatives of the subbiome, refer to Appendix D.1 (SI, p. 1).

- Climate:
  - Perpetually wet and warm, a phenomenon due to (i) its intertropical position, with an incidence of the solar rays quasi-perpendicular to the ground and (ii) evapotranspiration rejecting the excess of water pumped by trees from their roots to their direct air environment.
  - Low temperature variability per year and rainfall above 2000 mm annually.
- Heavy rainfall in the evening almost every day
- Plant composition: (i) Mainly evergreen and semi-deciduous species. Very high trees growing very fast. (ii) The canopy is large and multi-layered (up to six layers with several emergent trees higher than the general canopy and piercing it) and displays a complex structure. (iii) The canopy is closed (very high LAI); (iv) Evergreen trees have a wide range of leaf styles, which enables plants in the shade to capture light, even if the latter is dim. (v) Next to giant trees forming a continuous and closed canopy, bamboos, epiphytes, lianas and ferns, but grasses and hydrophytes are absent or rare. (vi) Gap phase (canopy disruption) initiated by tree fall or by tropical storms.
- Biodiversity:
  (i) Animal, as well as plant species richness, is higher than any other subbiomes, teaming with insects; large number of fungi and microbes. The diversity is great, especially in the canopy. This subbiome shelters half of the world's species. (ii) The biodiversity is highest in the forest canopy. (iii) Trophic networks complex and stratified. (iv) Mammals and birds play critical roles in plant diaspore dispersal and pollination.
- Key ecological drivers (quoted from [36, p. 39]):
  (i) Precipitation exceeds evapotranspiration with low intra- and interannual variability, while closed tree canopies maintain a humid microclimate and shade. (ii) Low diurnal and seasonal variation of temperature. (iii) Soil moist, but neither inundated nor peaty. (iv) Nutrients sequestered in vegetation or cycled through dynamic litter layer, no leaching, no run-off. (v) Decadal regimes of tropical storms drive cycles of canopy destruction and renewal of plants that were in the shadow of former higher trees, thus a possibility to reach sunlight.

SB_2: The tropical-subtropical dry forests and thickets subbiome:

- Distribution:
  This subbiome is a relatively continuous and broad zone around the equator, between the Tropics of Cancer and Capricorn, but sometimes projecting slightly outside the tropical zone, notably in South America and Madagascar and northwestern Mexico. Much broader than subbiome B_1. For the locations of the representatives of the subbiome, refer to Appendix D.2 (SI, p. 1).

- Climate:
  (i) The climate is warm year-round but is marked by dry and wet seasons lasting globally 4–7 months per year (one or two wet seasons per year). (ii) Compared to subbiome SB_1, the variability of temperature per year is greater due to the seasonal contrast; annual rainfall height is between 500 and 2000 mm.
- Plant composition:
  (i) Mainly broadleaf deciduous species that shed their leaves at the beginning and remain leafless at the rest of the dry season. Grow leaves at the beginning of the wet season in frequent heavy rains. Many trees have cuticle leaves retaining water in their tissues during the wet season (as is the case of teck and mountain ebony). (ii) The canopy is large and multi-layered during the wet season; during the leafless dry period, the sunlight reaches the soil and facilitates the growth of saplings and thick brushwood. (iii) The canopy is closed during the wet season (high seasonal LAI). (iv) Many plant seeds are dispersed by winds, rarely by vertebrates. Nutrients sequestered in litter layers, which is abundant and contain many decomposers and mineralisers (saprophagous fungi and diverse microorganisms). (v) Grasses rare or absent. Epiphytes, ferns and bryophytes limited in the wet period. (vi) Contrary to subbiome SB_1, taxonomic diversity is low in the canopy. (vii) Gap phase initiated by tree fall.
- Biodiversity:
  (i) Animal and plant species richness is high, lower than in subbiome SB_1 but with many herbivorous, frugivorous and carnivorous vertebrates, building up a complex trophic network.
- Key ecological drivers (quoted from [36, p. 39]):
  (i) Water surplus or deficit are limiting. (ii) Warm temperatures (rarely below 10 °C) varying little daily or seasonally are good drivers. (iii) Soil moist, but neither inundated nor peaty. (iv) Nutrients sequestered in vegetation or cycled through dynamic litter layer, no leaching, no run-off. (v) Tropical storms may be important disturbances, but flammability is low.

SB_3: The tropical-subtropical montane forests subbiome:

- Distribution:
  (i) The tropical-subtropical montane forests, also called cloud forests, are scattered among the intertropical mountains, at the montane altitudinal zone (between 800–1500 and 2000–3500 m) [51, 73]. The subbiome spreads above the rain- or semi-deciduous lowland forests, showing a gradient of diverse habitats adapted to increasingly more temperate climates when altitude increases (reaching even cold temperatures for the highest mountains). (ii) They cover about 2.5% of all tropical forests [73]. (iii) They are scattered in 59 countries [41% in Mexico, Central and South America, and the Caribbeans; 16% in East and Central Africa and 43% in Asia and Oceania (India, Sri Lanka, South-Eastern Asia, New Guinea and Hawaii)] [32, 87].

- Climate:
  (i) Gradient of cloud condensation increasingly persistent with increasing altitude; correlatively, temperature decreasing from 20 to 8 °C. The solar radiation is low. (ii) Annual rainfall height is very variable in this subbiome, ranging from 500 to 10,000 mm [73]. Permanent moisture. (iii) Due to these climatic properties, important forest tracts encompass peatlands, rendering the soil very acidic with low rates of decomposition and mineralisation. (iv) From the upper areas of the cloud forest, where clouds permanently cover the soil, fog drips adhere to the needles or leaves of trees and coalesce into larger drops that fall to the ground; these drops can foster water run-offs that contribute to the general water cycle.
- Plant composition:
  (i) Compared with subbiomes SB_1 and SB_2, typically, the trees of subbiome SB_3 are lower (between 5 and 20 m high) and more heavily branched and stemmed, with gnarled trunks and branches. Not remarkable at the lowest parts of the subbiome, these properties reinforce fast with increasing altitude. The trees grow small, hard, with cuticle-covered leaves, and are evergreen. Like SB_1 and SB_2, the canopy is closed but, contrary to the latter, single-layered. These traits typically characterise elfinwoods. (ii) The constant humidity in the air and soil favours great biomass and biodiversity of bryophytes (notably bulky hanging mosses), lichens, ferns, bromeliads and orchids. There are a lot of endemic plants. (iii) Regularly in the lower layers and permanently in the higher layer of the subbiome, the trees are in contact with clouds and mists, so their leaves are always soaked with water.
- Biodiversity:
  (i) The climatic and botanical traits mentioned above have favoured great species richness and high level of endemism. Taxonomic diversity though decreases and trophic structure becomes simpler with increasing altitude. (ii) The productivity is moderate, fuelled by autochthonous energy. (iii) Many animal and plant taxa are endemic, most of them in danger of extinction (e.g. the quetzal *Pharomachrus mocinno* in Central America, the mountain gorilla *Gorilla gorilla beringei* in central Africa; the spectacled bear *Tremarctos ornatus* in northern and western South America) [32]. (iv) Trophic webs are as complex as in subbiomes SB_1 and SB_2, with trophic chain length being at least 4. (v) High local endemism, especially among plants, invertebrates, amphibians and birds at the highest level of the subbiome.
- Key ecological drivers (quoted from [36, p. 39]):
  (i) Cloud moisture, controlling water circulation and evapotranspiration, is unlimiting; cloud cover from orographic uplift and closed tree-canopies helps moisture and shade to be constant. (ii) Altitudinal gradients in temperature drive the aspect of the forest from a quasi-lowland aspect up to a typical cloud forest aspect (see description above). (iii) Cool to cold temperatures and occasional frost impose the frailer structure, with thicker bark and more stemmy aspect. (iv) Large pools of inorganic nitrogen are efficient drivers, enhancing primary production [22].

SB_4: The tropical heath forests subbiome:

- Distribution:
  Tropical heath forests subbiomes are scarce and localised, only known from two areas: (i) the Rio Negro Campinarana region of the Amazon Basin (NE Brazil, E Columbia), and (ii) the Sundaland heath forests in Borneo and neighbouring islands. These forests are adapted to a strong nitrogen deficiency and strong soil acidity, which have important consequences for the vegetation.
- Climate:
  Equatorial, thus (i) warm and moist, weakly seasonal; (ii) drenched with water due to the equatorial regime (abundant rainfall every evening) that is little absorbed in the ground, resulting in leaching and run-offs.
- Plant composition:
  (i) Trees are smaller than in the other tropical forest formations (height 5 to 20 m), forming diverse types of canopy, from completely closed to open. Leaf sizes are generally smaller than in the subbiomes SB_1 and SB_2. (ii) Generally, in comparison with the other subbiomes of biome B_1, the tree stems are much slenderer (diameter < 20 cm) and, not uncommonly, are twisted. (iii) The LAI is generally lower than in the subbiomes previously mentioned. (iv) Due to the very acidic soil and paucity of nitrogen in the soils, the vegetation is highly adapted, and endemic [71].
  In the Sundaland heath forests, for example, a number of plants, e.g. *Gymnostoma*, developed, within root nodules, symbiotic relationships with nitrogen-fixing bacteria (e.g. *Rhizobia*); other plants, e.g. *Hydnophytum*, get their nitrogen from associations with ants; a third group consisting of carnivorous plants, e.g. sundews *Drosera*, are partially heterotrophic, trapping and digesting small arthropods.
- Biodiversity:
  Compared with the other forests mentioned: (i) Plant and animal biodiversity is much lower, but the rate of endemism is proportionally higher. (ii) Epiphytes are common, but lianas are rare, and ground vegetation is less dense; (iii) The trophic network is simpler; (iv) There exist proportionally larger numbers of insects feeding the carnivorous plants
- Key ecological drivers (quoted from [36, p. 39]):
  (i) Podzol type typical of hyperacidic soil, which is a limiting factor, prevent the soil from retaining nutrients, in particular water surplus, which is leached and runs off, and nitrogen. (ii) As a result, the ecosystems are subject to frequent drought, worsened by frequent floods occasioned by the many run-offs. (iii) This occasions periodic anoxia among the roots in relatively large areas, juxtaposing waterlogged peat forest tracts and palustrine wetlands.

3.7.1.2 The Temperate-Boreal Forests and Woodlands (B_2)

The temperate-boreal forests and woodlands biome covers a vast climatic zone from warm-temperate (e.g. Mediterranean type) to boreal areas. Compared to biome B_1, two features can be observed here: (i) B_2 is the home for fewer species, but each is characterised by more individuals per surface unit and occupies a more extensive distribution area. (ii) Its productivity is far more variable between the seasons [very cold to cold in winters, with weak productivity and unfavourable to reproduction, generally accompanied by dormancy, hibernation or migration out of the zone; chilly to hot (sometimes torrid) (summers) with high productivity and usually favourable to the return of the migrating animals, active life and reproduction] [36, 89]. The biome subsumes six subbiomes.

SB_5: **Boreal and temperate montane forests and woodlands subbiome:**

- Distribution:
  (i) Boreal forests (or taiga) span at the high latitudes of the Northern hemisphere (Northern America and Eurasia, e.g. Canada, Scandinavia, Russia) between the Arctic tundra in the north and the deciduous temperate forests or grasslands in the south. (ii) Temperate montane forests grow in the montane zone of mountains of the temperate latitudinal zone (e.g. Alps or Pyrenees or Carpathian Mountains in Europe, Rocky Mountains in N America), corresponding to the mid-elevation lying below the tree-line, above which treeless zones stretch up to the snow zone, and above the foothill zone which belongs to subbiome SB_6.
- Climate:
  (i) The taiga, as well as the montane temperate forests, share a number of features that cause relatively similar types of trees to grow in these two subbiomes, hence their grouping in the same entry. They are subject to very cold temperatures during relatively long to very long winters, have brief springs and autumns, and short summers that can vary from cool to moderately hot. These features are reminiscent of the idea that altitude recapitulates latitude. However, this idea needs to be nuanced. (ii) There are notable differences. During the four months around the winter solstice, in the taiga above the polar circle, the sun never rises, and the subbiome is plunged into darkness. Slightly below it, in the same period, a slight sunrise at the horizon level exists for one or two hours, thus a very dim light. This phenomenon does not occur with the montane temperate forests which are at midlatitudes, thus with much slighter seasonal daylight variation between the two solstices. This naturally impacts the biocoenoses in terms of species.
- Plant composition:
  (i) The taiga and the montane temperate forests are both composed of a majority of needle-leaved evergreen conifers and a minority of broad-leaved trees. (ii) However, due to the characteristic climatic difference mentioned above, the boreal forests are far less diverse characterised by dominant trees

as we shall see in the point below. (iii) Locally, due to the drivers described below, local conditions can favour peat formation, thus forested bogs.

- Biodiversity: (i) In comparison with all the other subbiomes of biome B_2, the species diversity of the taiga is much lower. Coniferous trees are dominant but represented by a very small number of genera, namely *Picea* (spruces), *Abies* (firs), *Pinus* (pines), *Larix* (larches), *Tsuga* (hemlocks) and a species of cedar, *Cupressus nootkatensis* (the Alaskan cedar);[18] and angiosperm trees are mainly represented by aspens (*Populus*) and birches (*Betula*). (ii) But the vegetation of the taiga, cold resistant, supports a larger number of migratory animals than, for example, the temperate montane forest. The boreal forests are home of emblematic animals like the moose (*Alces alces*) or the grey wolf (*Canis lupus*), but these animals can also be found in a number of temperate montane forests like in the Rocky Mountains in the US. (iii) In both subbiomes, small and large herbivores and omnivores (hare, bear, deer and insects) are found during summer. (iv) Montane forests, though phenotypically similar, are more diverse, with a greater variety of conifers and angiosperm trees, the same genera of taiga, but represented by fewer species, to which other genera can be added e.g. junipers, true cedars or alders.
- Key ecological drivers (quoted from [36, p. 39]):
  (i) Large seasonal temperature ranges; (ii) cold winters with large number of severely freezing days, with prolonged winter snow; (iii) short growing seasons (1–3 months with mean temperature > 10 °C); (iv) severe post-thaw frost; (v) overall water surplus and little annual precipitation + soil acidity in several localities favouring peat.

SB_6: Deciduous temperate forests subbiome:

- Distribution:
  The area of the deciduous temperate forests subbiome is very extensive in the northern hemisphere and comprises three main disjoint areas situated in (i) Europe (and two eastward extensions), (ii) East Asia and (iii) Eastern North America.

  (1) In Europe, subbiome SB_6 occupies most of the temperate part of the continent south of the boreal forest subbiome (SB_5) and north of a set of heterogeneous and rugged Mediterranean subbiomes (SB_{10}, SB_{12}, SB_{14}, SB_{18} and SB_{19}); this subbiome is bounded by
  (i) Southern Scandinavia and the Gulf of Finland in the north; (ii) Northern Iberia, the Mediterranean hinterland foothills (with the exception of the Alpine Arch) and a line in the south of the Balkan peninsula corresponding to northern Greece and European Turkey in the south; (iii) the British Isles and the Atlantic coast of France and Iberia in the West; (iv) the central part of the Russian plain in the East.

[18] Another "cedar", the so-called Siberian cedar, is an emblematic tree that despite its cedar aspect is actually a pine with its scientific name *Pinus sibirica*.

Eastward extensions are observed: one through Asia Minor to the foothills of the Zagros and Caucasus mountains, and another wedge-shaped one further north penetrating deeply into Central Russia, as far as western Siberia.

(2) In Asia, the main part of subbiome SB_6 spans in East Asia (Northeastern China, southeastern Siberia, Korea and Japan). Deciduous temperate forest tracts can also be found in the foothills of the Himalayas.

(3) In eastern North America, subbiome SB_6 covers a territory between the Atlantic Ocean and roughly a line between Minnesota and Luisiana[19] [18, 19].

- Climate:
  (i) Characterised by four seasons of approximately equal lengths: spring, summer (the hottest), autumn and winter (the coldest). A climatic cline due to the relative influences of oceanic and continental climatic drivers (different regimes between the three disjoint areas of the subbiome). In continental climates, the seasons are more marked, less buffered, than in the oceanic climates. (ii) Cold winters but not as severe ones as those characterising the boreal forests (coldest peaks averaging −30 °C, temperatures around or slightly above 0 °C frequent); hot summers but less torrid or scorching than in tropical or subtropical subbiomes (hottest maxima averaging 30 °C, generally between 15 and 25 °C). Average annual temperature ca 10 °C; annual rainfall between 750 and 1500 mm (data from NASA earth observatory, site: https://earthobservatory.nasa.gov/biome/biotemperate.php).
- Plant composition:
  (i) The subbiome is dominated by broad-leaved trees that periodically shed their leaves in autumn, being thus leafless in winter, grow their new foliage in early spring and are leafy the rest of the year. (ii) The leafy period lasts, on average, six months between May and October. This duration is variable, decreasing with increasing latitude (between 120 and 250 days). (iii) According to latitude and continentality influences, and to the continent where the subbiome is situated, vegetation formations vary in their species composition. But one finds the same genera or subfamilies of trees, e.g. maples, oaks, birches, beeches, or ashes.
- Biodiversity:
  - This deciduous temperate forest is highly productive in summer though not as much as the tropical-subtropical subbiomes.

[19] The deciduous temperate forests of North America used to cover 2,500,000 km^2 before the colonisation by Europeans; but since the eighteenth century, these ecosystems have been widely anthropised. However, an effort is made to conserve large samples of this habitat, still well preserved, in the frame of the parks of National Capital Region Network (NCRN, site: https://www.nps.gov/im/ncrn/eastern-deciduous-forest.htm).

- This high production though limited in time does not prevent the subbiome from sustaining a relatively rich and persistent fauna (even in the coldest part of the year):
  - The trophic network is multistorey: (i) For example, restricted to mammals, it contains as primary consumers large herbivores like cervids or small ones like rodents or lagomorphs, as secondary consumers omnivores like bears or the wild boar, small insectivores like shrews, or carnivores like canids (wolves, foxes, etc.) or felids (lynxes, wildcats). The latter can occasionally be tertiary consumers when they feed on insectivores that are frequently small-sized. (ii) If we add the birds to the mammalian fauna, we can count, as consumers 1, granivores like many passerines (buntings, sparrows); as consumers 2, for example, gulls, strigiformes and accipitriformes that have a strong impact on the ecosystem (notably by efficiently regulating the populations of rodents). If we add all the other vertebrates and invertebrates of the subbiome, the food web becomes immensely intricate.
  - All these animal species have developed strategies to overcome the climatically unfavourable seasons, and thus remain in the subbiome. (i) Most invertebrates are diapausing in the coldest part of the year (spring and winter). Diapause-inducing temperatures, thus the onset of this phenomenon, vary from one species to another [24]. (ii) Vertebrates have different strategies: hibernation for a number of mammals; migration for migratory animals [in particular birds, a number of fishes, rarely terrestrial mammals (unlike in the tropical subbiomes)]; dense pale-coloured winter fur or plumage for the warm-blooded vertebrates; food caching and winter foraging are common; etc. This diversity of strategy is specific to the subbiome.
- Key ecological drivers:
  The main drivers are:
  - large seasonal temperature range, from ca −1 °C (in winter) to 22 °C (in summer) (mean values);
  - substantial winter snow;
  - limiting growing season lasting 4–6 months;
  - constancy of water disposal;
  - great deciduous leaf turnover fostering the formation of humus, the basis of fertile and relatively deep soils in favour of broad and solid tree roots.

SB$_7$: Oceanic cool temperate rainforests subbiome:

- Distribution:
  Disconnected narrow areas in diverse continents. There are fundamentally three kinds of oceanic cool rainforests: two in the temperate climatic zone, at low elevation and limited to a long and narrow strip along oceanic shorelines, and one in the tropical climatic zone in mountainous areas at high altitudes.

In the temperate climatic zone:

– the Northern Hemisphere, mainly in North America, the Pacific Northwest coastline from Central California to Southern Alaska;
– the Southern hemisphere, namely in (i) South America with the hinterland of the Pacific coastline from central Chile to southern Patagonia, and (ii) Australasia, with the eastern coastline of Australia, Tasmania and New Zealand.

In the tropical climatic zone:
Localised in the New Guinea Highlands, where a central range montane rainforest extends above 1500 m elevation.

- Climate:
Cool temperate temperature with the coldest month, on average, below −3 °C and the warmest, on average, above 10 °C (between 4 and 12 °C) [1,43]. Great diversity of annual rain precipitation depending on local conditions, maximal in winter months; mean precipitation height on average above 1400 mm [1].
- Plant composition:
(i) This subbiome is composed of evergreen and semi-deciduous trees and shrubs. (ii) The rainforests are a mix of broad-leaved and needle-leaved trees (the latter being dominant). (iii) Canopy relatively closed (high LAI, up to 95%) during the whole year. (iv) High SLA but lower than in the temperate deciduous forests (subbiome SB_6).
- Biodiversity:

  – Tree diversity is low; endemism is strong ; many trees shelter abundant epiphytes and are covered by many types of bryophytes. Grasses are rare.
  – The subbiome in question is identifiable by its physiognomy and the commonness of its drivers, but each part of it, depicted above, is characterised by very specific taxonomic compositions:

    ◦ In the temperate North Hemisphere:
    (i) A number of emblematic conifers characterise the subbiome, e.g. the redwoods (genera *Sequoia, Sequoiadendron* and *Metasequoia*),[20] the Douglas fir *Pseudotsuga menziesii*, the Sitka spruce *Picea sitchensis*, the Pacific yew *Taxis brevifolia*, the western hemlock *Tsuga heterophylla*, etc. (ii) These conifers are associated with bigleaf maples, and with alders along stream banks. These angiosperms contribute to the fixation of nitrogen within the ecosystem. (iii) Nurse logs constitute a phenomenon typical of this habitat: it results from the natural fall of trees, whose dead organic matter fosters the growth of seedlings in the cleared areas.

[20] In particular, the coast redwood *Sequoia sempervirens*, the world's tallest tree.

  - In the temperate South Hemisphere:
    (i) One angiosperm tree genus, *Nothofagus*, is exclusive and common, never found north of the equator. This genus can be deciduous, semi-deciduous or evergreen. (ii) In addition, the conifers found there are very specific, different from those found in the Northern Hemisphere, and also more varied than their northern counterparts [21]. They are mainly represented by three very ancient families that are traced back to the Triassic, the Araucariaceae, Cupressaceae and Podocarpaceae [41, 63].[21]
  - In the tropical South Hemisphere:
    (i) The upper montane forests, extending from 1500 to 2500 m in elevation, are dominated by moss-covered *Nothofagus*; above 2500 m, different conifers are found that are typically austral like *Podocarpus* or *Araucaria* for example. This part of the subbiome shows high animal endemicity.

- The food web is less complex, with large herbivores and predators less diverse and abundant than in subbiomes SB_5 or SB_6. Vertebrates are mostly sedentary and show low diversity.
- Gap-phase dynamics are driven by individual tree falls, lightning strikes and storms.

• Key ecological drivers:
(i) Climate is largely regulated by oceanic influences, which brings climatic moisture compensating for relatively low precipitation. (ii) Frequent cloud cover limits light in winter, mainly at high latitudes. (iii) Soils are moderately fertile, hence frequent accumulation of peat. (iv) Fires are rare.

SB_8: Warm temperate laurophyll forests subbiome:

• Distribution:
Patchy distribution in different disconnected regions in the warm temperate climatic zone between 26° and 43° latitudes of both hemispheres [36, p.47]:
In the Northern Hemisphere:

  - North America:
    (i) *Mexico*: Sierra Madre Occidental, Sierra Madre del Sur, Eje Volcánico Transversal; (ii) *United States:* from Virginia southward to Florida, along the coast of Atlantic Ocean and Gulf of Mexico, eastward to Texas and along the southern banks of the Mississippi [7].
  - Macaronesia:
    The woody parts of the Azores, Canary Islands and Madeira.
  - Africa:

[21] ... which has provided arguments to the prior existence of the Gondwana.

South Africa: (i) the Knysna-Amatole montane forests within the Cape provinces and (ii) the Southern Mistbelt forest within the Drakensberg mountains [59].

- Asia:
  (i) *India, Nepal and northwestern Myanmar*: the subtropical montane forests of the Himalayas, Meghalaya and Chin Hills-Arakan Yoma area [96]; (ii) *China:* the Chiangjian plain evergreen forests,[22] the Guizhou broadleaf and mixed forests,[23] the Jiang Nan subtropical evergreen forests[24] and the South China subtropical evergreen forests; (iii) *Vietnam, Laos:* the northern Annamites rainforests (with a mean elevation of 453 m)[25] (iv) *Taiwan:* a subtropical evergreen forest covering the entire island (20% of the area protected), (v) *South Korea:* evergreen forests at the extreme south of the peninsula, (vi) *Japan:* the Tauheiyo evergreen forests on the Pacific side.[26]

In the Southern Hemisphere:

- South America:
  (i) *Chile:* two components between the Pacific coastline and the foothills of the southern Andes, a cooler one i.e. the *Valdivian temperate rainforests* between 37°S and 48°S latitude [29]; and a warmer one, of the Mediterranean type, i.e. the *Chilean Matorral* between 32°S and 37°S [30]. (ii) *Brazil:* Two contiguous areas in the South: the Serro do Mar along the Atlantic coast [60, 70] and the Paraná Plateau [77].

- Climate:
  (i) Warm temperate—mediterranean to wet subtropical. Summers are dry and winters wet. (i) Annual precipitation height between 1200 and 2500 mm.
- Plant composition:
  (i) A mix of evergreen and deciduous trees. The foliage is often leathery and glossy (laurophyll). The leaf sizes range from microphyll to notophyll (cf. Definition 3.8). (ii) Tree canopies have fewer storeys and are more uniform than the main tropical rainforest. They are relatively closed (high LAI). (iii) The SLA values are intermediate.
- Biodiversity:
  (i) The energy source is autochthonous and supports high primary productivity. (ii) The food web is more simple than in tropical forests, with moderate

[22] ...largely converted to rice paddies, now shrunk to minimal surfaces still sheltering the rarefied oaks *Cyclobanalopsis* and *Castanopsis*, and endangered emblematic animal species such as the Chinese sturgeon *Acipenser chinensis* and the Chinese alligator *Alligator chinensis*.

[23] ...almost gone, replaced by agriculture and secondary forests; remaining tracts in protected areas and inaccessible hills.

[24] ...of which only 0.5% is protected.

[25] Which explains why this formation is not tropical.

[26] Extant in remnant areas, close to temples, on steep slopes and in gorges.

local plant diversity and endemism; liana and epiphytes are uncommon or absent. (iii) Rich flying animal diversity (bats, birds, insects) depends on it. Low mammalian diversity but often endemic.

- Key ecological drivers:
  (i) Summer drought is not a limiting factor due to a modest water surplus that is an efficient driver. (ii) Moderate water deficit is relatively rapidly replenished. (iii) Moderate seasonality, relatively long growth period (6–8 months) and a limited number of frost days (temperature never decreasing much below 0 °C) are also an efficient drivers. (iv) Plants are shade-tolerant. (v) Regularly, canopy are locally broken down by tree falls due to old age, storms; which allows clearings to appear favouring young shoots; (vi) Acidic soils, not uncommon, are limiting. (vii) Moist microclimate does not favour fire, the latter being thus a poor driver.

SB_9: Temperate pyric humid forests:

- Distribution:
  This subbiome is restricted to three areas in Australia: (i) a long coastal rim in East and Southeast Australia along the Pacific shoreline, (ii) Tasmania, and (iii) a very localised zone in extreme southwest Australia (forests close to Cape Leeuwin, e.g. Blackwood River, Hilliger, Milyeannnup national parks).
- Climate:
  Wet temperate (cool and humid winters; relatively dry and warm summers); cyclic appearance of droughts within generally moist years.
- Plant composition:
  (i) Forests counting the highest flowering trees on the Earth (40–90 m tall); (ii) moist forests, high LAI; (iii) the canopy is open, letting light penetrate deeply onto the ground, which favours a multi-layered structure and a complex spatial structure under the canopy; (iv) many trees whose foliage compose the canopy are *sclerophyllous*, i.e. composed of hard, leathery and persistent leaves oriented parallel or oblique to direct sunlight;
- understorey generally not sclerophyllous; (v) ground flora shade-tolerant containing ferns, grasses and bryophytes, generally C_3 (i.e. obtaining their carbon via the most common photosynthetic pathway, see Sect. 9.4.3).[27]
- Biodiversity:
  - Plant diversity: these forests contain plant species that have two survival strategies against fire respectively caused by two types of plants [6, 44, 67]: (i) the *resprouters* that "can resprout post-fire from protected epicormic or basal buds and (ii) the *obligate seeders* "killed by fire and rely on soil- or canopy-stored seed banks for persistence" [44]. Due to the openness of the canopy, this forest is highly productive. The tallest trees representing this

[27] Other plants are C_4 (obtaining a proportion of their carbon via osmotrophy from water).

flora are eucalyptuses (e.g. *Eucalytus grandis, E. pilularis, E. saligna*).[28]

- Animal diversity:
  - The trophic networks are complex and regularly distributed between the different storeys of the forests, between the floor and the canopy. They comprise mammals, birds, reptiles, amphibians, and diverse invertebrates. Several animal species move between the strata.
  - We will mention several characteristic animals of this subbiome:[28] (i) an apex predator, the powerful owl (*Ninox strenua*), the largest owl in Australia mainly feeding on medium-sized marsupials; (ii) the eastern bristlebird (*Dasyornis brachypterus*), a small semi-flightless endemic passerine bird (listed as a threatened species)—mainly insectivorous but eating sometimes plant seeds [23]; (iii) a nocturnal small marsupial with the superficial look of a gerbil, the northern bettong (*Bettongia tropica*)—a herbivore listened as an endangered species; (iv) the yellow-bellied glider (*Petaurus australis*), a relatively large arboreal, nocturnal marsupial that has an omnivorous diet, notably eating nectar, pollen and honeydew as well as insects; (v) a tiny endemic rodent, the Hastings river mouse (*Pseudomys oralis*) strictly granivorous [9]; (vi) the Fleays barred frog (*Mixophyes fleayi*), a large Anuran species associated with flowing streams.

- Key ecological drivers:
  (i) Annual water surplus are favourable factors; (ii) fire is an important factor driving this type of habitat, a factor favoured by the heatwaves occurring during the droughts; (iii) growing seasons last 6–8 months; snow is very rare.

SB_{10}: Temperate pyric sclerophyll forests and woodlands subbiome:

- Distribution:
  This subbiome characterises areas subject to a Mediterranean climate. It is disconnected in three distinct parts of the world, i.e. North America, the Mediterranean Basin, and Australia.
  North America:
  Southern California, a coastal rim from Conception Point to the north of Baja California.
  Mediterranean basin:
  (i) Continental Spain except for the northern coastal plain along the Atlantic Ocean and mountainous hinterland, (ii) Central and Southern Portugal, (iii) the Balearic Islands, coastal Corsica, Sardinia and Sicily, (iv) the

[28] Species referred to in the website: www.qld.gov.au/environment/plants-animals/habitats/habitat/wet-sclerophyll-forest.

shores of mainland southern Italy and Croatia, and (v) the Mediterranean zone of Southeastern France.

Australia:

(i) Eastern and southeastern Australia; (ii) South Australia; (iii) Southwestern Australia.

These territories often lack continuity. In the Mediterranean areas, for example, the subbiome is interspersed with habitats belonging to biome B_3 (shrublands and shrubby woodlands).

- Climate:
  Fire-prone temperate region: either hot Mediterranean with warm and dry summers, and mild and wet winters or hot oceanic with dry summers and humid winters. Regular water deficit.
- Plant composition:
  (i) Forests or woodlands with trees much lower than subbiome SB_{10} (tree heights 10–30 m); (ii) LAI low to moderate, microphyll foliage, low SLA; (iii) canopy as well as shrubs and understorey sclerophyllous; plants are tolerant to water deficit thanks to physio-anatomical characteristics (e.g. ability to invaginate their stomata, possession of tubers, ability to constitute seedbanks) (iv) ground vegetation is sparse with a mix of C_3 and C_4 [see footnote 27] tussock grasses and forbs
- Biodiversity:
  (i) Due to the paucity of nitrogen in the soils, many plants possess nodules in the roots, that shelter *diazotrophs*, i.e. nitrogen-fixing bacteria or archaea [65]. (ii) Plants and animals are adapted to a habitat frequently subjected to fires due to the type of climate and types of vegetation; plants have thick barks protecting vital organs, *serotinous seedbanks* are held in the canopies or in the soils. (iii) Relatively high endemism and diversity among plants, birds and insects (dipterans, hemipterans) are observed, but food chains are relatively short. Many consumers have specialised diets, like the consumption of highly fibrous plants (e.g. koalas eating eucalyptus leaves). (iv) The trophic networks keep their fundamental functional units but are generally limited to three levels: primary producers, primary consumers, and secondary consumers, to which are added the decomposers/mineralisers
- Key ecological drivers:
  (i) Marked summer water deficits constitute ecological inhibitors limiting the lengths of food chain, as are acidic, sandy or loamy soils frequently encountered in those habitats; (ii) The concentrations of phosphorus and nitrogen are low, which is not favourable to the establishment of important food webs. (iii) Though still belonging to the B_2 biome, it is subbiome with the least mature climax, hence the relative short food chains.

3.7.1.3 The Shrublands and Shrubby Woodlands (B_3)

This biome has a climax that has not reached the forest stage but a more 'juvenile' one in the sense of Clements. That is due to relatively poor nutrient content compared to biomes B_1 and B_2, with consequently lower productivity in tropical, subtropical and Mediterranean climatic zones, mainly due to seasonal water deficits[29] with many types of leaves and roots. Many plants (e.g. legumes) practise osmotrophy via symbiotic relationships with soil microorganisms. Trophic networks are simpler but still composed of their main functional components, i.e. the primary producers (the photoautotrophs), the primary consumers (the herbivores), the secondary consumers (the omnivores and carnivores) and the decomposers/mineralisers. Fires and storms are the main disturbance events that cause the reallocation of resources and thus evolve the species distribution. Resprouting organs and seedbanks are common and adapt to fire-prone habitats. Four subbiomes are represented in this biome.

SB_{11}: Seasonally dry tropical shrublands subbiome:

- Distribution:
  This subbiome is divided into distinct intertropical areas: (i) Localised grassy habitats in the extreme north of Australia, i.e. within the northern tip of Cape York Peninsula, Arnhem Land and the Kimberley Plateau; (ii) South America: localised within the Brazilian sertão and cerrado, and portion of the Amazon forests that has already turned into grassland (campos ruprestres) [57]; Venezuelan tepui plateau; and the Güeppí table-top mountains in the extreme north of Peru; (iii) North America: the Florida scrub in sandy regions of southern Florida, a remaining remnants of grassland interspersed within subbiome SB_8 [61]; (iv) grasslands in different islands: in the Indian Ocean (La Réunion, Mauritius), in the Pacific Ocean (New Caledonia, French Polynesia).
- Climate:
  Identical to that prevailing in subbiome SB_2, one or two wet seasons and one or two dry seasons per year; which entails periodical seasonal drought
- Plant composition:
  (i) A mix of evergreen sclerophyllous shrubs with small leaf sizes and grassland; low open forests with trees whose height does not exceed 6 m. (ii) Dispersed within other subbiomes [SB_2, SB_{16}]; (iii) Conspicuous C_4 grasses [see footnote 27]
- Biodiversity:
  - The productivity is low, constrained by seasonal drought and poverty of nutrients, in particular nitrogen. The result of the last point is that (i) a large number of plants carry out a C_4 type of photosynthesis, (ii) a non-negligible

[29] ...during the dry seasons in the tropical/subtropical, and summers in the Mediterranean and warm-temperate oceanic areas.

number of plants are carnivorous, (iii) many species carry sclerophyllous leaves, and (iv) others have developed cluster roots or roots containing nodules favouring symbiotic interactions with fungi or prokaryotes (in particular nitrogen-fixing prokaryotes).
- Shrubs are the dominant primary producers.
- The trophic network is simple with limited food chain lengths: most primary consumers are arthropods, herbivorous vertebrates are rare (generally nectarivorous birds or granivorous small mammals); most secondary consumers are insectivorous; large vertebrates are lacking.

• Key ecological drivers:
(i) Recurrent water deficit during dry seasons, sandy soils and shallow rocky substrates are common features of these subbiomes. They are severely inhibiting factors. (ii) Other inhibiting factors are the paucity of nitrogen, phosphorus and other nutrients, easily explained by the acidity of the underlying bedrock. (iii) The largest pool of nutrients is held by the plants themselves; in a certain way, they are their own ecological drivers; thus the main drivers of the rest of the ecosystems. (iv) Recurrent fire on long-time scales is often a driver but rare in table-top mountains like the Venezuelan Tepui plateau.

SB_{12}: Seasonally dry temperate heaths and shrublands subbiome:

• Distribution:
This subbiome is located in the warm-temperate climatic zone.
- Mainly Mediterranean, strongly overlaps subbiome SB_{10} in Australia and in the western Mediterranean Basin. In addition, coastal areas of North Africa, Dalmatia, Greece, Turkey, Near East, Northern Chile, California and South Africa (Cape provinces).
- In addition, Macaronesia, mainly (Canary archipelago), and several areas in North America (heaths within the Appalaches and the Texas Edwards Plateau).

• Climate:
Generally in hot mediterranean climates (hot and dry summers, mild and wet winters).
• Plant composition:
(i) Evergreen shrubs with sclerophyllous small leaves that are often *ericoid*, i.e. with thick cuticles and are crowded on the branchlets (heathers are typical ericoid plants). (ii) On the ground, typically, geophytes and sclerophyll grasses or grass-likes. (iii) These plants have low SLA. As a reminder, consequently, they have a low relative growth rate.
• Biodiversity:
(i) Productivity is low. Plants and animals are resilient to drought and fire, and adapted to a paucity of nutrients (plants thanks to stomatal regulation and root structure). (ii) Trophic webs are simple, composed of the primary producers already mentioned (heath or heath-like shrubs, graminoids, geophytes, etc.),

primary consumers composed of vertebrate and arthropod herbivores in low densities (often pollinators or granivores, all adapted to consumption of plant parts containing little nutrients), secondary consumers essentially composed of small insectivores (large mammalian predators are rare or even absent); (iii) Endemism is common among vascular plants and invertebrates.

- Key ecological drivers:
  (i) Water deficit and sandy soils are strong inhibitor (ii) In many area, soils that are typically acidic with very low content in phosphorus, nitrogen and mineral cations are also inhibitors.

SB_{13}: Cool temperate heathlands subbiome:

- Distribution:
  This subbiome comprises (i) Atlantic coastal regions of western Europe from northern Portugal to northern Jutland, also comprising the British Isles; (ii) the Açores archipelago; (iii) in northern America, the Newfoundland and the Aleutian Islands; (iv) in southern America, the southern parts of Chili and Argentina including Tierra de Fuego.
- Climate:
  It is determined by maritime influence in relatively high latitudes, which entails two trends: (i) a dampening effect on the temperature: winters are moderately cold, and summers mildly hot; (ii) a constancy of air humidity and the relatively equal distribution of the rainfall height, moderate to high, during the whole year.
- Plant composition:
  (i) This subbiome is covered by more than 70% of low shrubs whose height is generally less than 1 m, growing small, tough, evergreen and ericoid leaves. (ii) This shrub cover is dense and interspersed with grasses [C_3 graminoids, see cf. footnote 27] and ferns. (iii) The SLA of the leaves is low, thus also the relatively growth rate of the vegetation.
- Biodiversity:
  (i) Primary productivity is low as a consequence of different ecological drivers (see below); (ii) The trophic web is limited. Due to the relative poverty of the abiotic factor system and the low productivity of the primary producers (plants), the ecosystem hardly sustains the primary consumers [e.g. cervids, lagomorphs, camelids (in South America)] that, for this reason, consume in addition plants of adjacent subbiomes (forests, anthropogenic ecosystems). Next to these mammals, among the smaller vertebrates, one can find ground-nesting birds and rodents. (iii) Canids and raptors count among the main secondary consumers.
- Key ecological drivers:
  (i) Soils are often acidic, entailing a paucity of nitrogen and phosphorus, which is a limiting factor. (ii) Cold temperatures are also limiting, but constant humidity or moisture constitute facilitating drivers. (iii) Water surplus is a facilitating driver. (iv) Fires are not strong drivers, due to the humidity of air,

though they can occur during infrequent hot and dry summers. (v) Seasonally low light is an inhibiting factor at the highest latitudes.

SB_{14}: Young rocky pavements, lava flows and screes:

- Distribution:
  Scattered worldwide in volcanic areas and major geological faults, e.g.
  - in seismic and volcanic regions of the Mediterranean basin (Italian peninsula, Greece and Turkey, Northern Africa);
  - in the African rift valley and extension: from the banks of Lake Tanganyika north-northeastwards across Kenya, the Horn of Africa, Yemen and the Hejaz, the Arava Valley, the Dead Sea and Jordan valley, the northernmost part of the rift reaching the foothills of the Lebanon mountain range;
  - in a very long seismic zone along the Persian Gulf, the Osman Sea and the Indian Ocean stretching from southeastern Iraq across southern Iran and southern Pakistan eastward nearly to the border of India;
  - in continuation, along the Aravalli mountains between Pakistan and India and the Himalayas;
  - in diverse volcanic islands: the Macaronesian archipelagos, Ascension island, Comoro Islands, Reunion, Mauritius and Rodrigues; in continuity, Sumatra, Java, Sulawezi, the Maluku Islands, the Philippines, the different islands of Japan and the Kuril islands (on the Asiatic continent, a narrow coastal rim in the extreme south of South Korea);
  - in Australasia: New Guinea, Solomon archipelago, New Zealand, Tonga Islands, Niue;
  - in the subarctic zone: Iceland, Aleutian islands, Kamchatka Peninsula;
  - in the Americas: California and Rocky Mountains in the United States, the Sierra Madre Occidental in Mexico; scattered areas in the mountains of Central America; northern part of the Andes down to Central Chile.
- Climate:
  Since the subbiome depends on telluric movements (seismic and volcanic phenomena) that can occur at any latitude and altitude, it does not depend on climatic factors (see distribution just above).
- Plant composition:
  (i) No trees, very rare shrubs scattered in a landscape where lichens and bryophytes dominate on rocky substrate with very little soil. (i) Thus, a very low, almost negligible LAI.
- Biodiversity:
  - Plant biodiversity, endemism and productivity are very low due to the fact that nitrogen is scarce, provided by an infrequent density of lichens and cyanobacteria (see the paragraph on ecological drivers below). Hence, an overwhelming percentage of the primary producers are either bryophytes, lichens, and photosynthetic microorganisms (e.g. cyanobacteria, microalgae) or chemosynthetic microorganisms (bacteria, archaea) [4]. The rare

vascular plants (e.g. ferns, gymnosperms and angiosperms) are of very modest sizes and are pioneering species.

- The food web is truncated. Among animals strictly linked to subbiome SB_{14}, the primary consumers are thus in limited quantity. Arthropods are very rare during a long period (several years) after an eruption. Recent studies showed that ant colonies are among the most resilient insects, and laboratory studies explained why generally, most arthropods could not survive an eruption, and thus are absent in this subbiome [20]; "reptiles and ground-nesting vertebrates are among the few resident vertebrates", but no mention is made about their foodweb status [36, p. 54]. The few primary consumers (e.g. larids or birds of prey) prey on species adjacent subbiomes. If the subbiome is continental, secondary consumers are not strictly linked to the subbiome. An interesting article authored by N. Philip and Myrtle J. Ashmole [3, p. 563] about an insular habitat belonging to subbiome SB_{14}, namely Ascension Island, describes the trophic web as having the following characteristics, I quote:

> (1) chemolithotrophic bacterial activity; (2) input of marine bacteria and other organic material from the sea-salt aerosol assisted by sea breezes; (3) transfer of organic material, by wind or movement of animals, from the productive intertidal zone towards the interior; (4) input of marine biomass by nesting seabirds; (5) input of biomass in the form of carrion and flotsam deposited on the beaches and utilised by terrestrial animals.

- Key ecological drivers:
  - Limited moisture and nutrient scarcity are inhibiting factors of productivity.
  - The substrate, composed of lava pavements, scree and rock outcrops covered by very thin soil, very poor in nutrients, constitutes a limiting factor.
  - Climate is a facilitating driver as well as an inhibitor. (i) High summer temperature and exposure to solar energy, if the air is sufficiently humid, can be favourable to the development of ecological succession. (ii) Conversely, if the temperature is exceedingly high and dry, or the cold very severe as it can occur in high altitudes and latitudes, the succession can be much slower, almost at a standstill.
  - Other inhibitors are lava flow, strong geothermal activities like the sprouting of geysers or mass movements, i.e. "bulk movements of soil and/or rock debris down slope or vertically downwards in response to gravity" [36, p. xvii].

3.7.1.4 The Savannahs and Grasslands (B_4)

Biome B_4 is characterised by a continuous ground layer of grasses (graminoids) that can be interspersed, or not, by trees or shrubs, sometimes constituting woodlands, in which case, forbs separate grasses and trees/shrubs. This biome exists at low and moderate latitudes (<55°N or S). Productivity is highest when the temperatures are hot and simultaneously, the precipitation heights or ambient humidity are high; this happens in the wet seasons in the tropical/subtropical and in spring in the temperate climatic zone. Trees/shrubs and grasses compete, with an advantage for the trees; sometimes, savannahs and grasslands turn into forests. An equilibrium is obtained between these two types of vegetation by (i) the consumption by herbivores and (ii) fire, two major drivers. That comforts the theory of the spontaneous maturation of ecological successions—tending to an adult forest climax but blocked here in the juvenile stage by biotic and abiotic agents. The trophic web is complex, favouring a great diversity of large-bodied mammalian herbivores, themselves prey to a non-negligible number of large mammalian or avian carnivores or scavengers. This biome counts five subbiomes.

3.7.1.5 Deserts and Semi-deserts (B_5)

Biome B_5 is characterised by low to very low biomass due to the lack of water in the atmosphere and the earth's surface (composed of very little soil tending to concentrate in oases and much sand or bedrock outcrop). It includes hot or cold arid to very arid areas. Due to the climatic conditions, primary productivity is weak to very weak. Plants are very sparse and generally small-sized, adapting their morphologies to resist desiccation. Rainfalls are (i) rare due to temperature extremes and dehydrating winds and (ii) erratic—with bursts of plant growths after long periods of dormancy. The trophic web is complete (with prey species and predators) but limited and sporadic, with a small number of animals—all adapted to acquire and conserve water. Some species can travel long distances to obtain their water and food. Their life history, relative to the harshest phases, is one of the following:

(1) Long-lived tolerance to drought with low metabolic rate,
(2) High mobility enabling to escape drought,
(3) Short favourable periods of activities followed by long unfavourable periods of dormancy.

That is made possible by the landscape heterogeneity. The biome has five sub-biomes.

3.7.1.6 Polar Alpine (Cryogenic) Biome (B_6)

Biome B_6 comprises extremely cold regions, i.e. Arctic and Antarctic areas, or high mountainous elevations. Covered with snow and ice caps during the longest periods of the year, and permanently in the most extreme ones. Seasonal with very short summers (season for growth and reproduction) and very long winters (unfavourable season). The primary productivity is comparable to deserts and semi-deserts, low to very low. Insolation is weak, the atmosphere very dry, and winds harsh and desiccating. For all these reasons, the metabolic rate of the organism is low, with a life strategy consisting of minimising energy loss. Microbial decomposition is slow, which leads to continuous peat formation. In the coldest area, microorganisms dominate, forming microfilms. In relatively milder ones, more elaborate organisms manage to grow: primary producers like lichens, bryophytes, algae, forbs, grasses, and dwarf shrubs. The trophic web is limited, with few large vertebrates migrating from the biome during winter or hibernating. Smaller vertebrates and invertebrates tend to hibernate or be dormant under snow cover, adopting different survival strategies (for example, by developing antifreeze substances in their internal cell water to preclude lethal ice formation). This biome subsumes five subbiomes.

3.7.1.7 Intensive Land-Use Biome (B_7)

This extensive anthropogenic biome results from cropping, pastoralism, plantation farming or urbanisation, where, for example, clearing, drainage, earthworks, ploughing, introducing fertilisers and chemical inputs in soils, and irrigation are involved. Besides, this activity introduces alien animals and plants voluntarily or involuntarily. The result is an artificial biome that is disequilibrated and impoverished in its biodiversity (low endemism, a weak number of species, trophic relationships impaired by genetic manipulation, the emergence of sporadic "cosmopolitan opportunists with short cycles" [36, p. 73]), excessive hunting and destruction of predators. It replaces natural ecosystems in all climatic zones, exploits a large quantity of water for irrigation, and depletes the natural water and soil system with all these practices. The increasing extension of biome B_7 enhances climate change by altering the water cycle and releasing greenhouse gases via fossil fuels and livestock. Five distinct subbiomes represent the biome.

3.7.2 Within the Subterranean Realm R_2

3.7.2.1 Subterranean Lithic Biome (B_8)

Biome B_8 consists of non-aquatic lithic systems, i.e. dry caves and bare rocks (sediments) sheltering in its interstices prokaryotic life. This ensemble shelters ecosystems mainly composed of microbial biofilm and some tiny symbiotic invertebrates and their direct environment. Sunlight hardly reaches this habitat, and there is no standing water inside; thus, the trophic web is minimal: all the primary producers are chemolithoautotrophic prokaryotes, feeding heterotrophic lithotrophic prokaryotes and symbiotic protists and invertebrates. The atmosphere is very mineral and was thought to be lifeless only several decades ago before modern technology discovered ecosystems teaming of life in improbable areas, and this biome is one of these. This biome is found in all the parts of the Globe and is represented by two subbiomes.

3.7.2.2 Anthropogenic Subterranean Voids (B_9)

Mines, tunnels, tombs, and underground military installations (in activity or abandoned), for example, constitute this anthropogenic biome, characterised by extremely low productivity and wide connection with the surface (generally a terrestrial biome) via sipping waters or aerial currents. Sunlight is rare or absent, privileging a redox chemical energy source. However, in some voids still in use and artificially lit, electric light can provide enough photic energy to sustain, among primary producers, algal species. Nevertheless, most primary producers are autochthonous chemolithoautotrophic microorganisms that form biofilms with heterotrophic unicells that constitute their consumers. Trophic webs are simple and opportunistic, dominated by allochthonous organisms that venture into these voids, finding shelter and breeding opportunities (invertebrates, namely arthropods; small mammals - generally insectivorous) whose primary energy can be the phototrophic algae mentioned above. This biome contains one subbiome, itself.

3.7.3 *Within the Freshwater Realm* R_3

3.7.3.1 Rivers and Streams Biome (B_{10})

It is found anywhere on the terrestrial part of the planet except in Antarctica. Includes lotic ecosystems, i.e. ecosystems within running waters, structured in arteries, that flow downwards from a source located on top of a mountain, a hill or a plateau to the sea or a lake at a minimally reachable altitude. It is constituted in a treelike system where the trunk is the main artery, the river itself from source to outlet, and the branches the tributaries (from secondary sources to the main river). Generally, flow regimes widely depend on rainfall catchments and vary from permanent to episodic, tributary to climatic and edaphic conditions. Productivity increases with decreasing altitude unless the lotic system at low altitudes leaves a climatic zone. Energy entering the heterotrophic levels of the trophic web can have two origins: (i) an external one—generally from the surrounding terrestrial biome, and (ii) an internal one either from internal biofilms or from photoautotrophs (phytoplankton, aquatic plants); these different situations depend on external parameters that can influence light penetration, with a strong effect on the complexity level of the trophic web. Seven very different subbiomes exemplify these situations.

3.7.3.2 Lakes Biome (B_{11})

Biome B_{11} contains lentic ecosystems, i.e. ecosystems within standing waters spreading in basins, structured in semi-closed bodies of different surfaces and depths, generally with an inlet (river(s) or any current(s) of water feeding it) and an outlet (river(s) or any current(s) running out of it). Biome B_{11} is thus a system in dynamic equilibrium submitted to the regime of biome B_{10} to which it is connected. Lakes can vary from permanent to episodic and can even dry under the influence of climatic or/and edaphic factors or the river regime they depend on. Energy is photic in the upper layers and redox-chemical in the lower ones when the lake is deep enough. Gradient climatic influences on the lake biome are observed but are sometimes strongly tempered by the geomorphology and groundwater around it and at the level of its bottom. Productivity depends (i) on the importation of nutrients and organisms from connected feeding rivers and (ii) on autochthonous organisms (phytoplankton, periphyton,[30] floating plants). Consumers of higher degrees feed on this organic matter, structured in the classical upper level of the food web (e.g. predators 1 on prey, predators 2 on predators 1, ..., predators k on predators $k-1$ for a food chain of length k). Trophic webs tend to increase in size and complexity with increasing lake volume. These traits are very general and shared to a large degree by its ten subbiomes.

3.7.3.3 Artificial Wetlands (B_{12})

This anthropogenic biome occurs in structures built to hold and transfer water or produce electricity for human use. It includes, for example, "large storage reservoirs, farm dams or ponds, recreational or ornamental wetlands, rice paddies, freshwater aquafarms, wastewater storages and treatment ponds, and canals, ditches and drains" [36, p. 121]. Water is generally standing except in canals and drains. Energy comes mainly from surrounding biomes via run-off, groundwater or human activities. It can also enter from autochthonous sources via in situ photoautotrophs that secondarily colonised the biome (green algae and plants), notably in artificial water bodies. Artificial wetlands are less variable and productive than natural lakes. However, they can shelter some native plant and animal species that survived the destruction needed to implement the structures of biome B_{12}, as well as several opportunistic colonists. As expressed by the IUCN report of 2020 [36, p. 121], "the simplest artificial wetlands support only microbial biota, while the most diverse can include submerged or emergent plant communities, which promote complex habitats for invertebrates, fish, waterbirds, amphibians, reptiles and, sometimes, amphibious mammals." The biome subsumes five subbiomes.

---

[30] For the definition of this concept, I quote a passage of [36, p. xvi]: "complex layer of algae, cyanobacteria and heterotrophic microbes embedded in a mucopolysaccharide matrix cohering to submerged aquatic surfaces."

3.7.4 *Within the Marine Realm* R_4

3.7.4.1 Marine Shelf (B_{13})

Located between the shoreline (B_{18}) and the deep sea-floor biome (B_{15}), thus above the continental shelf, it contains many habitats (e.g. "seagrass meadows, kelp forests, oyster beds and coral reefs" [36, p.131]) whose substrates can be rocky reefs, sandy bottoms or muddy bottoms. These substrates release essential mineral salts for the organisms of the ecosystems just mentioned; the latter mutually exchange resource subsidies (e.g. kelp detritus and plankton), the core primary production. Light penetrates this biome and conveys solar energy as the primary energy available to the organisms. Increasing water clarity conveys higher productivity. This parameter must have a value higher than a given threshold to allow aquatic plants, macroscopic algae (e.g. kelp), and animals depending on them, to establish and persist. Productivity also depends on upwellings that can deliver nutrients from the deep ocean bottom and resource subsidies from adjacent terrestrial and freshwater biomes. Biome B_{13} contains nine subbiomes.

3.7.4.2 Pelagic Ocean Waters (B_{14})

It is the largest biome on Earth, located in the open ocean called the pelagic zone, in a horizontal gradient from the shore to the horizon, beyond the marine shelf biome. Vertically and following a gradient from surface to sea bottom in case the ocean is not deeper than ca 4000 m, the water column is composed of three storeys, the (i) epipelagic (>200 m), (ii) mesopelagic (200–1000 m) and (iii) bathypelagic (1000–4000 m) zones. Vertically, from the surface to the bottom limit of the bathypelagic zone, light intensity decreases with increasing depth, and correlatively, productivity decreases. The upper layers forming the epipelagic zone are photic, thus concentrate primary production (via photosynthesis mainly carried out by phytoplankton) and dilute in their water the carbon dioxide (CO_2) adsorbed on the ocean's surface from the adjacent atmosphere. Then, a carbon flux is dragged downward via a food chain. This flux decreases with increasing depths, which explains the decrease in productivity. A specificity of biome B_{14} lies in the Redford ratio (cf. Sect. 3.6.4, p. 47), stipulating, as a reminder, that C: N: P is constant throughout all oceans. The pluricellular organisms in the deep zones (mesopelagic and bathypelagic)—exclusively animals, and essentially crustaceans and fishes—are adapted to a habitat where the pressure is intense and the light almost absent. Common adaptations for these animals are described in a publication by Eric J. Warrant and colleagues [94]: (i) bioluminescence or (ii) specific eyes with, I quote, a "*typical* form, with lateral placement on the head, large fields of view, and restricted binocular overlap"; or "*tubular* in form, with restricted fields of view and large binocular overlap". The pelagic ocean waters biome is diversified into five subbiomes.

3.7.4.3 Deep Sea Floors Biome (B_{15})

The deep-sea biome encompasses the ecosystems of the sea floors below 250 m depth. Here, the weak light prohibits photosynthesis, leading to the formation of biocoenoses in two distinct ways.

- One way is by gathering only heterotrophic species, with the consumers of the least degree solely depending, for their needed energy and matter, on the ecological layers above them, notably, the gradient of nutrients (detritus of different origins, venturing preys from above) falling from the top layers of the water column.
- Another way biocoenoses form is by utilising energy and matter from the sea bottom. These chemosynthetically based ecosystems derive their primary energy from the fluids emitted by hydrothermal vents and cold seeps.

Diluted oxygen is generally at the disposal of the eukaryotes and aerobic prokaryotes of the deep seas from the global currents transporting them across the oceans. All the species of the biome are detritivores or carnivores. The food chain they generally form goes from the smallest to the largest species in a classical prey-predation relationship. Depending on the downward flux of nutrients, classical deep-sea benthic biodiversity is high, mainly composed of medium-sized to large-sized animals. A well-studied ecosystem of this type lies on the extensive soft sediments of the abyssal plains. Other ecosystems that depend on hydrothermal vents or cold seeps show, low diversity but high biomass and high endemism. Biome B_{15} counts seven subbiomes.

3.7.4.4 Anthropogenic Marine Biome (B_{16})

Anthropogenic biome formed from human installations (active or abandoned) dumped on the sea floor (e.g. shipwrecks, pipelines, rubble piles, aquaculture infrastructures). The sea bottoms where these installations lie are generally in the epibenthic zone, i.e., sea floors located above 200 m depth. These objects form a substrate for sessile benthic organisms (animals dwelling on sea floors, such as, for example, sponges, corals, sessile molluscs and annelids) and *demersal* or *pelagic* mobile organisms (fishes, invertebrates) depending on them. Thus, a food chain spontaneously forms if the surface of the substrate is not too smooth, but this food chain has a weakness. The primary energy supply of such a food chain is allochthonous, stemming either from planktonic marine algae or phytoplankton drifting passively in currents reaching these artificial biocoenoses or actively by humans. This assisted biome is highly productive, but its visibility attracts large pelagic predators. It can be a source of microbial infection due to the concentration of waste products and decaying bodies, particularly in the dumped aquafarms. It is thus feared that this biome will be fragile and not sustainable in the long run. It counts two subbiomes.

3.7.5 *Within the Transitional Terrestrial-Freshwater Realm* R_{13}

The transitional terrestrial-freshwater realm contains only one biome.

3.7.5.1 Palustrine Wetlands Biome (B_{17}) [36, p. 94]

This biome represents varied habitats found on all continents, generally lands soaked with water to such a degree that they are often deprived of oxygen, resulting in an accumulation of peat. The resources available to organisms and the ecosystem's productivity depend on the water regime, namely the water flow, the catchment, the topography, the run-off, infiltration, percolation and local evapotranspiration. These habitats are productive but with restricted ecological niches: the organisms are specifically adapted to this type of habitat, the species are thus endemic. Amphibious plants are the dominant autotrophs, adapted to low oxygen concentrations. The heterotrophs that depend on this type of vegetation range from microbial decomposers to large vertebrates adapted to this type of ecosystem. The food web is complete. This biome contains seven subbiomes.

3.7.6 *Within the Transitional Terrestrial-Marine Realm*

3.7.6.1 Shorelines Biome (B_{18})

The shoreline biome [36, p. 158] comprises different habitats at the boundary between land and ocean, and it is found on all continents.

Productivity can vary continuously between almost nil to high, depending on several factors:

(1) It depends on the energy source, exterior to the biome, from sea or land.
(2) It is tributary to the vegetation coverage ranging from nonexistent to luxurious.
(3) It is influenced by coastal upwelling and the nutrient content of adjacent terrestrial sediments.

Biocoenoses are diversely structured according to the size of the primary particles (detritus, phytoplankton) in the intertidal zone, to the action of tides, cyclically exposing the habitat to air, and the strength of waves, tending to increase in the bays at the approach to adjacent headlands. Many organisms have morphological and behavioural adaptations against desiccation when the tide retreats and to resist wave thrusts. Biodiversity is high in the intertidal zone, with microorganisms as dominant elements when the primary organic particles of the environment are small. This biome subsumes four subbiomes.

3.7.6.2 Supralittoral Coastal Biome (B_{19})

This biome [36, p. 163] stretches just above the shoreline biome, between the latter and the adjacent terrestrial biome. It is regularly splashed but not submerged. Its width can vary between tens of metres and a few kilometres. The biota is specialised, with relatively low biomass. The diversity is low. The relief is characterised by sandy headlands and coastal dunes, which are very sandy and subject to winds and sea sprays. It is generally covered with low-sized bushy, often wind-carved vegetation highly adapted to desiccation, high salinity and substrate instability and is often partially flooded. The fauna, depending on this vegetation, is mobile and sea-dependent. It is the home for gastropods and isopods, which can notably feed sea birds assembled in vast colonies. The biome is represented by one subbiome, itself.

3.7.6.3 Anthropogenic Shoreline Biome (B_{20})

This anthropogenic biome within human constructions, such as artificial islands, harbour basins, and piers, is a clear example of human influence on natural environments [36, p. 165]. It is inhabited by benthic biota, with bacterial biofilm and algae as the main primary producers. The consumers are composed of filter-feeders like ascidians, barnacles, or planktivorous fishes. The second group of consumers comprises demersal predators, such as teleosts and sharks, generally hunted or deterred by humans, thereby weakening the trophic webs. The ecosystem's fragility is further exacerbated by the strictly allochthonous character of energy and nutrient matters, i.e. their dependence on external sources, which impairs its stability. The biome is represented by one subbiome, itself.

3.7.7 *Within the Transitional Terrestrial-Freshwater-Marine Realm* R_{134}

3.7.7.1 Brackish Tidal Biome (B_{21})

This biome [36, p. 168] occurs in estuaries where fresh and sea waters mix to form brackish waters that induce the formation of fluvial deltas, marine-dominated intertidal forests (e.g. mangroves), and coastal saltmarshes. It is widely distributed in tropical and temperate areas. Energy and nutrients can be allochthonous from the rivers that flow into this biome; energy can also be autochthonous, as aquatic flowering plants directly absorb sunlight. The productivity is high, and the trophic webs are complex.

(1) Autotrophs, mainly phototrophic, comprise (i) a rich aquatic flora (notably large flowering plants), but also (ii) planktonic algae, cyanobacteria, and all sorts of plant or algal detritus forming particles of dead biomass.

(2) These primary producers feed numerous invertebrates (crustaceans, molluscs, arthropods);
(3) The primary and second producers mentioned above feed rich fish nurseries, a mix of herbivorous and carnivorous fishes that, in turn, constitute food to vertebrates (reptiles, wading birds, mammals).

Three subbiomes are subsumed by this biome.

3.7.8 Within the Transitional Subterranean-Freshwater Realm $\mathbf{R_{23}}$

3.7.8.1 Subterranean Freshwaters Biome (B_{22})

This biome [36, p. 84] is composed of underground water bodies within caves or carved in rocks, forming the largest volume of water in the Planet. It is mainly laterally connected to the subterranean biomes (B_8 and B_9) and, above, to freshwater biomes (B_{10}, B_{11} and B_{12}). B_{22} is scattered worldwide. No sunlight reaches the ecosystems; the autochthonous primary energy source of the biocoenoses is only chemical via the chemoautotrophic prokaryotes. Water depends on the connection with external biomes. Planktonic species and aquatic invertebrates, fishes and amphibians live in the water bodies, but they depend on the exterior for their primary energy source. They form simple trophic chains: (i) bacteria and aquatic fungi that are food for protozoans and invertebrates and (ii) the latter being eaten by larger invertebrates and vertebrates (fishes, neotenic amphibians like axolotls or olms, bats). This biome comprises two subbiomes.

3.7.8.2 Anthropogenic Subterranean Biome (B_{23})

It is a biome settled within subterranean constructions, such as underground canals, water pipes, and abandoned flooded tunnels or mines [36, p. 87]. This biome is far more open to the connected freshwater biome (B_3) than biome B_{22}. Water directly flows from the external water surfaces, or like biome B_{22}, from seeping waters across and along the rocks and sediments. Sunlight is dimly diffusing through diverse openings (e.g. portals, vents) and is generally too weak to enable the prevailingness of photosynthesis, hence a chemical origin of most of the autochthonous primary energy. This energy source is limited to a restricted microorganismal biofilm. Most primary energy and nutrients (e.g. carbon, oxygen, organic chemicals) are allochthonous and transported to the biome by flowing water. All these factors foster the main trophic web: (i) The primary producers are composed of photosynthetic algae. (ii) The secondary producers (or consumer 1) comprise heterotrophic bacteria and protists. (iii) Consumers 2, 3, and following support successive levels of invertebrates of increasing sizes, small vertebrates (e.g. fish, rodents) adapted to dim light. This biome subsumes two subbiomes.

3.7.9 *Within the Transitional Subterranean-Marine Realm*

3.7.9.1 Subterranean Tidal Biome (B_{24})

The biome [36, p. 90] gathers (i) submarine caves, (ii) subterranean aquifers and (iii) deep marine pools isolated behind the shoreline and connected underneath to the sea. Except near the caves' entrance or the pools' surface, sunlight does not penetrate these habitats. The primary energy of the organisms inhabiting biome B_{24} and nearly all their nutrients (carbon, oxygen and other mineral or organic chemicals) are allochthonous. They are washed in from the ocean at the disposal of the biocoenoses in situ. However, this does not preclude autochthonous chemical energy sources via chemolithoautotrophs. A characteristic trait of this biome, in habitats surrounded by carbonate or volcanic walls, is a decreasing gradient of productivity of the ecosystems as one steps inwards from the opening. Among producers, chemoautotrophs and phototrophic algae coexist. Consumers 1 comprise heterotrophic microorganisms consumed by filter-feeders. The latter is food to predators, a cave-dwelling exclusively carnivorous fauna, with typical adaptations like, e.g., depigmentation, reduced vision, enhanced tact and chemical sensitivity, and K-reproductive strategy (e.g. bats). Three subbiomes represent biome B_{24}.

3.7.10 *Within the Transitional Freshwater-Marine Realm*

3.7.10.1 Semi-confined Transitional Waters Biome (B_{25})

This biome [36, p. 127] is located where the freshwater and marine biomes meet and dynamically equilibrate, far from land influence. It includes coastal inlets constantly humid under the pressure of waves and sprays, fjords at the highest latitudes, and estuaries and lagoons worldwide. It is an area of brackish water characterised by a decreasing upstream salinity. The regime of these waters varies according to climatic zones and local seasons, periodically balancing the dominance of the sea and that of the freshwater currents. Like in all aquatic habitats, the nutrients (oxygen, carbon, and other vital elements) are diluted in the waters, varying concentrations according to location. Primary energy is overwhelmingly solar. The primary producers carry out oxygenic photosynthesis and comprise cyanobacteria, phytoplankton, macroalgae and plants. The trophic web is complete and complex, with food chains containing animals of increasing sizes according to increasing trophic levels, e.g. vertebrates (fish, reptiles, birds and mammals), all adapted to the aquatic habitat. This biome contains three subbiomes.

Quizzes and Problems

3.1 Prove that for any plant $\mathcal{P}$, if $I = [0, t]$ is the time interval it needs to have a size corresponding to its dry mass $M(t)$ and if the initial dry mass $M(0) = M_0$, then the relative growth rate of $\mathcal{P}$ is

$$\mathrm{RGR} = \frac{\ln M(t) - \ln M_0}{t}$$

References

1. Alaback PB (1991) Comparative ecology of temperate rainforests of the Americas along analogous climatic gradients. Rev Chil Hist Nat 64:399–412
2. Allen TFH, Starr TB (1982) Hierarchy: perspectives for ecological complexity. Univ. of Chicago Press.
3. Ashmole NP, Ashmole MJ (1997) The land fauna of Ascension Island: new data from caves and lava flows, and a reconstruction of the prehistoric ecosystem. J Biogeogr 24:549–589
4. Bay SK et al (2021) Chemosynthetic and photosynthetic bacteria contribute differentially to primary production across a steep desert aridity gradient. ISME J 15:3339–3356, https://doi.org/10.1038/s41396-021-01001-0
5. Blackman (1919) The compound interest law and plant growth. Ann Bot 33:353–360.
6. Bond WJ, Midgley JJ (2001) Ecology of sprouting in woody plants: the persistence niche. Trends Ecol Evol 16:45–51
7. Box EO, Fujiwara K (1988) Evergreen broad-leaved forests of the southeastern United States: preliminary description. Bull Inst Environ Sci Technol Yohohama Natn Univ 15:71–93
8. Bradley R (1718) New improvements of planting and gardening, both philosophical and practical, W. Mears, London, 3 Vols.
9. Breed B, Ford F (2007) Native mice and rats. In: Australian Natural History series, CSIRO Publishing, Collingwood Victoria, AUS.
10. Cabello-Yeves PJ, Zemskaya TI et al (2019) Microbiome of the deep Lake Baikal, a unique oxic bathypelagic habitat. Limnol Oceanogr 65:1471–1488, https://doi.org/10.1002/lno.11401.
11. Cael BB et al (2016) The volume and mean depth of Earth's lakes, Geophys Res Lett 44:209–218 https://doi.org/10.1002/2016GL071378.
12. Cáliz J et al (2018) A long-term survey unveils strong seasonal patterns in the airborne microbiome coupled to general and regional atmospheric circulations. Proc Natl Acad Sci USA 115:12229–12234, https://doi.org/10.1073/pnas.1812826115.
13. Castellan G et al (2022) Drawing the borders of the mesophotic zone of the Mediterranean Sea using satellite data. Sci Rep 12:5585, https://doi.org/10.1038/s41598-022-09413-4
14. Chen JM, Black TA (1992) Defining leaf area index for non-flat leaves. Plant Cell Environ 15:421–429.
15. Cherif M (2012) Biological stoichiometry: the elements at the heart of biological interactions. In: Stoichiometry and research-the importance of quantity in biomedicine, IntechOpen, ch 16, pp 357–376
16. Childress JJ, Fisher CR (1992) The biology of hydrothermal vent animals: physiology, biochemistry and autotrophic symbioses. Oceanogr Mar Biol Annu Rev 30:337–341.
17. Clements FE (1916) Plant succession: an analysis of the development of vegetation. Carnegie Institution of Washington.

18. Delcourt HR, Delcourt PA (2000) Eastern deciduous forests In: Barbour MG, Billings WD (eds) North American terrestrial vegetation, 2nd edn, p. 357–395, Cambridge Univ Press, Cambridge (UK).
19. Dyer J (2006) Revisiting the Deciduous Forests of Eastern North America. BioScience 56:341–352, https://doi.org/10.1641/0006--3568(2006)56[341:RTDFOE]2.0.CO;2.
20. Elizalde L (2014) Volcanism and arthropods: a review. Ecol Austral 24:3–16, https://doi.org/10.25260/EA.15.24.1.0.32.
21. Fragnière Y et al (2015) Fighting their last stand? A global analysis of the distribution and conservation status of gymnosperms. J Biogeogr 42:809–820, https://doi.org/10.1111/jbi.12480.
22. Gay JD et al (2022) Global distribution and climate sensitivity of the tropical montane forest nitrogen cycle. Nat Commun 13:7364, https://doi.org/10.1038/s41467-022-35170-z.
23. Gibson L, Baker J (2004) Diet of the eastern bristle bird *Dasyornis brachypterus* in New South Wales. Corella 28:79–81
24. Gill HK et al (2017) Insect diapause: a review. J Agric Sci Technology 7:454–473, https://doi.org/10.17265/2161-6256/2017.07.002
25. Grant PR (2001) Reconstructing the evolution of birds on islands: 100 years of research. Oikos 92:385–403
26. Grisebach AHR (1838) Über den Einfluß des Climas auf die Begränzung der natürlichen Floren. Linnaea 12:159–200
27. Hall BK, Kerney R (2011) Levels of biological organization and the origin of novelty. J Exp Zool B: Mol Dev Evol 318:428–437, https://doi.org/10.1002/jez.b.21425
28. Heer O (1835) Beiträge zur Pflanzengeographie. Orell, Füssli & Co, Zürich.
29. Hogan CM (2014) Chilean matorral. In: Saundry P (ed) Ecoregions, Encyclopedia of Earth.
30. Hogan CM (2014) Valdivian temperate forests. In: Saundry P (ed) Ecoregions, Encyclopedia of Earth.
31. Holdridge LR (1947) Determination of world plant formations from simple climatic data. Science 105(2727):367–368
32. Hostettler S (2002) Tropical Montane Cloud Forests: A Challenge for Conservation. Bois et forêts des tropiques 274:19–31.
33. Hult R (1881) Försök till Analytisk Behandling av Växtformationerna. Soc Fauna Fl Fenn Meddeland 8:1–155
34. von Humboldt A & Bonpland A (1805) Essai sur la géographie des plantes. In: Voyage de MM Alexandre de Humboldt et Aimé Bonpland, pp. 2–35 (in French). Levrault, Schoell et Co, Paris
35. von Humboldt A & Bonpland A (1805) Tableau physique des régions équatoriales. In: Voyage de MM Alexandre de Humboldt et Aimé Bonpland, pp. 37–155 Levrault, Schoell et Co, Paris
36. IUCN (2020) IUCN global ecosystem typology 2.0. IUCN, Gland (CH).
37. Jørgensen SE (2009) Ecosystem Ecology. Elsevier
38. Keith DA et al (2013) Scientific foundations for an IUCN red list of ecosystems. PLos One 8:e62111, https://doi.org/10.1371/journal.pone.0062111
39. Kenoyer L (1929) Plant physiognomy. Ecology 10:409–414.
40. Kerner von Marilaun A (1863) Das Pflanzenleben der Donauländer. Verlag Wagner'schen Univ, Innsbruck.
41. Kershaw P, Wagstaff B (2001) The southern conifer family Araucariaceae: history, status, and value for paleoenvironmental reconstruction. Ann Rev Ecol Syst 32:397–414.
42. Kolasa J, Pickett, STA (1989) Ecological systems and the concept of biological organization. Proc Natl Acad Sci USA 86:8837–8841.
43. Köppen W (1900) Versuch einer Klassifikation der Klimate, vorzugsweise nach ihren Beziehungen zur Pflanzenwelt. Geogr Z 6(12H):657–679.
44. Le Breton T et al (2023) Fire frequency and severity mediate recruitment response of a threatened shrub following severe megafire. Fire Ecol 19:67, https://doi.org/10.1186/s42408-023-00217-z
45. Le Layec V, Hourdez S (2021) Oxygen consumption rates in deep-sea hydrothermal vent scale worms: effect of life-style, oxygen concentration, and temperature sensitivity. Deep-sea res I,

Oceanogr Res Pap, 172 (103531), pp. 103531. https://doi.org/10.1016/j.dsr.2021.103531

46. Lecoq H (1855) Etudes sur la géographie botanique de l'Europe. vol 4, Baillière, Paris.
47. Lee Z et al (2007) Euphotic zone depth: Its derivation and implication to ocean-color remote sensing. J Geophys Res 112:C3, https://doi.org/10.1029/2006JC003802, 2007.
48. Linnaeus C (1758) Systema naturae. Laurentii Salvii, Stockholm.
49. Lorenz JR (1858) Allgemeine Resultate aus der pflanzengeographischen und genetischen Untersuchung der Moore im präalpinem Hügellande Salsburgs. Flora 14:209–221.
50. Lyons TW et al (2012) Geobiology of the proterozoic eon. In: Knoll AH et al (eds) Fundamentals of Geobiology, Wiley-Blackwell
51. Martin PH et al (2011) Vegetation zonation in a neotropical montane forest: environment, disturbance and ecotone. Biotropica 43:533–543, https://doi.org/10.1111/j.1744-7429.2010.00735.x
52. Marshall SM et al (2021) Identifying molecules as biosignatures with assembly theory and mass spectrometry. Nat Commun 12:3033, https://doi.org/10.1038/s41467-021-23258-x.
53. Merriam CH (1894) Laws of temperature control of the geographic distribution of terrestrial animals and plants. Nat Geog Mag VI:229–243.
54. Meyen FJ (1836) Grundriss der Pflanzengeographie. Haude und Spenersche Buchhandlung. (In Geman), Berlin.
55. Monteith JL, Unsworth MH (1973) Principles of Environmental Physics, 2nd edn. Edward Arnold, London
56. Monteith JL, Unsworth MH (2013) Principles of Environmental Physics, 4th edn. Elsevier Science, Burlington
57. Morellato LPC, Silveira FAO (2018) Plant life in campo rupestre: New lessons from an ancient biodiversity hotspot. Flora 238:1–10, https://doi.org/10.1016/j.flora.2017.12.001
58. Mucina L (2019) Biome: evolution of a crucial ecological and biogeographical concept. New Phytol 222:97–114, https://doi.org/10.1111/nph.15609.
59. Mucina L, Gelldenhuys CJ (2006) Afrotemperate, subtropical and azonal forests. In: Mucina L, Rutherford MC (eds) The vegetation of South Africa, Lesotho and Swaziland. Strelitzia 19:585–615.
60. Murray-Smith C et al (2008) Plant diversity hotspots in the Atlantic coastal forests of Brazil. Conserv Biol 23:151–163, https://doi.org/10.1111/j.1523-1739.2008.01075.x
61. Myers RL (1990) Scrub and high pine. In: Myers RL, Ewel JL (eds) Ecosystems of Florida, p. 150–193, Univ Central Florida Press, Orlando (FL).
62. Nicolson, M. (2013). Community concepts in plant ecology: from Humboldtian plant geography to the superorganism and beyond. Web Ecol 13: 95–102, https://doi.org/10.5194/we-13-95-2013.
63. Ogden J, Stewart GH (1996) Community dynamics of the New Zealand conifers. In: Enright NJ, Hill RS (eds) Ecology of southern conifers, p. 81 119, Smithsonian Institution Press.
64. O'Neill RV, DeAngelis DL et al (1986) A hierarchical concept of ecosystems. Princeton Univ Press, Princeton, NJ.
65. Ormeño-Orrillo E et al (2013) Dinitrogen-fixing prokaryotes. In: Rosenberg E et al (eds) The prokaryotes - prokaryotic physiology and biochemistry, p. 427–452, Springer-Verlag, Berlin, Heidelberg, https://doi.org/10.1007/978-3-642-30141-4_72.
66. Osone Y et al (2008) Correlation between relative growth rate and specific leaf area requires associations of specific leaf area with nitrogen absorption rate of roots. New Phytol 179:417–427, https://doi.org/10.1111/j.1469-8137.2008.02476.x
67. Pausas JG, Keeley JE (2014) Evolutionary ecology of resprouting and seeding in fire-prone ecosystems. New Phytol 204:55–65, http://doi.org/10.1111/nhp.12921
68. Pérez-Harguindeguy N et al (2013) New handbook for standardised measurement of plant functional traits worldwide. Aust J Bot 61:167–234, https://doi.org/10.1071/BT12225
69. Philippi RA (1836) Ueber die Flora Siciliens im Vergleiche zu den Floren anderer Länder. Arch Naturgesch 2:337–366
70. Pitman RL (2013) Serro do Mar coastal forests. In: Howarth RW, Mohan J (eds) Biomes and ecosystems, p. 1127–1129, Salem Press

71. Proctor J (1999) Heath forests and acid soils. Bot J Scotl 51:1–14, https://doi.org/10.1080/03746609908684920/
72. Raunkiær C (1934) The life forms of plants and statistical plant geography, being the collected papers of C. Raunkiær. Oxford (UK), Clarendon Press.
73. Ray DK (2013) Tropical montane cloud forests. In: Pielke RA (ed) Climate Vulnerability, p. 79–85, Elsevier Inc., Academic Press.
74. Redfield AC (1934) On the proportions of organic derivatives in seawater and their relation to the composition of plankton. In: Daniel RJ (ed) James Johnstone Memorial Volume, p. 177–192, UP of Liverpool (UK).
75. Ricklefs RE (2008) The Economy of Nature. 6th edn, Freeman & co, New York.
76. Rynearson TA, Menden-Deuer S (2016) Drivers that structure biodiversity in the plankton. In: Glibert PM, Kana TM (eds). Aquatic microbial ecology and biogeochemistry: A dual perspective, pp. 13–24, Springer, https://doi.org/10.1007/978-3-319-30259-1_2.
77. Schipper J (2023) Araucaria moist forests. One earth site web: https://www.oneearth.org/ecoregions/araucaria-moist-forests/.
78. Schouw JF (1823) Grundzüge einer allgemeine Pflanzengeographie (German translation from Danish). G. Reimer, Berlin.
79. Senterre B et al (2021) Ecosystemology: a new approach toward a taxonomy of ecosystems. Ecol Complex 47:100945, https://doi.org/10.1016/j.ecocom.2021.100945
80. Shelford VE (1945) The relative merits of life zone and biome concepts. Wilson Bull 57:248–252.
81. Spalding MD et al (2007) Marine ecoregions of the World: a bioregionalization of coastal and shelf areas. BioScience 56(7):573–583
82. Tansley AG (1935) The use and abuse of vegetational concepts and terms. Ecology 16:284–307, https://doi.org/10.2307/1930070.
83. Terrasson F (1988) La peur de la nature. (In French) Sang de la terre, Paris
84. Terrasson F (1994) La civilisation anti-nature. (In French) Sang de la terre, Paris
85. Terrasson F (2002) En finir avec la nature. (In French) Edition du rocher, Paris
86. Thurmann J (1849) Essai de phytostatique appliqué à la chaîne du Jura et aux contrées voisines. (In French) Jent et Gassmann, Bern.
87. Toledo-Aceves et al (2011) Tropical montane cloud forests: Current threats and opportunities for their conservation and sustainable management in Mexico. J Environ Manage 92:974–981, https://doi.org/10.1016/j.jenvman.2010.11.007.
88. Uroz S et al (2015) The mineralosphere concept: mineralogical control of the distribution and function of mineral-associated bacterial communities. Trends Microbiol 23:751–762, https://doi.org/10.1016/j.tim.2015.10.004
89. Vankat JL (2002) Boreal and temperate forests. In Encyclopedia of Life Sciences, Macmillan Ltd.
90. Van der Gast CJ et al (2008) Temporal scaling of bacterial taxa is influenced by both stochastic and deterministic ecological factors. Environ Microbiol 10:1411–1418.
91. Von Post H (1862) Försök till en systematik uppställning af växtställena i mellersta Sverige. (In Swedish). Bonnier, Stockholm.
92. Walker JKM et al (2008) Long-term experimental warming alters nitrogen-cycling communities but site factors remain the primary drivers of community structure in high arctic tundra soils. ISME J 2:982–995
93. Walker LR, del Moral R (2003) Primary succession and ecosystem rehabilitation. Cambridge University Press
94. Warrant EJ et al (2003) Eye design and vision in deep-sea fishes. In: Collin SP, Marshall HJ (eds) Sensory processing in aquatic environment, p. 303–322. Springer.
95. Whittaker RH (1970) Communities and ecosystems. Macmillan, London.
96. Wikramanayake E et al (2002) Terrestrial ecoregions of the Indo-Pacific: a conservation assessment. Island Press, Washington, DC.
97. Zainordin AF, Hamid SA (2017) Determination of trophic structure in selected freshwater ecosystems by using stable isotope analysis. Trop Life Sci Res 28:9–29, https://doi.org/10.21315/tlsr2017.28.2.2.

Chapter 4
The Interactions

4.1 Interactions Between the Heterospecific Populations of a Biocoenosis

The interactions between the species forming a biocoenosis consist of their mutual relationships. Those existing between a biocoenosis and the abiotic factors of its environment are, to a certain extent, shared by all the species of the biocoenosis (e.g. oxygen, water, mineral salts, or standard interval of physical variables like temperature, humidity or pH for example) at disposal in the environment. Last but not least, the interactions between the abiotic factors supporting the ecosystem significantly affect the taxonomic composition of the biocoenosis. We will now specifically focus on the interactions between the species forming a biocoenosis.

Species can evolve together across time due to interactions that also change with time. Hence, the temptation of some authors to call *coevolution* all the interactions of a biocoenosis, i.e. trophism, mutualism and competition[1] [11, 31].

I am afraid to disagree with that concept since trophism is a set of relationships accompanying the pathway of matter and energy through the ecosystems, and as such fuel the ecosystems. The *red queen theory* [35] explains a kind of arms race forcing the predator and its prey to develop an increasingly efficient survival strategy with time (for example, the development of increasingly efficient organs and capacities), which sometimes has, as a side effect, the parallel evolution of the involved species. This parallel evolution is not obligatory, even if spectacular

Supplementary Information The online version contains supplementary material available at https://doi.org/10.1007/978-3-031-90656-5_4.

[1] On the contrary, others restrict the term *coevolution* to a category of interactions that we prefer to call *phenotypic mutualism*. However, I also disagree with this point of view which I find too restrictive.

M. Salomon, *Integrative Ecology*, https://doi.org/10.1007/978-3-031-90656-5_4

examples corroborate the theory [7, 20]. Thus, evolutionary processes are more correlated to, and enhanced by trophism than actually caused by it.

When I refer to coevolution, I am describing a process in which one partner species depends on the other for its well-being or survival, at least to a certain degree. This dependence is often reciprocal, meaning that both species benefit from the relationship at the individual level, with automatic implications at the population level. As a result, the process can change the ecophenotypes of the species involved across time, for their own sakes and not as a means for fuelling the ecosystem.

In addition to trophism and coevolution, two other interactions within biocoenoses will be mentioned: (1) *competition* and (2) the chains of relationships involved in *decomposition/mineralisation*. As a reminder (see Sect. 2.3), this interaction is the process of releasing the chemical elements and simple compounds of the organic matter of the biocoenoses into simple compounds in the abiotic factor system; the result of this process is the disposal of these chemical bodies as nutrients to the primary producers (e.g. plants or other autotrophic producers), ensuring the cycle perpetuating the ecosystem.

4.2 Interactions in Biocoenoses

4.2.1 Trophic Relationships

4.2.1.1 Directionality of the Trophic Components of a Biocoenosis

We already saw the concept of trophic components in Chap. 2 (Axiom 2.1, p. 16). These components must be seen as stages in a weblike scaffolding structure called a *food web*, which models the relays of energy flow and matter cycling through biocoenoses. A food web is directional, thus reflecting the pathway of energy and matter through biocoenosis, and is composed of food chains whose successive stages are the trophic components.

Definition 4.1 The mode of nutrition of the species composing a given trophic component can be scaled from 1 to 4, figures called trophic levels. Trophic levels can be greater than 4 (though infrequently) but are always a finite integer. The trophic components of an ecosystem (see Axiom 2.1) are listed below, corresponding to their trophic level:

- Trophic level 1: The primary producers,
- Trophic level 2: The primary consumers or consumers 1,[2]
- Trophic level 3: The secondary consumers or consumers 2,
- Trophic level 4: The tertiary consumers or consumers 3.

[2] ... in certain ecosystems, called secondary producers.

An apex consumer (or apex predator) is a species occupying the highest trophic level in a food chain. According Footnote 15 of Chap. 2, if the highest trophic level of a food chain is 4, i.e an apex predator being thus a consumer 3, then the length of its member food chain is 3.

4.2.1.2 Consumers Level

Predator-prey relationship:

Predation is the relationship between two species based on consumption, i.e. one species, A, consumes the other one, B [A being the predating species (or predator) and B, the prey species (or prey)]. When the species involved are animals in very simple terrestrial habitats like in the example of Fig. 2.1, generally, the predator is carnivorous and the prey species herbivorous, in which case predation is necessary to preserve a high abundance level among the plant species on which the prey species feeds. An equilibrium is thus attained between the three trophic components of the limited biocoenosis studied. This relationship is a dynamic process beneficial to the predators and prey, even if it is lethal for a number of individuals of the latter, doomed to be captured, killed and eaten by the former. The limitation of the number of prey individuals by the predators prevents a far larger number of prey individuals from dying from starvation as a consequence of their exhausting the plant resources.

Parasite-host relationship:

Parasitism is another type of trophic interaction. Here, contrary to the predator, the parasite does not kill its victim immediately (or at all) but diverts the host's metabolic apparatus for its own benefit. Generally, the parasite keeps the host alive, but the latter is severely weakened or even seriously ill in many cases. A variant of parasitism is parasitoidism, in which the parasitoid diverts the metabolic apparatus of the host like a parasite in the first stage but kills and feeds on the host in the second stage. Such a relationship was observed notably between the larvae of wasp species (the parasitoids) and the larvae of aphids (the prey): (1) the parasitoid lays its eggs in the tissue of its victims, (2) the larvae, hatched from the parasitoid's eggs penetrate the tissues of the hosts' larvae. The parasite-host relationship benefits the parasite as long as the latter is not numerous. However, the host populations do not benefit from that relationship because they are globally weakened. When the host species is reduced to a small population, parasitism can lead to stochastic extinction of both species when the parasite is species-specific [6].

Phytophagy:

This type of consumption can be considered predation in some cases (when the consumed plant is totally or partly destroyed, e.g. herbivory) and not in others (when the plant is not destroyed but just a nonvital part of it eaten, e.g. frugivory when the embryo in the seed is not destroyed). Phytophagy enables the basic transfer of energy and circulation of organic carbonic compounds from the *primary producers* to the *consumers*.

4.2.2 Coevolution

Coevolution is a relationship where the involved species become interdependent for their survival across time at the individual level, i.e. the relationship takes place between two individuals of different species either for the benefit of both or for the benefit of one and a neutral effect for the other. This relationship has the same effect collectively at the group level. Coevolution complements the ecophenotype of these species, linking their evolutionary destinies. Coevolution establishes relationships between partner species with different tightness, from very loose and not obligate to extremely tight and obligate. By increasing order of tightness, we can enumerate the coevolutionary relationships: (1) commensalism, (2) ecological mutualism, (3) phenotypic coevolution, (4) symbiosis, and (5) endosymbiosis.

Commensalism [34]:

This form of coevolution is the loosest one. Species A and B share the same habitat, B benefiting from A, but conversely, A obtaining neither advantage nor disadvantage from B.

This case is exemplified by:

1. the house sparrow *Passer domesticus* and Humans,
2. the ancestors of modern dog and Humans in prehistorical times,
3. several non-pathological bacteria and their hosts,
4. epiphytic orchids and the trees on which they cling or
5. birds and the trees providing holes to the former.

In that coevolution type, species B's disappearance would not impair species A.

Ecological mutualism [34]:

Here, the related species A and B mutually benefit from one another. This relationship is tighter than in the case of commensalism. The disappearance of one partner necessarily impairs the other, though not necessarily fatally. In the tightest relationships falling in this category of coevolution, the partner species are obligate. Hereunder are several examples of mutualism:

- The vital cohabitation of passerine birds, the oxpeckers (*Buphagus*) and several large herbivorous mammals of the African savannah: the former rid the latter of itchy and often unhealthy ectoparasites, which, in turn, feed the birds.
- Many plant seeds are disseminated by frugivorous birds and bats in tropical regions. When the disseminator has a species-specific relationship with the plant species, the species involved are obligate partners. A dramatic case is the relationship in Mauritius between a tree, the Tambalacoque or dodo tree, and an extinct bird species, the dodo.

  The quasi-disappearance of the Tambalacoque (*Syderxylon grandiflorum*) is explained by the disappearance of the Dodo (*Raphus cucullatus*). This bird, related to pigeons, exterminated in the eighteenth century, was the only species whose gastric juices could corrode the very hard shell of its seed-carrying fruits. The animal also disseminated the seeds. At present, Dodo trees are revived and seem saved thanks to the artificial corrosion of the fruit shell via chemical and mechanical means. However, sadly, these plants are now strictly dependent on humans for survival.

Phenotypic mutualism [37]:
This type of interaction, often called "coevolution" or included in mutualism, is exemplified by the relationship existing between several pollinating insects and plant flowers (angiosperms), often species-specific and based on the convergent evolution of certain features (shape, colour, *pheromones*[3]). This type of coevolution, known for long but studied in detail for only ca 20 years has coerced the phenotypes of the partner species in such a way that they are functionally necessary to each other.

Symbiosis [34]:
Symbiosis superficially resembles a parasite-host relationship. It involves two partners whose sizes differ in two or more orders of magnitude, the giant organism hosting the tiny one. For example, the larger species is a vertebrate or a large invertebrate, and the smaller one is a microorganism. However, these two similar interactions are radically different because symbiosis benefits both partners, contrary to parasitism—a trophic relation.
An interesting case of symbiosis is a review, reported in the book chapter of Ann K. Dunn [10], on the association between luminous bacteria of the Vibrionaceae family and marine fishes or squid living in the abyssal zones, thus places not reached by sunlight. The bacterium provides light to the host, the latter being a source of glucose or iron to the microbe. The microorganisms are generally housed in a specialised organ of the host, generating a system lighting the surroundings of the latter. The mentioned author exemplified the case with *Vibrio fischeri* and a cephalopod species, the Hawaiian bobtail squid *Euprymna scolopes*. *Vibrio fischeri* is not an obligate symbiont but can live freely or as a probiotic species of the mollusc. The bacterium produces light to enable the vision of the squid; in return, the mollusc is a source of glucose, carbon and iron for the microbe. The association takes place significantly after the hatching of the squid, the bacterium being horizontally acquired by the host from seawater.

Endosymbiosis [34]:
Endosymbiosis is a highly tight hypothetical symbiosis that would have occurred about 2.7 billion years ago within the cell of the ancestors (they were unicells) of all the present eukaryotes. We owe Lynn Margulis (1938–2011) [27][4] an explanation of the origin of the eukaryotes involving endosymbiosis, which she defended

[3] Pheromones are volatile, hormone-like chemicals generally used by members of the same species (insects, mammals) to elicit social responses between themselves, e.g. attraction of reproductive mate.

[4] The cited article refers to Sagan, the second married name of Lynn Margulis, née Lynn Alexander. This brilliant Americam evolutionary biologist is known under the name Margulis, her first married name.

with so many brilliant arguments taken from comparative cytology and organic chemistry, that her theory, after a decade of scepticism, had became broadly recognised as standard. We will now remind the reader of some biological facts that will help him follow the explanations of the functioning of endosymbiosis. The Living World is divided into two types of cellular organisms:[5]

- The prokaryotes (bacteria and archaea): they are the simplest and smallest organisms. They are all unicellular. The cytoplasm of their cell does not contain any organelle, and has no nucleus. The only visible structures are the ribosomes. They generally contain one single circular chromosome.
- The eukaryotes (all the rest of the Living World): they are the most complex and by far the largest organisms. They gather unicellular and pluricellular organisms. The cytoplasm of their cell(s) includes, next to their ribosomes, a nucleus containing several linear chromosomes, and organelles. Importantly, among them, several organelles contain nucleic acid sequences, i.e. the mitochondria, the basal bodies connected to flagella, and plastids. All eukaryotes possess mitochondria and basal bodies. Only phototrophic eukaryotes contain plastids.

It was remarked that mitochondria and plastids had a striking resemblance with free-living bacteria [18, 25, 27, 33, 36]. This led these scientists and others to argue that these types of organelles had, in the Aechean era (−4 to −2.4 Ga), free-living bacterial ancestors that were engulfed and integrated into a greater cell, probably during the *Great Oxidation Event* 2.4 billion years ago. By the mid-1900s, alpha-protebacteria were supposed to be the ancestors of the mitochondria, and Cyanobacteria, those of the plastids. The earlier authors supposed these first host cells to be proto-eukaryotic amitochondriate amoeboids. But more recent authors, in the light of novel findings, have been considering that the first host cells were probably relatively large archaea that evolved to the first amitochondriate eukaryotes; and that later, their descendants acquired the one or the two types of DNA- or RNA-bearing organelles [16, 17]. These organelles have become such obligate partners to the eukaryotic host that they entirely became part of the latter.

4.2.3 Competition

This type of interaction was initially proposed by Darwin in 1859 as a component of the *struggle of life* [8], opposing the deleterious effect of Malthusian population growth, i.e., the natural tendency of each species to grow exponentially without bounds when the individuals are not subject to density-dependent limiting factors.

[5] This divison only concerns the cellular organisms, and thus excludes the viruses, considered at the edge of life [26].

The Darwinian struggle of life comprised the trophic relationships we already saw in Sect. 4.2.1 and competition, which is the subject of this subsection.

More than three-quarters of a century later, the Russian biologist Gregory F. Gause (1910–1986) pointed out that the struggle for life, as mentioned above, needed scientific validation [14, ch. 1]. He proposed that it be done (1) by decomposing the complex behavioural patterns into elementary processes and (2) by applying protocols to these elementary processes, based on experimental methodology, and mathematical modelling inspired by the works of many contemporaneous scientists.[6] He gave an overview of the models of these authors, and then mainly dedicated his work to the concept of competition.

Contrary to the types of interactions seen previously, which necessarily take place between different species of a biocoenosis, competition mutually involves any living beings within or between species. In the first case, we will speak of intraspecific competition; in the second, we will speak of interspecific competition.

Individuals and groups of organisms compete for space, resources (energy, nutrients), sexual and social partners, and all vital factors ensuring their survival and health. These eco-ethological characteristics infer a level of agonistic behaviour that, with other behaviour patterns like those seen previously (see Sects. 4.2.1 and 4.2.2), governs the occupation of space by individuals. These features do not occur randomly but are organised to ensure the sustainability of the whole.

4.2.3.1 Intraspecific Competition

Intraspecific competition, like trophic relations, is necessary for organisms to counter Malthusian population growth. For example, assume a theoretical population of bacteria starting initially with one individual, with discrete generations, a mean generation of 20 mn, and individuals reproducing by binary fission. At each generation g_i, a bacterium is replaced by two identical offspring. Figure 4.1 illustrates the development of this population when no density-limiting factors hamper it, e.g. no mortality, no competition, no selection, no mutation, no migration and no generation overlap.

If at generation g_0, the initial number of bacteria is $N_0 = 1$, the total number of bacteria N_k at generation g_k is:

$$N_k = 2^k \quad \text{(see legend of Fig. 4.1)} \tag{4.1}$$

In this population, after 288 generations (4 days), we would obtain about 5×10^{86} offspring, more than the total number of atoms in the universe![7] Imagine such a colony enclosed in a Petri dish containing limited nutrients; that shows the

[6] E.g. Ivan Pavlov (1849–1936), Ludwig Plate (1862–1937), Ronald Ross (1857–1932), Vito Volterra (1860–1940) and Alfred Lotka (1880–1949).

[7] This number being estimated between 10^{78} and 10^{82}.

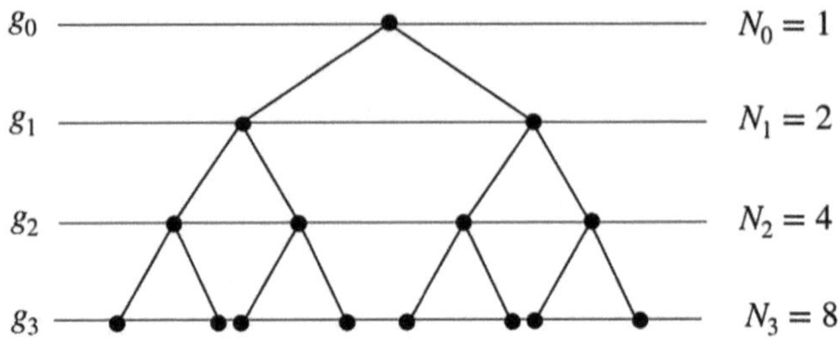

Fig. 4.1 Tree-like diagram representing the descendants of a bacterial population across the first four generations, stemming from one bacterium. The microbe splits into two new equal individuals at each generation;each bullet corresponds to an individual before it divides. $N_i = 2^i$ is the total number of the population at generation g_i

Malthusian growth cannot continue forever and that competition must arise at a given instant. In real life, the numbers of offspring vary between the generations, which overlap and can sometimes be almost continuous. The developments are then more complex to analyse mathematically, but the ideas remain identical. We will see different mathematical models used to describe intraspecific competition in Chap. 13.

Intraspecific competition is a process within species opposing the exhaustion of the resources that are not unlimited.

4.2.3.2 Interspecific Competition

This notion has already been glimpsed by Darwin [8]:

> "As the species of the same genus usually have, though by no means invariably, much similarity in habits and constitution, and always in structure, the struggle will generally be more severe between them, if they come into competition with each other, than between the species of distinct genera."

Interspecific competition between species A and B occurs when (1) they are sympatric, (2) their ecological needs and trophic levels are similar, and (3) their nutrient and energy sources are scarce. These conditions induce A and B to compete for the same resources and territory. In such a definition, many parameters are involved, e.g., mass and size of individuals, energy expenditure, life history, behaviour and lifestyle. [Successive formalisations of this state of affairs have been called ecological niche, e.g. sensu Elton [12] or Hutchinson [15] (cf. Chap. 13)]. Frequently, between species A and B, the more demographically dynamic and aggressive taxon supplants the other one. If one species arrives in the sympatry zone later than the other one, the most successful competitor is often the newcomer.

The species mentioned above, A and B, can be either (1) related species that did not evolve enough to attain niche differentiation or (2) ecological equivalents—unrelated species that, by parallel evolution, share identical feeding habits and habitat exploitation. Generally, two competing species cannot share the same habitat; they will exclude one another territorially or eliminate each other—a phenomenon that has overwhelmingly been observed and stressed by Gause [14], hence its being termed “Gause exclusion principle”. This phenomenon can be more or less pronounced according to food disponibility and seasons [14, 21].

Cases of competition where one species supplanted another are unfortunately countless, often due to voluntary or involuntary introduction by humans for ages. In many cases, the native species are replaced mainly by the introduced ones, to the point that the supplanted species almost disappears. Hereunder, we give a few examples reported in the literature [14, 21, 30]:

1. the intertidal barnacles sharing a substrate of limited surface,
2. the sympatric Russian crayfishes *Potamobius astacus* and *P. leptodactylus*, two related taxa that used to be allopatric and eventually overlapped in an area where they competed (at the expanse of *Potamobius astacus*),
3. the native fish *Schizothorax intermedius* and the voluntarily introduced carp *Cyprinus albus* in lakes of South Central Asia,
4. the native trout *Salvelinus fontinalis* and the voluntarily introduced perch *Perca flavescens* in lakes of northeastern United States,
5. reptiles like skinks of the genus *Emoia* in the Marshall Islands, and the lizards *Lacerta melisellensis* and *L. sicula* on 18 islands of the Dalmatian coast,
6. birds like the hawk eagles *Spizaetus* on the western Sumatran islands, the babblers of the genus *Malacopteron* in Indonesia and the two species of nightingale, i.e. the thrush nightingale (*Luscinia luscinia*) and the common nightingale (*L. megarhynchos*) in Poland.

Globally, there are two schools of thought regarding the generality of interspecific competition. The first one, considering that competition is an interaction which plays a crucial role in the formation of biocoenoses, mainly gathers the neo-darwinian school of thought represented by one of the leading geneticists and evolutionary biologists, the Ukraino-American Theodosius Dobzhansky (1900–1975). The latter showed, in an article published in 1937, that incipient interspecific competition existed at F_1, between the hybrid products A × B of crossings between two speciating taxa A and B of the fruit fly *Drosophila obscura* and the pure products of each taxon (A or B). The hybrids displayed lowered fitness compared to the pure taxa, being less viable and partially sterile, which induced the progressive disappearance of the hybrids. For this scientist, this finding could be a general rule applicable to similar microevolutionary cases [9].

The other school of thought was represented by Warder C. Allee (1885–1955) and colleagues who had a more altruistic point of view concerning the formation of the animal part of the biocoenoses. These scientists considered competition, though important, to play a lesser role, contrary to other major interactions in the elaboration of communities [1]. These colleagues pointed out that competition mainly plays an important role among plants (1) for access to space and light, with a crucial consequence for the formation of vegetation layers and (2) for pollination when the pollinators get scarce. In animals, though there are spectacular cases of competition, as seen in the example above, competition is less general and relatively infrequent compared to other types of interactions (cooperative and trophic). Interestingly, Ernst Mayr thoroughly treated the problem of interspecific competition in Chapter 4 of his masterwork "Animal Species and Evolution" [21, ch. 4], showing, among examples of higher vertebrates, notably birds, that many species had mechanisms avoiding competition. When food is abundant, different potentially competing species living in sympatry will specialise according to slight preferences. Several examples are presented in the literature:

- Four sympatric species of triclads (free-living flatworms) belonging to two related genera living in British lakes, *Polycelis nigra*, *P. tenuis*, *Dugesia lugubris* and *D. lacteum* feed on the same oligochaetes and arthropodes taxa, but avoid competition thanks to distinct preferences: *D. lacteum* prefers crustacea, *D. lugubris* molluscs, and each of *Polycelis* species different types of oligochaetes [23].
- The bustards (order Otidiformes) are large terrestrial birds adapted to dry grasslands and steppes; they are generally omnivorous, their diet being extensive and composed of leaves, fruit and small animals (invertebrates, little rodents). Two sympatric (and endangered) species of this order, the little bustard (*Tetrax tetrax*) and the great bustard (*Otis tarda*), compete relatively fiercely in European and Asiatic cereal farmlands, but thanks to distinctive preferences, they manage to share the same habitat: the little bustard breeding preferentially in the centre of the agrarian habitat where they periodically form leks prior to their reproduction, whereas the great bustard are more mobile in their habitat choice, as they do not practise leks [32].

4.3 Interactions Between Biocoenosis and Abiotic Factor System

4.3.1 Decomposition and Mineralisation

4.3.1.1 Presentation of the Process

The process of decomposition/mineralisation is a unidirectional relationship between a *biocoenosis* and its *abiotic factor system*, precisely an action from the former to the latter. Its functioning, which I will sketch an overview hereunder, was recently described in detail in two review articles [3, 4]. *Decomposition* is the process where organic material is progressively broken down to its tiniest elements, which can be assessed by estimating the flux of released organic carbon between two states. One must remember that intact organic molecules almost exclusively exist in live organisms. Decomposition and mineralisation occur in organisms' decaying dead bodies and excreta. The chain of heterotrophic organisms feeding on the same decaying matter during the successive stages of this process eventually releases portions of the organic matter decreasing in size. The final result of this process is *mineralisation*, whereby the once-organic material is reduced to ions or simple chemical compounds (e.g. mineral salts)—nutrients to the primary producers (*photoautotrophs* or *chemoautotrophs*).

- In continental ecosystems, this phenomenon occurs in plant litters, detritus of unicellular organisms and fungi, and carrion when death is non-consumptive (i.e. not resulting from predation), accumulated in the soils. It is thus a local phenomenon; the dead organic matter remains mainly circumscribed in a given territory unless it falls in a watercourse where it can be drifted or is actively transported by a predator, sometimes to long distances. But these events are relatively rare.
  The article by Noah Fierer and colleagues published in 2009 [13] was one of the first comprehensive cross-biome comparisons performed on belowground biocoenoses, where the process of decomposition/mineralisation takes place in terrestrial habitats. Microbial biomass data were collected from 400 individual soils distributed between seven terrestrial biomes, namely (1) the tundra, (2) boreal forest, (3) temperate coniferous forest, (4) temperate deciduous forest, (5) temperate grassland, (6) desert and (7) tropical moist forest. Biomasses, measured in $\mathrm{g\,C\,m^{-2}}$ of different decomposer/mineralisers were compared between these biomes: Acari, Collembola, Enchytraeids, Nematodes, Earthworms, Microbes (cf. Table 4.1[8]).
  The table shows discrepancies between the different biomasses. Across all the biomes, microbial biomass is the most abundant, followed by the biomass of the Annelids, the other fossorial animals (Acari, Collembola, Nematodes) being less represented. We can see in this sample that the chains of decomposers/mineralisers are highly structured, and though differently organised from one biome to the others, follow a similar pattern.
- In the aqueous habitats, decomposition and mineralisation happen:

Table 4.1 Biomasses of categories of decomposers and mineralisers in terrestrial biomes. Mean biomasses (in parentheses, *SEM* stard error of the means), *ND* no data

	Acari	Collembola	Enchytraeids	Nematatodes	Earthworms	Microbes
Tundra	0.13 (0.02)	0.10 (0.02)	0.99 (0.16)	0.18 (0.05)	1.4 (0.80)	136 (20)
Boreal for.	0.20 (0.02)	0.06 (0.01)	0.32 (0.06)	0.08 (0.02)	0.28 (0.18)	57 (6)
Temp. conif. for.	0.35 (0.12)	0.24 (0.06)	0.80 (0.10)	0.10 (0.01)	1.2 (0.6)	175 (26)
Temp. decid. for.	0.23 (0.03)	0.12 (0.03)	0.64 (0.16)	0.25 (0.11)	2.0 (0.4)	116 (9.4)
Temp. grassl.	0.18 (0.03)	0.16 (0.05)	0.31 (0.07)	0.36 (0.08)	3.8 (1.9)	131 (10)
Deserts	0.01 (0.00)	0.01 (0.00)	ND	0.01 (0.00)	ND	43 (8)
Trop. moist. for.	0.16 (0.05)	0.02 (0.01)	ND	0.01 (0.00)	4.9 (1.78)	203 (20.84)

[8] Data extracted from [13, Table 1].

- within drifting dead organic matter in the form of *marine snows*, i.e. aggregates of decomposed carrion of light-weighted animals and particulate organic matter (POM), i.e. detritus of autotrophic organisms (phytoplankton, macroalgae, plants) drifting in the water columns;
- within sinking dead organic matter, generally corpses of dead sharks or mammals, in particular whales—a phenomenon called falls. An impressive example of this phenomenon is the whale falls—the fall of whale carcasses to the bottom after their death—. Indeed, the sediments beneath whale carcasses would receive a pulse of labile organic matter equivalent to almost 2000 years of organic-carbon flux [29].

Necromass, i.e. the decaying organic matter found in dead organisms and excreta, is differentiated between autotrophically and heterotrophically derived necromass (see Definition 4.2), a distinction that has a biological meaning [3].

Definition 4.2 (Typology of the Necromasses) Autrophically derived necromass is the necromass derived from dead autotrophs (e.g. plants, algae and phytoplankton).
Heterotrophically derived necromass is the necromass derived from dead heterotrophs (mainly animals).

A very important notion to understand the destiny of the necromass in an aqueous environment is that of resource subsidies, as explained in the article by Richardson and colleagues [24] (see Definition 4.3). As described at the beginning of the introduction of the article, ecosystems were seen (until the cited paper) as closed entities as far as the circulation of matter and energy was concerned. This circulation was seen as an autochthonous phenomenon. The authors cited above pointed out that in aqueous habitats, the source of nutrients and energy has a major allochthonous component: the nutrients senso latu circulate across the boundaries of the ecosystems, even across different biomes.

Definition 4.3 (Resource Subsidies) Resource subsidies are flows of matter and energy (nutrients) circulating from one ecosystem to another, as is the rule in aqueous habitats. These nutrients can be chemicals at the disposal of autotrophs, plant or alga detritus, plankton live or dead, and can include nekton [2]. The foraging active animals secondarily adapted to the aqueous habitat (amniotes) are not considered resource subsidies [24].

Dead matter is brought into aqueous ecosystems as resource subsidies from the terrestrial, freshwater and marine realms.

4.3.2 Genesis of Biogenic Molecules from Metabolism

The two best-known biogenic molecules of this category are gaseous dioxygen and limestone.

Gaseous dioxygen (O_2) is a waste product of photosynthesis, a chemical reaction that has continuously occurred within the photoautotrophs since their origin, 2.5 Ga ago. The ancestors of the green lineage (or Viridiplantae) that comprises the green algae and the land plants, and of the other autotrophs (brown and red algae), were very probably heterotrophic unicellular eukaryotes that incorporated a cyanobacterium (or cyanobacterium-like cell) through a process of endosymbiosis (the ancestors of the plastids) [19]. These events marked the beginning of the Proterozoic aeon, when the first cyanobacteria appeared on an Earth with an atmosphere where gaseous oxygen was absent. It is the continuity of the photosynthetic activities since that remote epoch, which explains the accumulation of O_2 in the atmosphere and hydrosphere until the present, at the disposal of the aerobic organisms, the living beings that breathe with oxygen acting as a terminal electron acceptor (almost all the multicellular organisms and the aerobic microorganisms).

Limestone ($CaCO_3$) is a precipitate of calcium previously dissolved in water. Though this chemical reaction can be obtained via non-biological chemical reactions, most of the limestone outcrops found presently at sea levels (cliffs) as well as at high altitudes are due to the lengthy deposit of foraminifera tests, coral calcareous skeletons and mollusc shells accumulated during the last 540 million years. These carbonate sediments are thus formed from a mixture of calcium and carbon atoms previously processed by the metabolism of the organisms and piled after their death. Limestone deposits form most of the carbonate rocks, accounting for about one fourth of all the sedimentary rocks. The limestone sediments frequently embed fossils that give invaluable paleontological information [5].

4.4 Interaction within the Abiotic Factor System

Chemical agents:

Some basics of chemistry are prerequisites for this book (see Sect. 10.1). To have a general idea of the existing elements composing matter in the universe, the reader can refer to Mendeleev's table (Fig. 10.1). Elements play predominant roles in the constitution of living beings, thus of biocoenoses. There are six of them that are central in the constitution of the vital molecules: (1) *carbon* (C), (2) *hydrogen* (H), (3) *nitrogen* (N), (4) *sulphur* (S), (5) *carbon* (O), and (6) *phosphorus* (P). The first five mentioned elements are the most abundant throughout the universe [22]. These elements are suited for developing the complex structures and functions of living systems (see Table 2.1).

Physical agents:

Other abiotic agents are crucial. Among the most significant, we can enumerate:

1. Temperature,
2. Humidity,
3. Solar and other electromagnetic radiation,

4. Wind,
5. pH (in aqueous environments).

These items vary proportionally according to various physical and geographical parameters. (1) Available sunlight, (2) water, (3) carbon dioxide and (4) gaseous dioxygen vary significantly, namely, with altitude, latitude, decreasing atmospheric pressure, maritime or continental influence, exposition to winds or wind intensity. These variations and the resulting relationships between the available hydrogen, carbon, and oxygen atoms will influence vegetation type, hence the biocoenosis [28].

References

1. Allee WC et al (1949) Principles of animal ecology. Saunders Company, Philadelphia, Pennsylvania, USA
2. Allen D (2013) Nutrient subsidies from nekton in salt marsh intertidal creeks. Limnol Oceanogr 58:1048–1060, https://doi.org/10.4319/lo.2013.58.3.1048
3. Benbow E et al (2019) Necrobiome framework for bridging decomposition ecology of autotrophically and heterotrophically derived organic matter. Ecol Monogr 89(1):e01331, https://doi.org/10.1002/ecm.1331
4. Benbow E et al (2020) Death and decomposition in aquatic ecosystems. Front Ecol Evol 8:17, https://doi.org/10.3389/fevo.2020.00017.
5. Boggs S Jr (2006) Principles of sedimentology and stratigraphy. 4th edn, Pearson Prentice Hall, Upper Saddle River (NJ).
6. Boots M, Sasaki A (2002) Parasite-driven extinction in spatially explicit host-parasite systems. Am Nat 34:706–713, https://doi.org/10.1086/339996
7. Brockhurst MA et al (2014) Running with the Red Queen: the role of biotic conflicts in evolution. Proc R Soc B: Biol Sci 281:20141382, https://doi.org/10.1098/rspb.2014.1382.
8. Darwin C (1859) On the origin of species by means of natural selection, or the preservation of favoured races in the struggle for life. Murray, London
9. Dobzhansky, T. (1937) Genetic nature of species differences. Am Nat 71:404–420.
10. Dunn, A. (2012) *Vibrio fischeri* Metabolism: symbiosis and beyond. Adv Microb Phys 61:37–68, https://doi.org/10.1016/B978-0-12-394423-8.00002-0
11. Ehrlich PR, Raven PH (1964) Butterflies and plants: a study in coevolution. Evolution 18:586–608, https://doi.org/10.1111/j.1558-5646.1964.tb01674.x
12. Elton CS (1927) Animal ecology. McMillan Co, New York.
13. Fierer N et al (2009) Global patterns in belowground communities. Ecol Lett 12:1238–1249, https://doi.org/10.1111/j.1461-0248.2009.01360.x
14. Gause GF (1934) The struggle for existence. Williams & Wilkins, Baltimore
15. Hutchinson GE (1957) Concluding remarks. Cold Spring Harbor Symp Quant Biol 22:415–427
16. Keeling PJ, McCutcheon JP. (2017) Endosymbiosis: the feeling is not mutual. J. Theor. Biol 434:75–79, https://doi.org/10.1016/j.jtbi.2017.06.008.
17. Lane N (2017) Serial endosymbiosis or singular event at the origin of the eukaryotes? J Theor Biol 434:58–67, https://doi.org/10.1016/j.jtbi.2017.04.031
18. Lederberg J (1952) Cell genetics and heredity symbiosis. Phys Rev 32:403–430, https://doi.org/10.1152/physrev.1952.32.4.403.
19. Leliaert F et al (2012) Phylogeny and molecular evolution of the green algae. Crit Rev Plant Sci 31:1–46, https://doi.org/10.1080/07352689.2011.615705
20. Marshall CR (2017) Five palaeobiological laws needed to understand the evolution of the living biota. Nature Ecol Evol 1:0165, https://doi.org/10.1038/s41559-017-0165
21. Mayr E (1963) Animal species and evolution. Belknap Press of Harvard UP, Cambridge (MA).

22. National Research Council (US) Committee on Planetary Biology and Chemical Evolution - Space Study Board (1990) The search for life's origin. National Academic Press, Wahington DC.
23. Reynoldson TB (1966) The distribution and abundance of lake-dwelling triclads - towards a hypothesis. Advan Ecol Res 3:1–71, https://doi.org/10.1016/S0065-2504(08)60309-8
24. Richardson JS et al (2010) Resource subsidies across the land-freshwater interface and responses in recipient communities. River Res Appl 26:55–66, https://doi.org/10.1002/rra.1283
25. Ris H, Plaut W (1962) Ultrastructure of DNA-containing areas in the chloroplast of *Chlamydomonas*. J Cell Biol 13(3):383–391, https://doi.org/10.1083/jcb.13.3.383
26. Rybicki E (1990) The classification of organisms at the edge of life, or problems with virus systematics. S Afr J Sci 86: 182–186.
27. Sagan L (1967) On the origin of mitosing cells. J Theoret Biol 14:225–274
28. Smith WK et al (2008) Alpine forest. In: Jørgensen SE, Fath BD (eds) Encyclopedia of Ecology, p. 144–153, Academic Press, Oxford (UK).
29. Smith CR, Baco AR (2003) Ecology of whale falls at the deep-sea floor. Oceanogr Mar Biol 41:311–354.
30. Sorjonen J (1986) Mixed singing and interspecific territoriality - consequences of secondary contact of two ecologically and morphologically similar nightingale species in Europe. Orn Scand 17:53–67
31. Strotz LC et al (2018) Getting somewhere with the Red Queen: chasing a biologically modern definition of the hypothesis. Biol Lett 14(5):20170734, https://doi.org/10.1098/rsbl.2017.0734
32. Tarjuelo R et al (2017) Intraspecific and interspecific competition induces density-dependent habitat niche shifts in an endangered steppe bird. Ecol Evol 7(22):9720–9730, https://doi.org/10.1002/ece3.3444
33. Thrash JC et al (2011) Phylogenomic evidence for a common ancestor of mitochondria and the SAR11 clade. Sci Rep 1(1):13, https://doi.org/10.1038/srep00013
34. Van Beneden P-J (1883) Animal parasites and messmates. Paul Kegan, Trench & Co, London.
35. Van Valen L (1973) An evolutionary law. Evol Theory 1:1–30.
36. Wallin IE (1927) Symbionticism and the origin of species. Williams & Wilkins Co, Baltimore.
37. Wöhrmann K, Jain SK (1990) Population Biology, Springer Verlag, Berlin.

Chapter 5
Example of an Ecological System: The Kelp Zone of Central California

This example concerns a complex ecosystem, the Pacific littoral area of *Central California*, mainly between Monterey Bay and Point Conception (Fig. 5.1). This habitat includes a wealthy and typical *subbiome*—the *kelp forest*—and the nearby continental shelf (above a depth of ca 100 m). I propose to call the ecosystem formed by these two interconnected habitats the kelp zone (KZ). Kelp forests grow on a narrow belt at shallow depth (generally above a depth of 40 m), forming rocky reefs in (I quote) "temperate oceans of both the southern or northern hemisphere" or "in some subtropical areas that experience considerable upwelling"[1] [10].

The Californian kelp forests are remarkable due the great *species richness* of their constitutive kelp, i.e. brown algae of different sizes, shapes and aspects belonging to a single order, the Laminariales [7], and to the fact that they are home to the Giant kelp (*Macrocystis pyrifera*), by far the largest and most complex representative of the group and key species of the total biocoenosis of the KZ. The *thallus* of the Giant kelp—a strong lignified stipe—can grow up to 60 m, developing *blades* at irregular intervals [51]. The Bull kelp (*Nereocystis luetkeana*)) is another giant brown alga, though smaller than the Giant kelp—its thallus can grow up to 36 m—with a holdfast of 40 cm. Its stipe is hollow and has many blades on its top. It is the second dominant species of the habitat. Two other kelp species are characteristic of this habitat, namely the Elk kelp (*Pelagophycus porra*) and the Feather boa kelp (*Egregia menziesii*) [10]. The kelp forest described here, as we shall see, shelters a rich biodiversity that has been thoroughly studied, thus making it possible to outline

Supplementary Information The online version contains supplementary material available at https://doi.org/10.1007/978-3-031-90656-5_5.

[1] *Upwellings* are vertical currents sprouting cold waters from the depths of an ocean up to its warmer surface. These water currents carry nutrients from bottom layers upward to upper layers of the water column.

M. Salomon, *Integrative Ecology*, https://doi.org/10.1007/978-3-031-90656-5_5

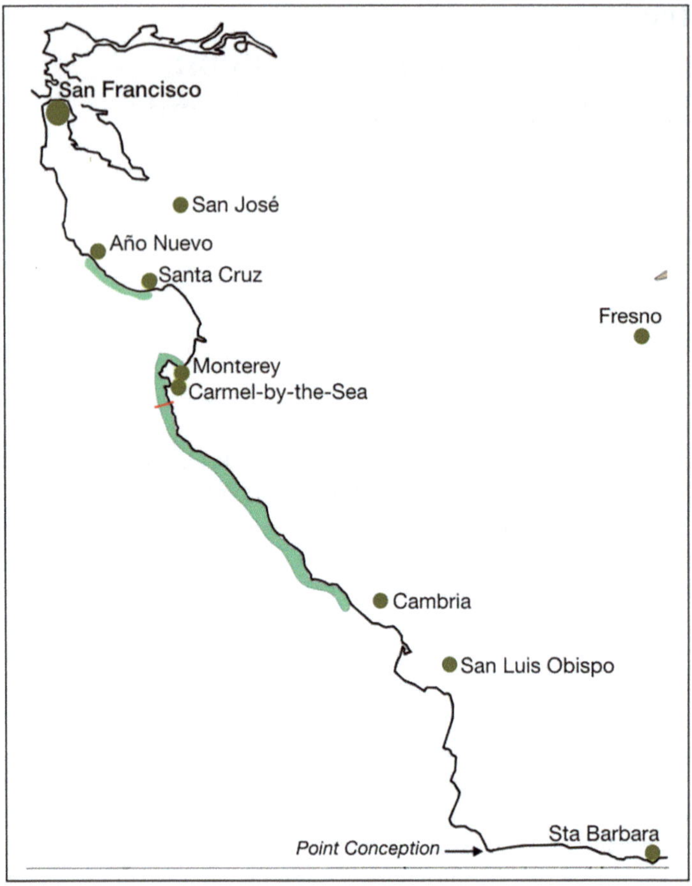

Fig. 5.1 Range of the Central Californian kelp forests (map performed with QGIS). The main species of this formation are the Common giant kelp *Macrocystis pyrifera* and the Bull kelp *Nereocystis luetkeana*; the range is represented by the two *greenblue* stretches along the coastline [23]. *Red line* segment under Carmel-by-the-Sea: transect studied by McLean [34]

a quasi-total ecosystem.[2] Besides, it contains an informative limited biocoenosis, involving (1) these kelps, (2) a herbivore—the purple sea urchin (*Strongylocentrotus purpuratus*), and (3) a carnivore—the sea otter (*Enhydra lutris*) , highlighting the functions of this ecosystem that was on the brink of disappearing because of a man-induced disequilibrium—the overhunting of the sea otter for its fur. That explains why this milieu has been extensively studied since the 1920s.

[2] No precise data exist for the KZ described here, but 130 species of algae and 800 species of animals were reported in the KZ of Baja California, a similar ecosystem as ours. The species richness is probably similar in the KZ here studied.

Thus, the study of this kelp zone provides us with (1) an almost total biocoenosis (TB) and its environment and (2) a limited biocoenosis (LB) based on the trophic relationships between kelps, sea urchins or abalones, and sea otters. The first type of research [10, 34] gives an extensive list of species, conveying a clear picture of the structure of the habitat, but these works needed to provide more information on how the ecosystem works. On the contrary, the second category of research [17] highlights the functions of this ecosystem but only within a small part of it, and the difficulty, after that, lies in the extrapolation of the findings to the total ecosystem. The study about the LB, mainly focusing on the trophic relationships between kelps (the producers), sea urchins and abalones (the primary consumers), and sea otters (the secondary consumers), explains how the TB equilibrium results from a limited portion of the entire habitat. These studies were very important because of their practical interest. They dramatically revealed the fragility of this ecological equilibrium when the quasi-extermination of the sea otters at the beginning of the twentieth century led to the destruction and quasi-disappearance of the coastal kelp forests because of the pullulation of sea urchins (owed to the quasi-absence of the otters, their sole predators).

The two narrow turquoise surfaces along the coast (Fig. 5.1) represent the studied *kelp zone*.

5.1 Description of an Almost Total Biocoenosis

An almost *total biocoenosis* (TB) has been surveyed with its producers, consumers and decomposers/mineralisers in the Monterey Bay area (central California) in several publications [10, 34].

5.1.1 The Primary Producers

They are constituted by different species, i.e. various taxa of *kelps*, coralline algae, marine flowering grasses and *phytoplankton* forming *submarine forests*.

5.1.1.1 Within the Intertidal Subzone

This subzone is characterised by a depth range of 0–4.50 m, extending to 0–100 m offshore. Kelps like *Postelsia palmaeformis* (Fig. 5.2a) are close to the top limit on rocky points, “receiving the full force of the surf”. The feather boa kelp (*Egregia menziesii*) (Fig. 5.2b) is found at about the mean low tide level in protected areas. Just below, at the lowest level of the intertidal zone and upper level of the subtidal area (down to a depth ≤ 3 m), the coralline algal species *Calliarthron sp.* (Fig. 5.2c) form dense beds of branching individuals.

Fig. 5.2 Photographs of kelps of the intertidal subzone-exposed areas. (**a**) The palm seaweed *Postelsia palmaeformis* (courtsey of Peter Pearsall); (**b**) The feather boa kelp *Egregia menziesii* (courtesy of Steve Lonhart, NOAA); (**c**) The coralline alga *Calliarthron sp.* (credit: Steve Lonhart)

The most common autotrophic species in protected areas is a flowering marine plant, the Scouler's surfgrass (*Phyllospadix scouleri*) (Fig. 5.3a). Other prevailing kelp species are mentioned, i.e. *Dictyoneurum californicum* (Fig. 5.3b) on level surfaces, *Cystoseira osmundacea* (Fig. 5.3c) forming thick layers at depths down to 300 m, and *Desmarestia herbacea* (Fig. 5.3d).

5.1.1.2 Within Subtidal Subzone I

The subzone is characterised by a depth range of 4.50–18 m, extending to 100–200 m offshore. The Bull kelp (*Nereocystis luetkana*) (Fig. 5.4a) "predominates in the exposed areas ... where depth rapidly increases"; whereas in the more sheltered areas, "large stands of the Giant kelp (*Macrocystis pyrifera*) (Fig. 5.4b) are found".

5.1.1.3 Within Subtidal Subzone II

This area has a depth range of 18–150 m and an extension >200 m offshore. Succeeding to subtidal zone I towards the horizon, this area comprises a steep wall diving into a continental shelf, slightly declining between 90 and 150 m deep. The algal species replacing the autotrophic community of subtitle zone I are:

- the Southern Stiff-stiped kelp (*Laminaria setchellii*) growing on the top of the wall mentioned above in dense forests (Fig. 5.5a),
- the Red alga *Fryeela gardneri* which "produces an iridescent horizontal blade" [34, p. 98] on a stratum at ca 60 m and below (Fig. 5.5b),

Fig. 5.3 Photographs of kelps of the intertidal subzone-protected areas. (**a**) Scouler's surfgrass *Phyllospadix scouleri* (Courtesy of Chad King, NOAA); (**b**) Brown alga *Dictyoneurum californicum* (credit: Steve Lonhart); (**c**) Chain bladder kelp *Cystoseira osmundacea* (credit: Steve Lonhart); (**d**) Acid kelp *Desmarestia* sp. (credit: Steve Lonhart)

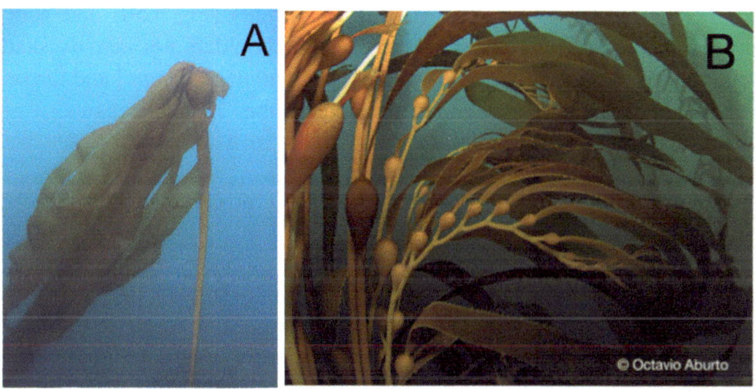

Fig. 5.4 Photographs of kelps of the subtidal I subzones. (**a**) Bull kelp (*Nereocystis luetkana*) (credit: Steve Lonhart); (**b**) Giant kelp (*Macrocystis pyrifera*) (courtesy of Octavio Aburto)

Fig. 5.5 Photographs of kelps and seaweeds of the subtidal II subzone. (**a**) Southern Stiff-stiped kelp (*Laminaria setchellii*) (credit: Steve Lonhart); (**b**) *Fryeela gardneri* (credit: Steve Lonhart); (**c**) Stalk kelp (*Pterygophora californica*) (credit: Steve Lonhart); (**d**) Streamer kelp (*Costaria costata*) (credit: Steve Lonhart); (**e**) *Polysiphonia paniculata* (Sandra Lindstrom)

- on the continental shelf, a dense forest of Stalk kelp (*Pterygophora californica*) (Fig. 5.5c), Streamer kelp (*Costaria costata*) (Fig. 5.5d) and acid weed *Desmarestia herbacea* (Fig. 5.3d). "On tops of boulders not heavily shaded by Pterygophora", other kelps [*Dictyota* sp., *Plocamium pacificum*] and the seaweed *Polysiphonia paniculata* sp. (Fig. 5.5e) are common.

5.1.1.4 About the Phytoplankton and the POM

Phytoplankton comprises microscopic oxygenic photosynthetic organisms that drift in suspension in the waters of the intertidal and subtidal zones of the *kelp zone*. It includes unicellular eukaryotes (cf Fig. 5.6) and cyanobacteria[3]. In the upper layers of the open ocean, as phototrophs (organisms taking in their energy from solar light to trigger their metabolism), many of the marine photosynthetic protists[4]

[3] Formerly called greenish-blue algae because of their colour and the fact that they often dye the waters surrounding them with a greenish-blue tinge prior to their true determination as bacteria via modern technologies.

[4] Though modern systematics do not recognise protists as a natural group (in the sense of phylogeny, they are paraphyletic), it is convenient for ecologists to maintain the term 'protist'

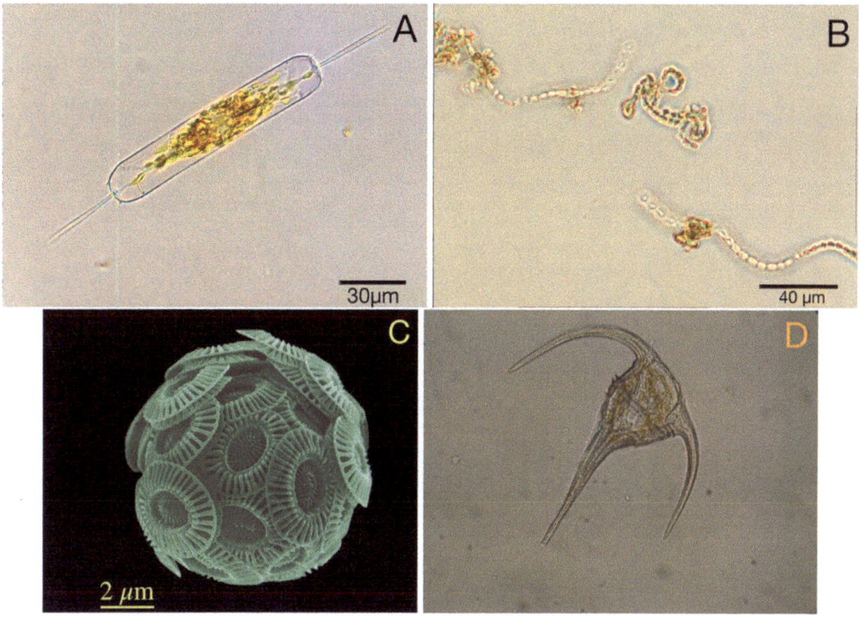

Fig. 5.6 Photographs of some photosynthetic unicellular eukaryotes. (**a**) a diatom: *Ditylum brightwellii* (Courtesy of Gwenaëlle Gain, IBPC, Paris, and of RCC, Roscoff, France); (**b**) a diatom: *Leptocylindrus danicus* (Credit:Gwenaëlle Gain, IBPC, Paris, and RCC, Roscoff, France); (**c**) a coccolithophore: *Emiliana huxleyi* (credit: Stig Bjarte Haugen, courtesy if the IMR); (**d**) a dinoflagellate *Ceratium* sp. (Credit: Lars-Johan Naustvoll, courtesy of the IMR)

functionally play the role of the plants and multicellular algae of the terrestrial and littoral ecosystems [18]. However, contrary to most vascular plants, they are not strictly autotrophic (i.e. taking all their carbon from the CO_2 dissolved in the atmosphere or hydrosphere). They are thus *mixotrophic* (i.e. partly autotrophic, partly heterotrophic). They are *photoautotrophic* when the environmental conditions are normal (when they can freely and abundantly take in solar light energy). But in rare stress situations, e.g. when sunlight energy is lacking, they can adopt a *chemoheterotrophic* mode of nutrition to survive [22]. With the detritus of kelps, frequently deriving from the withering blades of giant kelps (*Macrocystis pyrifera*), they form a bulk of organic matter called *particulate organic matter* (POM). This biomass is food for ca 70% of the animals of the ecosystem, mainly invertebrates [35, 38]. The POM is scattered through all the subzones of the kelp zone.

to denote the marine unicellular eukaryotes composing the microplankton (planktonic species of a size less than 0.2 mm).

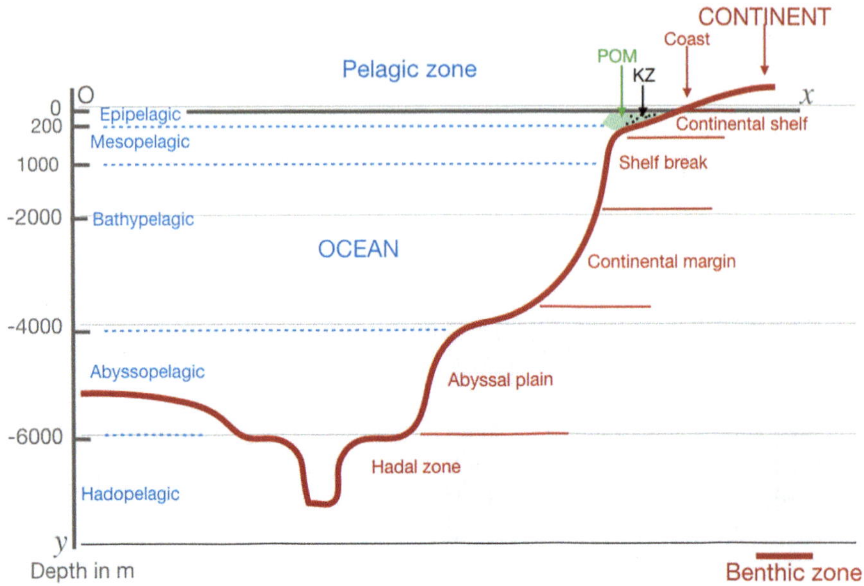

Fig. 5.7 Cross-section of the transition between continent and ocean from the littoral to the pelagic zone of the open ocean. In *grey*, the coordinate system with arbitrary origin O, x representing the earth's tangential lengths and y vertical depth. The benthic zone, in brown, corresponds to the habitat comprising the sea bed and its direct environment (sediments below and the immediate oceanic layer above). The pelagic zone is the water column of the open ocean divided into vertical subzones (epipelagic, ..., hadopelagic) corresponding to definite physicochemical and biological characteristics (see text). In *green*, the location of the POM. *Black-dotted* surface, the kelp zone (KZ)

- A census of eukaryotic unicellular phytoplankton species of the California current system[5] has been carried out [52]. This study reports 404 taxa, i.e. 179 diatoms,[6] 58 coccolithophores, 139 dinoflagellates or silicoflagellates, and 28 other categories.
  The microorganisms described above are members of the *holoplankton*, organisms remaining planktic for all their life. As a reminder, plankton (or planktic species) are living beings that live in the water column and cannot swim against a current[7] Their original habitat lies probably in the epipelagic subzone of the pelagic zone in the vicinity of the kelp zone (KZ) or within the KZ (cf. Fig. 5.7).

[5] The California current system is a" broad, slow, meandering current flowing south from the northern tip of Vancouver Island (50° N)" to the Mexican border (excerpt of a website of NOAA fisheries—www.fisheries.noaa.gov/west-coast/science-data/oceanography-northern-california-current-study-area).

[6] The most recent denomination of these unicellular algae by taxonomists is Bacillariophyceae. But this is a specialised term for systematists (related to their derived characters or synapomorphies). I will use the more classical denomination for ease of reading.

[7] The organisms that are planktic only during their larval or juvenile stage form the *meroplankton*. Most meroplankton organisms belong to the *zooplankton*, i.e. the heterotrophic plankton.

This seems likely because the KZ undergoes a very strong hydrodynamic drag (much stronger than the kelp-free areas of the Californian coast) preventing the particles afloat within and around this area to be washed away to long distance or penetrate it easily from far away; these organisms drift passively with the currents to the benthic zone of the KZ [30]. The movement of these trophic resources from one ecosystem to another, which is clearly the case here, is called a *cross-ecosystem subsidy*. Kelp forests shelter many filter-feeding invertebrates consuming plankton (see p. 108), partly arriving at this habitat via the cross-ecosystem subsidy just mentioned [57].

- Though the eukaryotic plankton were almost the only phytoplanktonic species investigated within the rich body of literature concerning the ecology of the western Pacific littoral zone, we have to bear in mind that a great part of the marine photosynthetic organisms found in the open ocean are bacteria belonging to the phylum Cyanobacteria[8] It has been suggested that marine cyanobacteria could be responsible for 25% of the ocean *net primary productivity*[9] [20]. Other authors assess that this percentage is even higher (about 66%) [33]. These bacteria form a very ancient phylum, having existed on our planet for ca 3.5 *Ga*, at a time when very probably all the organisms were marine (or aquatic in isolated pools on emerged land) and living in an atmosphere and a hydrosphere totally different from the present ones in their chemical composition. They were thus the first organisms to have practised oxydative photosynthesis with, as a direct consequence, the rejection of gaseous oxygen in the atmosphere as by-product of their metabolism, a chemical species then toxic for most organisms [42]. The cyanobacterial biomass probably increased enormously 2.3 Ma ago, entailing a primary rise of atmospheric dioxygen that very probably have caused the mass extinction of many organisms not adapted to this new situation and forced many organisms to adapt to the new atmosphere—a phenomenon that has been called the Great Oxidation Event (GOE). All the organisms existing on Earth descend from the organisms that have adapted to the GOE, either generally by getting aerobic or by remaining anaerobic in given restricted niches, which goes to assert the gigantic role played by ancient marine cyanobacteria for the existence of life as we know it on the Earth. This fact also explains that though Cyanobacteria is a phylum that has evolved in many types of habitats, terrestrial and aquatic, this group remained very present and in abundance in marine ecosystems [1].

  The Cyanobacteria constitute a phylum of about 8000 species (5185 formally identified) composing seven orders: Chroococcales, Gloeobacterales, Nostocales, Oscillatoriales, Pleurocapsales, Spirulinales, and Synechococcales. The phylum conquered many habitats—marine, freshwater and terrestrial ones. But within this group, four orders are well represented in the marine zones, the Chroococcales,

[8] Photosynthetic prokaryotes are distributed among six groups, namely (1) five bacterial phyla, i.e. the Cyanobacteria, Chlorobi, Chloroflexi, Proteobacteria and Firmicutes; and (2) one archaeal class, the Halobacteria. However, only the Cyanobacteria practise oxygenic photosynthesis.

[9] For the meaning and understanding of the concept, see Chap. 9.

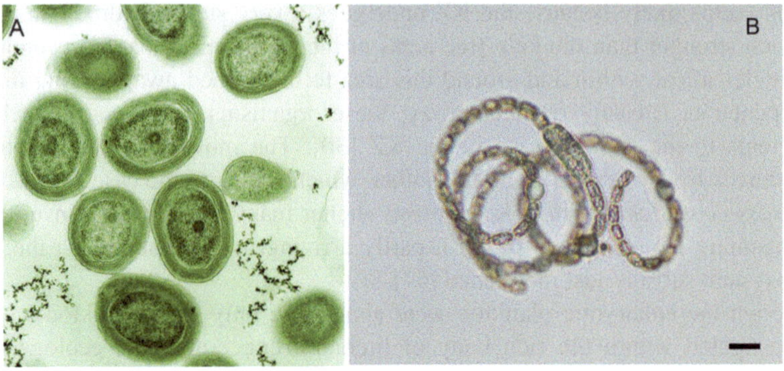

Fig. 5.8 Two cyanobacteria. (**a**) *Prochlorococcus marinus* (courtesy of Luke Thompson, NOAA); (**b**) *Anabaena sp.* (Courtesy of Charlotte Duval, MNHN, Paris)

Oscillatoriales, Pleurocapsales and Synechococcales. The last group is by far the most ancient and abundant cyanobacterial group in the oligotrophic open ocean. It is represented by two genera containing extremely small-sized coccal (spherical to cylindrical) bacteria, *Prochlorococcus* (Fig. 5.8a), with a diameter not exceeding 0.6 μm, and *Synechococcus*, with a diameter ranging between 0.8 and 1.5 μm [39]. These microorganisms do not form colonies; they appear completely unicellular, thus invisible to the naked eye. Among marine cyanobacteria playing a great role in the nitrogen fixation in the ocean and the heterotrophy in the marine carbon cycle, we can count the chroococcale *Crocosphaera watsonii* [5]. Other examples of abundant marine cyanobacteria, more colonial and building up filaments[10] and thus visible to the naked eye, are described in the literature, e.g. *Richelia intracellularis* (a symbiont of diatoms) or *Anabaena sp.* (Fig. 5.8b) [21, 24].

5.1.2 The Consumers

They consist of all the heterotrophs feeding on the producers. We will now outline the different categories of consumers in increasing order of trophic level.

5.1.2.1 The Zooplankton

Zooplanktonic species in any marine ecosystem are the heterotrophs constituting the trophic level just above that of the *phytoplankton*. They practise a mix of

[10] Filamentous cyanoacteria are often bacteria which fix nitrogen to plants or other producers thanks to a special cell called heterocyst.

primary, secondary and/or tertiary consumption (eating primarily phytoplankton but also adding to their diet zooplanktonic organisms and, for some species, practising cannibalism [13]). Thus, they are opportunist feeders as they lack specialised organs enabling them to be strictly herbivorous. But they constitute the trophic level that passes the organic matter of the phytoplankton to the rest of the biocoenosis. Zooplankton is divided into two categories according to size: (1) the *microzooplankton*, composed of protists (organisms with a size less than 0.2 mm) and (2) the *mesozooplankton* (also termed *macrozooplankton*), composed of minute multicellular animals (with a size between 0.2 and 50 mm).

Zooplankton, like phytoplankton, is mainly passively transported as an influx from the nearby pelagic ocean to the kelp zone (KZ). It comprises *holoplanktonic* as well as *meroplanktonic* species. Though they are consumers, this zooplankton constitutes trophic subsidies to the KZ. Their energy intake from phytoplankton, acquired in the pelagic zone, is at the disposal of many consumers of the KZ, once drifted in this area [57, p. 4]. According to the literature, research is still to be carried out to ascertain the number of zooplankton in the KZ and nearby pelagic areas. Articles treating microzooplankton in ecological studies are not numerous and quite recent despite the great role played by these microrganisms as trophic agents, and only a small number of publications talk about the composition of macrozooplankton in the Californian kelp zone.

A number of recent publications [8, 14] show that microzooplankton is a crucial functional group[11] in the marine ecosystem, as primary grazers of the phytoplankton as well as relays of the respiratory system of the oceans. Among these microorganisms, radiolaria, foraminifera or chlorophyta like *Volvox* [22] play an active role in marine ecosystems. The only article I know giving some information about representatives of the microzooplankton in the eastern Pacific kelp zone is a recent paper [43] where, within a very detailed list of planktic species mainly belonging to the mesozooplankton, two microzooplankton taxa were referred to, a phagotrophic dinoflagellate (*Noctiluca* sp.) and a foraminifera (*Ammonia tepida*, Fig. 5.9). This shows that the search for microzooplankton in general and in particular in the Pacific coast kelp forests is underway and represents a topical issue.

A recent review [57] described representatives of the macrozooplankton of the Californian kelp zone belonging to phyla Cnidaria, Ctenophora, Bryozoans, Annelids, Molluscs, Arthropods, Echinodermata and Chordates, mostly pelagic species at juvenile or egg stages, e.g. (1) the hydozoan (by-the-wind sailor) *Velella velella* (Fig. 5.10a), (2) the juvenile Dungeness crab (*Metacarcinus magister*) (Fig. 5.10b) and Pelagic red crab *Pleuroncodes planipes*, (3) tunicates like salps, doliolides or pyrosomes (Fig. 5.10c), (4) the rockfish (*Sebastes* sp.) (Fig. 5.10d), and (5) eggs of the herring *Clupea pallasii*.

[11] The microzooplankton gathers species that are very diversified phylogenetically, gathering taxa "no more related to one another than a human is to a house fly" [14, p. 2].

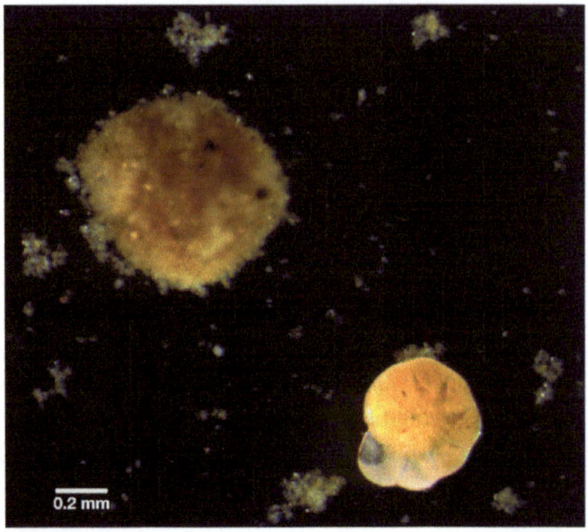

Fig. 5.9 A taxon of the microzooplankton of the kelp zone. The foraminifera *Ammonia tepida* [43] (courtesy of Olivier Maire)

Fig. 5.10 Four taxa belonging to the macrozooplankton of the kelp zone [57]. (**a**) a jellyfish, the by-the-wind sailor *Velella velella* (credit: Steve Lonhart); (**b**) a juvenile Dungeness crab *Metacarcinus magister* (credit: Steve Lonhart); (**c**) a tunicate: the doliolide *Dolioletta gegenbauri* (courtesy of Linda Ianniello); (**d**) a juvenile black rockfish *Sebastes melanops* (courtesy of Will Fennie)

5.1.2.2 The Upper Trophic Levels

This entry concerns marine animals whose trophic level is higher than the species composing the plankton. These organisms are distributed into two types of consumers:

(1) the strictly aquatic animals that can actively swim against the current of the water column, called the *nekton*, and
(2) the secondarily aquatic organisms (e.g. tetrapods, insects [53]) feeding either on the plankton or the nekton.

a-The Nekton

The species listed and analyzed here are borrowed from the very detailed chapter of Mark H. Carr and Daniel C. Reed, 'Shallow Rock Reefs and Kelp Forests' [10]. They are divided into different categories of consumers, though the limits of these categories are not always clear-cut: the *grazers* (thus *primary consumers*), *detritivores*, *planktivores*, *secondary consumers* and *tertiary consumers*.

a.1 The Grazers

Among the nektonic grazers, we can find echinodermata, molluscs and crustaceans of various sizes, ranging from minute amphipods to medium-sized decapods (crabs).

- **Sea urchins: three species**:
  The red sea urchin *Strongylocentrotus franciscanus*, purple sea urchin *S. purpuratus* (Fig. 5.11a), and white sea urchin *Lytechinus anamesus*.
  These species are detritivores, depending essentially on drifting POM when in small numbers, but voracious herbivores when in dense populations. In a relatively *climax* kelp ecosystem where the urchins are in limited numbers and in trophic equilibrium with the organic particles, as James H. McLean could have observed in Monterrey Bay near Carmel in the early 1960s (see map of Fig. 5.1, p. 100), the purple sea urchins were mainly restricted to the intertidal subzone of the KZ (see Sect. 5.1.1, p. 101) [34]. However, they become formidable grazers when they get numerous with, as a consequence, the detritus lacking due to their overharvesting or when their main natural predator, the sea otter, is absent. They then pullulate, form huge covers on the sea bed, and devour all the multicellular algae and kelps they encounter, even their holdfast, transforming the benthic zone supporting what was once a kelp forest into a marine desert called *urchin barren*.

Fig. 5.11 Within the nekton of the Central Californian KZ-Grazers. (**a**) Purple sea urchin *Strongylocentrotus purpuratus*; (**b**) Black turban snail *Tegula funebralis*; (**c**) Jewel top snail *Calliostoma annulatum*; (**d**) The Gumboot chiton *Cryptochiton stelleri*; (**e**) Whitecap limpet *Acmea mitra*; (**f**) Spider crab *Pugettia producta* (For all the photographs, credit: Steve Lonhart)

- **Herbivorous molluscs: characteristic snails, chitons and limpets**
  These species graze on the kelps and on red and green algae. If we concentrate strictly on our study zone—as a reminder, the kelp forest along the Central Californian coast[12]—we can list the following species:

  - Three strongly related species of turban snail (Fig. 5.11b) are present in the ecosystem. At an adult stage, all of them are generalist kelp feeders: *Tegula brunnea*, *T. funebralis* abundant in shallow water (0–6 m), *T. monterreyi* relatively rare at ca 6–7 m and *T. pulligo* common in deep water (7–12 m); they are distributed freely in the waters of the entire KZ, but the juvenile stages of these species are more restricted in their locations close to the sea bottom: young *T. brunnea* live at the limit of low-intertidal subzone and subtidal subzone I, while young *T. monterreyi.* and *T. pulligo* dwell at the limit of subtidal subzones I and II. [54]. Other turban snails are more specialised and graze on minute algae that grow on rocks (*T. aureotincta* and *T. eiseni*). Another snail, characteristic of the KZ, is the genus *Calliostoma* (Fig. 5.11c)—generalist algae feeders that can occasionaly be omnivorous.
  - A primitive group of molluscs, the polyplacophora, is represented by the Gumboot chiton *Cryptochiton stelleri* (Fig. 5.11d), mainly grazing on turf algae.
  - Limpets e.g. the whitecap limpet *Acmea mitra* (Fig. 5.11e), are efficient herbivores on macroalgae. These bivalves can inflict serious wounds to the algae, for example internal damage and tissue breaking [6].

[12] Area corresponding to the green ribbon along the coast on the map of Fig. 5.1, p. 100.

- **Marine crustaceans: amphipods and crabs** [10, 15, 41]
  - The minute and nearly planktonic amphipods have been shown to be efficient grazers, in particular, *Amphitoe longimana*, a species originating from the eastern coast of North America and reported to have extensively invaded the Pacific coast of the USA and Baja California.
  - Larger species feeding on kelp were reported to be abundant, notably spider crabs (*Pugettia producta*) (family Majidae) (cf. Fig. 5.11f).
  - A common species in the intertidal subzone of the kelp zone is the small herbivorous rockweed isopod *Idotea wosnesenskii* (cf. Fig. 5.15) that feeds mostly on seaweeds (e.g. *Ulva lactuca*, *Fucus gardneri*) [50].

a.2 Detritivores and Planktivores

The nektonic detritivores [i.e. animals feeding on POM (see p. 105)] and planktivores [i.e. animals feeding on phytoplankton or zooplankton] are either sessile or mobile animals, generally invertebrates.

- **Sessile detritivores and planktivores**
  They count among them immobile animals—sponges, cnidarians, sessile annelids in their calcareous tubes, barnacles (sessile crustaceans mimicking colonies of attached molluscs) or tunicates [25, 32, 36]. Representatives of these groups dwelling in the Californian kelp zone are presented here [10].

  Sponges

  These animals constitute one of the most primitive forms of the animal kingdom and constitute the phylum Porifera. They are all sessile and lack radial symmetry, tissues and organs; thus without a nervous system, though they possess isolated and diffused nervous cells enabling them to contract parts of their body. In the KZ, the group is represented by the numerous encrusting sponges, e.g. *Cliona* sp. (Fig. 5.12a)

  Cnidarians

  The species are characterised by radial symmetry and the possession of cnidocytes, specialised cells containing flagella used to eject venom to immobilise or kill their prey and capture them [31]. At adult stage, they can be sessile [sea firs (Fig. 5.12b), anemones (Fig. 5.12c)] or mobile (jellyfish).

  Annelids

  Some sessile polychaete worms are abundant albeit inconspicuous annelids, dwelling in calcareous tubes at sea bottoms. They are exemplified by *Dodecaceria* presented in (Fig. 5.12d).

Fig. 5.12 Within the nekton of the Central Californian KZ-Sessile detritivores and planktivores. (**a**) Boring sponge *Cliona celata* (credit: Steve Lonhart); (**b**) Sea fir *Abietinaria sp* (credit: Steve Lonhart); (**c**) Fish-eating anemone *Urticina piscivora* (credit: Chad King,) (**d**) Annelid *Dodecaceria fewkesi* (credit: Steve Lonhart)

- **Mobile detritivores and planktivores**
  The phrase "mobile animals" here means "permanently or occasionally mobile animals", contrasting with "sessile animals" treated just above. Mobile detritivores and planktivores are numerous and consist of crustaceans (crabs and shrimps), molluscs (mussels, abalone) and echinoderms (sea stars, sea urchins). I will present and illustrate several examples reported in the literature.

  - Within the intertidal subzone, emblematic examples of such animals are molluscs like the California mussel *Mytilus californianus* (Fig. 5.13a) or abalone species like black and red abalone[13] (resp. *Haliotis cracherodii* and *H. rufescens*) (Fig. 5.13b); or the echinoderm that we will focus on, the Purple sea urchin *Strongylocentrotus purpuratus* (Fig. 5.11a) [34]. As already mentioned, the latter is a detritivore-planktivore when POM is abundant, or their number limited by efficient predation, but otherwise, a very efficient grazer.

[13] Species that have rarefied for the last decades, which preoccupies conservationists at present (Pew conservation project, url: https://www.pewtrusts.org/en/research-and-analysis/articles/2020/04/22/kelp-collapse-in-northern-california-waters-has-far-reaching-impact).

Fig. 5.13 Within the nekton of the Central Californian KZ-Mobile detritivores and planktivores. (**a**) California mussel *Mytilus californianus*; (**b**) Red abalone *Haliotis rufescens*; (**c**) Warty sea cucumber *Apostichopus parvimensis*; (**d**) Bat sea star *Patiria miniata*; (**e**) Blacksmith *Chromis punctipinnis*; (**f**) Kelp rockfish *Sebastes atrovirens*. All photographs (credit: Steve Lonhart)

- Among other detritivorous echinoderms listed in the publication of Carr and Reed [10, p. 32] are mentioned sea cucumbers like *Apostichopus parvimensis* (Fig. 5.13c) or the bat sea star *Patiria miniata* (Fig. 5.13d).
- Several planktivorous fishes are found in the central Californian kelp forest, e.g. the blacksmith (*Chromis punctipinnis*) (Fig. 5.13e), gobies (four species), the Californian halfmoon (*Medialuna californiensis*), the señorita *Oxyjulis californica*, rockfishes of the genus *Sebastes* (Fig. 5.13f) [25].

a.3 Secondary Consumers

The grazers, together with the detritivores and planktivores studied above, constitute a constellation of species that we can grossly style as "primary consumers" albeit a part of mixotrophy has already been mentioned among some of the planktonic prey they feed on (see sub-subparagraph b.1.2). Therefore, we will use the typology of Carr and Reed [10] by calling *secondary consumers* the animals that feed on these "primary consumers". The examples below have been taken from this publication. Among the secondary consumers, we can count the following taxa.

Within the molluscs:

The carnivorous gastropods (sea snails and sea slugs) are abundant in species and individuals in the kelp zone along the Californian coast. An accurate estimation of their species richness (ca 200 species), carried out off Baja California, México, gives an idea of the biodiversity of the group [56]. Among these species, we can mention the leafy hornmouth *Cerastoma foliatum* that prey on barnacles, the Kellet's whelk *Kelletia kelletii* (Fig. 5.14a) that consumes polychaete worms, and a venomous snail, the California cone snail *Californiconus californicus* (Fig. 5.14b), predating herbivorous snails. Some sea slugs, like dorid nudibranchs (Fig. 5.14c), feed on sponges.

In addition, three octopus species that are very efficient secondary consumers are commonly found in the Californian KZ: the Giant Pacific octopus *Enteroctopus dofleini* [29], the ruby octopus *Octopus rubescens* [37] and the California two-spot octopus or bimac *Octopus bimaculatus* [2].

Within the crustaceans:

- The decapods[14]: an order comprising the large crustaceans exploited by fisheries (crabs, lobsters, shrimps, crayfish). 5 pairs of locomotor appendages characterise them, hence their names; the front pair of these appendages, in many species of the group, are pincers. 35 species are estimated to be potentially found in the kelp zone [55]. They are primarily secondary consumers (often scavengers) feeding on sessile and mobile primary consumers, e.g. the intertidal shrimp *Heptacarpus palpator* (Fig. 5.14d).
- The other crustacean orders are small invertebrates more or less belonging to the macrozooplankton or the small nekton and distributed into four classes, i.e. Malacostraca, Ostracoda, Maxillopoda and Branchiopoda (e.g. the green isopod, cf. Fig. 5.14e). We will not examine the biology of these species, but we will say that their biomass is enormous, although the number of species is unknown.

Within the echinoderms:

It comprises ca 50 species of sea stars [12, 28], almost all carnivorous [26]. Echinoderms constitute a group of very primitive and ancient bilaterians which have existed since the lower Cambrian era. They are at the basis of the deuterostomes, one of the main two branches of the animal kingdom from which the chordates, our phylum, have emerged. Among this group, sea stars are known

[14] From the greek δεκαποδ̀ος meaning ten legs.

Fig. 5.14 Within the nekton of the Central Californian KZ-Nonchordate secondary consumers. (**a**) Kellet's whelk *Kelletia kelletii* (credit: Steve Lonhart); (**b**) Californian cone snail *Californiconus californicus* (credit: Steve Lonhart); (**c**) Brown acanthodorid *Acanthodoris brunnea* (credit: Steve Lonhart); (**d**) Intertidal shrimp *Heptacarpus palpator* (credit: Steve Lonhart); (**e**) Green isopod *Penditodea wosnensenskii* (credit: Marlin Harms); (**f**) Sunflower sea star *Pycnopodia helianthoides* (credit: Steve Lonhart)

to be very efficient predators that have a primitive way of ingesting and digesting prey; they evert their two stomachs from the centre of their body, at the root of their rays, the one excreting digestive enzymes that break down the hard parts of the prey and the other eventually ingesting its flesh. One of the most spectacular representatives of this group, a giant with a diameter of up to 2 m and 20 arms, is the sunflower sea star *Pycnopodia helianthoides* (Fig. 5.14f) that feed on turban snails, abalone, sea urchins and other species of sea stars. One can see that this species is practising a very efficient camouflage!

Within the chordates

Present chordates constitute a phylum composed of three recognised clades, the cephalochordates (ca 25 species), tunicates (ca 3000 species) and vertebrates (the last being the overwhelming majority—according to the authors between 45,000 and 80,000 species—and the most evolved part of the phylum). The first two subphyla form a paraphyletic group classically called prochordates and contain very primitive, superficially wormlike or jellyfish-like invertebrates (several of them sessile at adult stage) (e.g. representatives in Sect. 5.1.2, p. 109). Most of them belong to the macrozooplankton and are POM feeders. However, there exist a group of carnivorous tunicates, represented in the Californian KZ by the *Hexacrobylus* and *Megalodicopia* genera, that can swim freely at a larval stage, thus belong to the nekton and the secondary consumers [46].

Within the prochordates:

The tunicate species *Megalodicopia hians* (Fig. 5.15a) is found in great numbers at the deepest level of the subtidal II subzone of the KZ. Though this species has not been thoroughly studied in the field, researches on specimens in aquariums suggest that at the larval stage, it is a free-swimming organism [27] and thus could belong to the nekton. At the adult stage, this is a very impressive sessile organism that has a siphon evolved in a huge mouth that can grab as large prey as small fish.

Within the fishes:

These animals are entirely nektonic. Many fish species are found along the Californian coast within the kelp forest. Their exact number, is not mentioned in the literature. However, a rough estimate, inferred from several sources [9, 36, 43, 45], would give 134 species of fish, among which 1 species belongs to the Cyclostomes, 13 to the Chondrichthyans, 1 to the Chondrostei and 118 to the Teleosts. The majority of the latter are nektonic secondary consumers.

- Teleosts [10, p. 322]:
  - The California sheephead *Semicossyphus pulcher* (Fig. 5.15b) is an efficient secondary consumer mainly feeding on red or purple sea urchins and California mussels
  - The wolf-eel *Anarrhichthys cellatus* (Fig. 5.15c) looks superficially like true eels (Anguilliformes) but is not related to them. It belongs to the order Scorpaeniformes. Wolf-eels feed on sea urchins and crabs.
  - Many small-sized cryptic reef fishes feed on other grazers and detritivores: clingfishes, eelpouts (Fig. 5.15d), blennies and clinids (Fig. 5.15e), pipefishes, poachers, snailfishes, pricklebacks (Fig. 5.15f) and gunnels, gobies and sculpins.
  - The painted and kelp greenlings (resp. *Oxylebius pictus* and *Hexagrammos decagrammus* (cf. Fig. 5.15g)) feed on small crustaceans, crabs and snails. Cabezons *Scorpaenichthys marmoratus* feed on crustaceans, molluscs (in particular abalone) and fish.

Fig. 5.15 Within the nekton of the Central Californian KZ-Chordate secondary consumers. (**a**) Predatory tunicate *Megalodicopia hians* (Courtesy of Kevin Stierhoff); (**b**) California sheephead *Semicossyphus pulcher* (credit: Steve Lonhart); (**c**) Wolf eel *Anarrhichthys ocellatus* (credit: Chad King); (**d**) Bigfin eelpout *Lycodes cortezianus* (credit: Kevin Stierhoff); (**e**) Crevice kelpfish *Gibbonsia montereyensis* (credit: Chad King); (**f**) Six spot-prickleback *Kasatkia seigeli* (credit: Steve Lonhart); (**g**) Kelp greenling *Hexagrammos decagrammus* (credit: Steve Lonhart), (**h**) Horn shark *Heterodontus francisci* (credit: Steve Lonhart)

Fig. 5.16 Within the nekton of the Central Californian—Invertebrate tertiary consumers. (**a**) Lewis' moonsnail *Euspira lewisii* (credit: Chad King), (**b**) Humboldt squid *Dosidicus gigas* (credit: Kevin Stierhoff), (**c**) Rock crab *Romaleon antennarium* (credit: Steve Lonhart), (**d**) Morning sun star *Solaster dawsoni* (credit: Chad King)

- Chondrichthyans [10, p. 322]:
  Horn sharks *Heterodontus francisci* (Fig. 5.15h) and bat rays *Myliobatis californica* are good examples of secondary consumers feeding on relatively large invertebrates difficult to ingest because they are protected by hard tests (exoskeletons or shells). Horn sharks feed mainly on hard-shelled molluscs and sea urchins; and bat rays, on molluscs and crustaceans in the seabeds or buried shells like clams (Fig. 5.16).

a.4 Tertiary Consumers

Within the invertebrates:

- Molluscs like moon snails (Fig. 5.16a) eat predatory snails "in the sand adjacent to rocky reefs" [10, p. 323], mainly bivalve molluscs. Among other tertiary consumers of the phylum, we can mention large cephalopods like the Humboldt squid (Fig. 5.16b).
- Among crustaceans, rock crabs, e.g. *Metacarcinus productus* and *Romaleon antennarium* (Fig. 5.16c) also feed on carnivorous snails.

Fig. 5.17 Within the nekton of the Central Californian—Tertiary and apex consumers. (**a**) Kelp bass *Paralabrax clathratus* (credit: Steve Lonhart); (**b**) Olive rockfish *Sebastes serranoides* (credit: Steve Lonhart); (**c**) Lingcod *Ophiodon elongatus* (credit: Steve Lonhart); (**d**) Buffalo sculpin *Enophrys bison* (Andy Murch, permission requested); (**e**) Pacific electric ray *Torpedo californica* (credit: Chad King); (**f**) Giant sea bass (Courtesy of Robert Schwemmer)

- The morning sun star *Solaster dawsoni* (Fig. 5.16d) is a relatively large echinoderm preying on smaller carnivorous or omnivorous sea stars and snails.

Within the fishes:

Larger piscivorous fishes, in fewer numbers, feed on carnivorous gastropods and octopuses, and smaller carnivorous or omnivorous fishes, e.g.

- the kelp bass *Paralabrax clathratus* (Fig. 5.17a),
- nine species of rockfish, e.g. the olive rockfish *Sebastes serranoides* (Fig. 5.17b),
- the lingcod *Ophiodon elongatus* (Fig. 5.17c)

- three large sculpins, the Buffalo sculpin *Enophrys bison* (Fig. 5.17d) and two species of irish lords, the red and the brown ones (resp. *Hemilepidotus hemilepidotus* and *H. spinosus*).

a.5 Apex Consumers

Apex predators belonging to the nekton are numerous (accurate number of species unknown) but transient [10], including invertebrates and large fishes. Due to the intricateness of the kelp forests, their impact on the abundance of the species of lower trophic levels is poorly known. Apex predation by invertebrates was recently documented in a paper [49] that pointed out that this phenomenon has been meagrely investigated, so more information is needed in the studied zone. Among the fishes, we can mention the white shark *Carcharodon carcharias* , the Pacific electric ray *Torpedo californica* (Fig. 5.17e), the California moray eel *Gymnothorax mordax* and the giant sea bass *Stereolepis gigas* (Fig. 5.17f).

b. Secondarily Aquatic Animals

b.1 Primary Consumers

The green sea turtle *Chelonia mydas* (Fig. 5.18a) is the only known primary consumer among this category of animals in the Central California kelp zone. Green sea turtles have the particularity to be carnivorous at a juvenile stage and become mainly herbivorous when becoming adults.

b.2 Secondary Consumers

The southern sea otter *Enhydra lutris nereis* (Fig. 5.18b) is the only known secondary consumer of the KZ, among its permanent residents. It is an effective predator of herbivorous invertebrates, principally sea urchins, but clams, abalone, and crabs enter its diet. As such, it has a powerful impact on the sustainability of the ecosystem. These animals, almost eradicated for their fur at the beginning of the twentieth century, were protected at the last minute and recovered to such a point that they are now extremely abundant locally, mainly in Central California. We will see in detail how this taxon is functionally a key species for the ecosystem in the study of a limited biocoenosis, hardly predated which nearly makes of this animal an apex consumer that has a strong impact on the ecosystem (see Sect. 5.2, p. 124).

b.3 Tertiary and Apex Consumers

Most other secondarily aquatic animals are resident birds and mammals that feed on kelp forest fishes, whose impacts on the ecosystem have not been fully studied but seem less dramatic than those of the sea otter. Among these animals, we can

Fig. 5.18 Secondarily aquatic animals of the Central Californian KZ. (**a**) Green sea turtle *Chelonya mydas* (credit: Chad King), a primary consumer; (**b**) Sea otter *Enhydra lutris* (credit: Steve Lonhart), a secondary consumer; (**c**) Snowy egret *Egretta thula* (credit: Steve Lonhart), an apex consumer; (**d**) Harbour seal *Phoca vitulina* (credit: Steve Lonhart), an apex consumer

mention Brandt's cormorant *Phalacrocorax penicillatus* , the snowy egret *Egretta thula* (Fig. 5.18c) and the harbour seal *Phoca vitulina* (Fig. 5.18d).

5.1.3 The Process of Decomposition and Mineralisation in the Kelp Zone (KZ)

Like in all marine habitats, the dead organic matter subject to decomposition and eventual mineralisation originates from corpses and excreta in the form of marine slow (mainly kelp or plant detritus, or corpses of dead planktonic organisms or minute invertebrates.) or marine animal falls (whale falls, shark falls, i.e. corpses of these heavy animals that sink to the sea bottom). Some of this material is brought into the kelp zone via cross-ecosystem subsidies (allochthonous material), whereas the remaining material is autochthonous.[15] The decomposition level is measured by

[15] Some of the organic carbon is autochthonous, originating from the local kelp; the rest is brought to the kelp forests via cross-ecosystem subsidies, e.g. transport by animals—guano from seabirds, excreta from fish like menhaden regularly migrating from rearing estuaries to littoral ecosystems—

the rate of dissolved inorganic carbon (DIC) and dissolved organic carbon (DOC) originating from a given decaying organic material and surrounding it. That can only be estimated in laboratory experiments in defaunated mesocosms. A few very recent publications gave some indirect or direct information on the process of decomposition/mineralisation in the kelp zones; however, up-and-coming research is presently carried out on the subject, mainly focused on the kelp.

Besides, kelp displayed differentiated degradation rates depending on environmental parameters in their natural aqueous habitats. It has been found that, in an aqueous environment, (1) the rate of DIC and DOC of kelp fragments, thus their rate of decomposition, increases positively with the rate of oxygen, (2) the rate of decomposition in anoxic environments is much reduced, even negligible and (3) temperature did not play a significant role [4]. Much of their organic carbon is recalcitrant to decomposition [19]. It has been shown that kelp decomposition is a crucial contributor to mineralisers, and that two opposite phenomena coexisted in kelp zones. (1) A portion of the POM biomass, in particular originated from kelp, is buried within anoxic sediments that embed a large amount of organic carbon in them, tantamount to slowing down decomposition, thus mineralisation. This phenomenon has been traditionally called *blue carbon* storage. (2) However, a number of marine fungi species exist in the kelp zones that, within a complex associating a whole bacterial chain, foster the remineralisation of the embedded organic carbon and other elements trapped in the sediment [40]. A recent publication [48] has shown that the fungi and bacteria of the aforementioned complex are involved in quorum sensing, i.e., a system of communication between cells that regulate their density, thus activity, with feedback on the gene expression controlling it. The decomposition of marine primary producers has been described by the following equation, with G_t biomass at time t, k the decrease constant, and R the refractory pool of organic matter [3]:

$$G_t = G_0 e^{-kt} + R \tag{5.1}$$

5.2 Description of a Crucial Limited Biocoenosis: Kelps, the Purple Sea Urchin, the Sea Otter

The different predators on sea urchins would not have the same impact on sea urchins in the kelp zone of Central California (see green belt on Map Fig. 5.1, p. 100) [10]. Sea otters are, by far, the most efficient predators on sea urchins, notably compared to predators that are more common in the kelp zone, namely predatory fishes and invertebrates. This fact has been shown by indirect evidence, for example, the survey of urchin barrens, a habitat at the sea bottom where the kelps have become

or diverse forms of live or dead organic material conveyed by marine currents either from the open sea or from the freshwater or land biomes [57].

very scarce or even inexistent due to their overgrazing by purple sea urchins. Urchin barrens were rare in the kelp zone of Central California, where the southern sea otter (*Enhydra lutris nereis*) was common and more frequent though patchily distributed in the other, more southern or northern portions of the Californian kelp zones, where this species is much rarer and where the most efficient predators are several fishes and invertebrates, notably the California spiny lobster *Panuliris interruptus* [10].

It is now firmly established that the sea otter is the only predator that efficiently controls the strongest herbivore impact on the kelp performed by sea urchins [11, 16]. It has been shown that in the areas where the sea otter has been removed, the following phenomenon is observed:

- The sea urchins are less efficiently controlled by the other predators (e.g. fishes like sheephead and cod, predatory crabs) since these species are overfished; thus, this situation cannot prevent the formation of urchin barrens.
- When sea urchins are removed by overfishing, the kelp forest recovers but becomes less stable and resilient.

These points explain why, when the sea otters were reintroduced (notably off Central and Southern California), the kelp forests sustainably resettled, whereas along the rest of the western Pacific coast, the kelp forest inexorably and lastingly dwindled [44, 47].

References

1. Allaf MM, Peerhossaini H (2022) Cyanobacteria: Model microorganisms and beyond. Microrganisms 10:696, https://doi.org/10.3390/microorganisms10040696.
2. Ambrose RF (1982) Shelter utilization by the molluscan cephalopod *Octopus bimaculatus*. Mar Ecol Prog Ser 7:67–73.
3. Banta GT et al. (2004) Decomposition of marine primary producers: consequences for nutrient recycling and retention in coastal ecosystems. In: Nielsen SL et al (eds) Estuarine nutrient cycling: the influence of primary producers; the fate of nutrients and biomass, pp. 187–216, Kluwer AP.
4. Boldreel EH et al (2023) Microbial degradation dynamics of farmed kelp deposits from *Saccharina latissima* and *Alaria esculenta*. Mal Ecol Prog Ser 709:1–15, https://doi.org/10.3354/meps14285
5. Bullerjahn GS, Post AF (2014) Physiology and molecular biology of aquatic cyanobacteria. Front Microbiol 5:359, https://doi.org/10.3389/fmicb.2014.00359.
6. Burnett NP et al (2021) Kelp morphology and herbivory are maintained across latitude despite geographic shift in kelp-wounding herbivores. Biol Bull 241:168–184, https://doi.org/10.1086/715039
7. Cain M et al (2014). Ecology, 3rd edn. Sinauer, Sunderland (MA).
8. Calbet A, Landry MR (2004) Phytoplankton growth, microzooplankton grazing, and carbon cycling in marine systems. Limnol Oceanogr 49:51–57, https://doi.org/10.4319/lo.2004.49.1.0051
9. California Department of Fish and Wildlife (2023) California Marine species portal Url site: https://marinespecies.wildlife.ca.gov/.
10. Carr MH, Reed DC (2016) Shallow rocky reefs and kelp forests (ch. 17). In: Mooney H, Zavaleta E (eds) Ecosystems of California, University of California Press, pp 311–336.

11. Cebrián J (2004) Grazing on benthic primary producers. In: Nielsen SL et al (eds) Estuarine nutrient cycling: the influence of primary producers; the fate of nutrients and biomass, pp. 153–185, Springer. Nature
12. Cintra-Buenrostro CE et al (2005) Oceanographic conditions and diversity of sea stars (Echinodermata: Asteroidea) in the Gulf of California, México. Rev Biol Trop 53(Suppl3):245–361.
13. Deserti ML et al (2017) Trophic ecology and diet of *Hydra vulgaris* (Cnidaria; Hydrozoa). Anim Biol 67:287–300, https://doi.org/10.1163/15707563-00002537.
14. Dolan J (2012) Microzooplankton: the microscopic (micro) animals (zoo) of the plankton. institut océanographique, Monaco. url: https://www.oceano.org/wp-content/uploads/2020/02/20.Microzooplankton_Dolan.pdf.
15. Duffy JE, Hay ME (2000) Strong impacts of grazing amphipods on the organization of a benthic community. Ecol Monogr 70:237–263.
16. Estes JA, Palmisano JF (1974) Sea otters: their role in structuring nearshore communities. Science 185:1058–1060.
17. Estes JA, Duggins DO (1995) Sea otter and kelp forests in Alaska: generality and variation in a community ecological paradigm. Ecol Monogr 65:75–100.
18. Falciatore A et al (2022) Light-driven processes: key players of the functional biodiversity in microalgae. C. R. Biol. 345:15–38, https://doi.org/10.5802/crbiol.80
19. Feng X et al (2022) Microbial-mediated contribution of kelp detritus to different forms of oceanic carbon sequestration. Ecol Indic 142:109186, https://doi.org/10.1016/j.ecolind.2022.109186
20. Flombaum P et al (2013) Present and future global distributions of the marine cyanobacteria *Prochlorococcus* and *Synecoccus*. Proc Natl Acad Sci 110:9824–9829, https://doi.org/10.1073/pnas.1307701110.
21. Flores E et al (2022) Adaptation to an intracellular lifestyle by a nitrogen-fixing, heterocyst-forming cyanobacterial endosymbiont of a diatom. Front Microbiol 13:799362, https://doi.org/10.3389/fmicb.2022.799362.
22. Flynn KJ et al. (2013) Misuse of the phytoplankton-zooplankton dichotomy: the need to assign organisms as mixotrophs within plankton functional types. J Plankton Res 35:3–11, https://doi.org/10.1093/plankt/fbs062.
23. Foster MS, Schiel D (1985) The ecology of giant kelp forests in California: a community profile. U.S. Fish Wildl. Serv. Biol. Rep. 85(7.2)
24. Golubic S et al (1999) Diversity of marine cyanobacteria. In: Charpy L, Larkum AWD (eds) Marine Cyanobacteria, Bull Insti Océanogr (Monaco), num. spéc 19, Monaco, pp 53–76.
25. Graham M et al (2008) Diversity and dynamics of California subtidal kelp forests. In: McClanahan T, Branch GM (eds) Food webs and the dynamics of marine reefs, p 103–134, Oxford Univ Press, https://doi.org/10.1093/acprof:oso/9780195319958.003.0005
26. Harrold C, Pearse JS (1987) The ecological role of echinoderms in kelp forests. Echinoderm Stud 2:137–233.
27. Havenhand JN et al (2006) *Megalodicopia hians* in the Monterey submarine canyon: distribution, larval development, and culture. Deap-Sea Res I: Oceanogr Res 53:215–222, https://doi.org/10.1016/j.dsr.2005.11.005.
28. Hemery LG et al (2016) Ecological niche and species distribution modelling of sea stars along the Pacific Northwest continental shelf. Diversity Distrib 22:1314–1327, https://doi.org/10.1111/ddi.12490
29. Hochberg FEG (1998) "*Enteroctopus; Enteroctopus dofleini* Wülker, 1910 new combination". In: Valentich Scott P, Blake JA (eds) Taxonomic Atlas of the Benthic Fauna of the Santa Maria Basin and the Western Santa Barbara Channel. Vol. 8, p. 203–208. Santa Barbara Museum of Natural History, Santa Barbara, CA
30. Jackson GA (1997) Currents in high drag environment of a coastal kelp stand off California. Cont Shelf Res 17:1913–1928, https://doi.org/10.1016/S0278-4343(97)00054-X
31. Jouiaei M et al (2015) Ancient venom systems: a review on cnidaria toxins. Toxins 7:2251–2271, https://doi.org/10.3390/toxins7062251.

32. Kenner MC, Tomoleoni, J (2019) Kelp forest monitoring at naval base Ventura County, San Nicolas Island, California: fall 2017 and spring 2018. Fourth Annual Report, open-file report 2019-1147, USGS, https://doi.org/10.3133/ofr20191147
33. Knoot CJ et al (2018) Cyanobacteria: Promising biocatalysts for sustainable chemical production. J Biol Chem 293:5044–5052, https://doi.org/10.1074/jbc.R117.815886
34. McLean JH (1962) Sublittoral ecology of kelp beds of the open coast area near Carmel, California. Biol Bull 122:95–114.
35. Miller RJ et al (2013) $\delta^{13}C$ and $\delta^{15}N$ of particulate organic matter in the Santa Barbara Channel: drivers and implications for trophic inference. Mar Ecol Prog Ser 474:53–66, https://doi.org/10.3354/meps10098
36. Nelson C. et al (2015) A field guide to kelp forests of the Santa Barbara Channel. Revised 5th edn, Santa Barbara Coastal Long Term Ecological Research Program.
37. Onthank (2008) Aerobic metabolism and dietary ecology of *Octopus rubescens*. MSc Thesis, WWU .
38. Page HM et al (2008) Assessing the importance of land and marine sources of organic matter to kelp forest food webs. Mar Ecol Prog Ser 360:47–62, https://doi.org/10.3354/meps07382
39. Partensky F et al (1999) *Prochlorococcus*, a marine photosynthetic prokaryote of global significance. Microbiol Mol Biol Rev 63:106–127.
40. Perkins AK et al (2023) Fungi increases kelp (*Ecklonia radiata*) remineralisation and dissolved organic carbon, alkalinity, and dimethylsulfoniopropionate (DMSP) production. Sci Total Environ 905:166957, https://doi.org/10.1016/j.scitotenv.2023.166957
41. Reish D, Barnard JL (1967) The benthic Polychaeta and Amphipoda of Morro Bay, California. Proc U S Natl Mus 120:1–26.
42. Sánchez-Baracaldo P et al (2022) Cyanobacteria and biogeochemical cycles through Earth history. Trends Microbiol 30:143–157, https://doi.org/10.1016/j.tim.2021.05.008
43. Shaffer JA et al (2020) Kelp forest zooplankton, forage fishes, and juvenile salmonids of the northeast Pacific nearshore. Mar Coast Fisher 12:4–20, https://doi.org/10.1002/mcf2.10103.
44. Steneck RS et al (2002) Kelp forest ecosystems:biodiversity, stability, resilience and future. Environ Conserv 29:436–459, https://doi.org/10.1017/S0376892902000322
45. Stephens JS et al (2006) Rocky reefs and kelp beds. In: Allen LG et al (eds) The ecology of marine fishes: California and adjacent waters, p. 227–252, University of California Press, Berkeley (CA), https://doi.org/10.1525/california/9780520246539.003.0009
46. Tatiàn M et al (2011) Molecular phylogeny endorses the relationship between carnivorous and filter-feeding tunicates (Tunicata, Asciiacea). Zool Scr 40:603–612, https://doi.org/10.1111/j.1463-6409.2011.00493.x
47. Tegner MJ, Dayton PK (2000) Ecosystem effects of fishing in kelp forest communities. ICES J Mar Sci 57:579–589, https://doi.org/10.1006/jmsc.2000.0715
48. Tourneroche A et al (2019) Bacterial-fungal interactions in the kelp endomicrobiota drive autoinducers-2 quorum sensing. Front Microbiol 10:1693, https://doi.org/10.3389/fmicb.2019.01693
49. Valdez JW (2020) Arthropods as vertebrate predators: a review of global patterns. Glob Ecol Biogeogr 29:1691–1703, https://doi.org/10.1111/geb.13157.
50. Van Alstyne KL et al (2006) Dopamine functions as an antiherbivore defense in the temperate green alga *Ulvaria obscura*. Oecologia 148:304–311, https://doi.org/10.1007/s00442-006-0378-3
51. Van den Hoek C et al (1995) Algae, an introduction to phycology. Cambridge University Press, Cambridge.
52. Venrick EL (2012) Phytoplankton in the California current system off southern California: Changes in a changing environment. Progr Oceanogr 104:46–58, https://doi.org/10.1016/j.pocean.2012.05.005
53. Wallace JB, Anderson NH (1995) Habitat, life history, and behavioral adaptations of aquatic insects. In: Merritt RW, Cummins KW (eds) An introduction to the aquatic insects of North America, 3rd edn, Kendall/Hunt, Dubuque (IA).

54. Watanabe JM (1984) The influence of recruitment, competition, and benthic predation on spatial distributions of three species of kelp forest gastropods (Trochidae: *Tegula*). Ecology 65:920–936.
55. Wicksten MK (1989) Ranges of offshore decapod crustaceans in the eastern Pacific Ocean. Trans San Diego Soc Nat Hist 21:291–316
56. Zamorano P, Hendrickx ME (2011) State of knowledge about the community of molluscs on both sides of the Baja California peninsula, Mexico: a comparative analysis. Cah Biol Mar 52:13–22
57. Zuercher R, Galloway AWE (2019) Coastal marine ecosystem connectivity: pelagic ocean to kelp forest subsidies. Ecosphere 10:e02602. https://doi.org/10.1002/ecs2.2602

Chapter 6
Ecological Models

6.1 The Scientific Nature and Methodology of Ecology

By the nature of its content and scientific methodology, ecology is situated at an intermediate position between the humanities and social sciences on the one hand and the experimental sciences on the other hand. Indeed,

- Ecology is an open discipline that has consistently promoted the elaboration of new and complex concepts, e.g. *ecological niche* [5, 6, 9, 11, 27, 33, 34], *biological diversity* or *biodiversity* [7, 10, 14, 17, 21, 25, 30, 31], *cline* [1, 12, 13, 15, 16, 24, 36], etc. The data which are valuable for this science come from two sources: field observations and scenarios simulated by *models*. In the most ideal cases, these two sources of information are compared, and when they match, conclusive but falsifiable hypotheses are derived.
- The experimental approach is limited in ecology. It is often integrated into constructing models or as a support of field data collection in case studies [3, 35]. Case studies in ecology, notably in behavioural ecology,[1] maintly rely on observation and historical reconstruction, and only partly on experimental approach.

Models are simplified, representations of realities that are too complex to be understood in their details as a whole. They allow us to analyse phenomena, that is, to decompose them into simplified, understandable parts, or principles. In order to be valid, models must be parsimonious, i.e. explain these parts in the simplest way

Supplementary Information The online version contains supplementary material available at https://doi.org/10.1007/978-3-031-90656-5_6.

[1] Behavioural ecology: branch of ecology interested in the behaviour patterns within studied species. The behaviours shed light on the social units and the impact on the environment of the studied species (1) to describe these phenomena in their own right, but also (2) to see whether behaviours have a major influence on the structure and functions of the community of which the species is a member.

M. Salomon, *Integrative Ecology*, https://doi.org/10.1007/978-3-031-90656-5_6

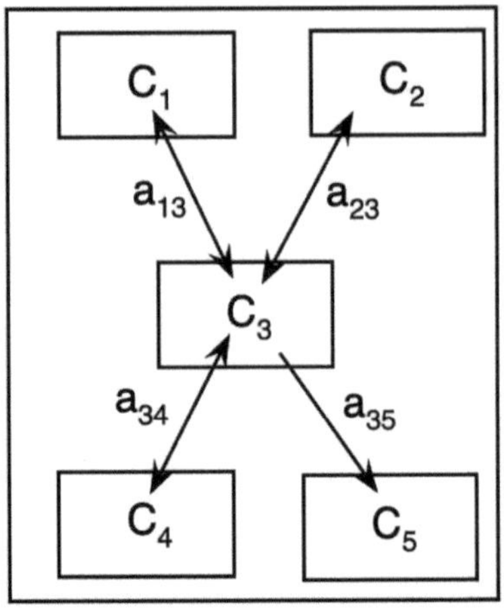

Fig. 6.1 Ecosystem compartmental model. The rectangles C_i, where $i \in \{1, 2, \cdots, 5\}$, are compartments: they may represent *a group of organisms* as well as *abiotic factors*—the agents of the system. The *arrows* represent relationships, the *single-headed* ones being one-way actions and the *double-headed* ones interactions, the heads pointing towards the target of the actions. For example, in this figure, arrow a_{35} is the action of C_3 upon C_5, but arrow a_{34} is the interaction between C_3 and C_4

(if two hypotheses compete for the same explanation, one must choose the simplest). Models must be falsifiable [18], i.e. refutable and reconstructed when new data or paradoxes contradict them. A model can be verbose, graphical or mathematical.

Graphical representations like Fig. 6.1 have the advantage of being straightforward and synthetic. Models also give clues for data collection management and enable us to perform mathematical processing and syntheses (i.e. reconstructing the entire phenomena from the analysed parts). This method applies not only to ecology but, in fact, to any science.

6.2 The Compartmental Model

An ecosystem can be described by a *compartmental model*—a representation that makes it possible to grasp efficiently the notion of interactions (cf. Fig. 2.1). A model has the advantage of lending itself to mathematical processing, making itself falsifiable and predictable. Biomathematical tools and *modelling* (the elaboration of models) play a significant role in describing and sometimes predicting ecosystems' structure, dynamics and functions. An *ecosystem compartmental model* identifies an ecosystem and all its components with *compartments*. Compartments included in a given compartment C are called subcompartments of C. Every subcompartment encompasses subcompartments similarly. This pattern is repeated an undefined but finite number of times, depending on the ecosystem described. Such a model is called recursive. The direct subcompartments of a given compartment (like for instance the populations forming a biocoenosis) constitute, by definition, a *level of integration* [22]. The compartments of the same level of integration endowed with similar properties (same structures and functions) interact to generate the structures and functions of the compartment that encompass them (i.e. *emergent properties*). The compartment endowed with emerging properties will be defined as the *emergent compartment*. An example where interactions between emergent compartments at the same level of integration target encompassing compartments is illustrated in Fig. 6.2.

In this figure, the single- and double-headed arrows correspond to one-way actions and interactions respectively. The direction of the arrow indicates the direction of the flow of matter, energy and information.

Arrows' colours and thickness:

- *black and thin:* interactions between the compartments at the lower level of integration;
- *yellow and thick: id.* between the emergent compartments. AF = abiotic factors;

Abbreviations used in the figure: LB = limited biocoenosis under study; OB =Other biocoenoses; TB = total biocoenosis; EC = ecosystem; EN = environment.

6.3 Illustration from Our Example

- Figure 6.2a shows that five compartments are the targets of *direct* or *indirect interactions*; e.g. C_1 interacts directly with C_3; and C_3, with C_2. However, several compartments do not interact directly with one another, e.g. C_1 and C_2; they do it indirectly, via C_3. Thus, the five compartments are interlinked, either directly or indirectly; altogether, they induce emergent properties on compartment C that encompass them, i.e. a compartment belonging to the level of integration just above it. For example, compartments C_i can represent *populations* and C a *community* containing them.
- In Fig. 6.2b, the limited biocoenosis under study (LB) and the environment (EN) are emergent compartments induced by the subcompartments they encompass, namely (i) the plant formation (Fp), the lynx population (Ly) and the hare population (Ha) populations for LB; and (ii) the other biocenoses (OB) and abiotic factors (AF) for EN. LB is made up of subcompartments (i.e. populations of different species) linked by trophic relationships—Fp to Ha, and Ha to Ly, the first element being plant resource of the second, which in turn is prey of the third. The trophic interactions induce the alteration of the number of individuals per compartment, thus the biomass of the latter, across time, launching a dynamic at the level of the emergent compartment. At every time t of the process, the states of the subcompartments of LB entail the state of the emergent compartment.

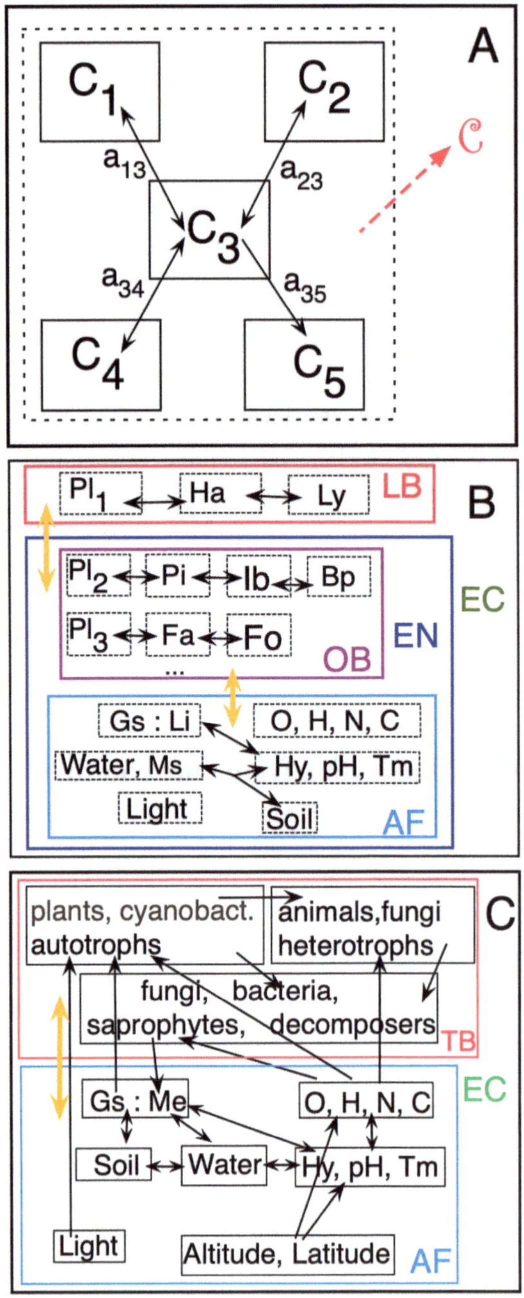

Fig. 6.2 Ecosystem compartmental model as illustrated in Fig. 2.1. (**a**) Compartmental model at a given level of integration: C_i = compartments; a_{ij} = interactions between C_i and C_j. *Large square* surrounded by *dotted line* = system composed of compartments whose interactions generate the emergent properties (symbolised by the *thick red dotted arrow* pointed towards emergent compartment C that encompass them). (**b**) Compartmental model of an ecosystem *sensu* the first definition: *Bp* birds of prey, *C* carbon, *Fa* frugivorous animals, *Fo* foxes, *Fp* flowering plants, *Gs* geological substrate, *H* hydrogen, *Ha* hares, *Hy* hygrometry, *Ib* insectivorous birds, *LB* limited biocoenosis under study, *Li* lithospheric elements, *Ly* lynxes, *Ms* mineral salts, *N* nitrogen, *O* Oxygen, *Pi* pollinating insects, *Tm* temperature. (**c**) Compartmental model of an ecosystem *sensu* the second definition: id. for the AF compartment

EN is the subcompartment of EC that contains OB and AF. Now, let us look OB and AF as emergent compartments resulting from the activities of some of their subcompartments.

- OB's subcompartments are of a similar nature to LB's, i.e. represent populations of distinct species interacting with one another. The interactions can be trophic, coevolutive, competitive or involved in mineralisation. Here, we deal with interaction links relating to compartments in parallel chains, with bridges between the latter. The chains altogether globally constitute the interaction network. As an example, Fig. 6.2b displays several chains:

  (1) Plants (Fp)—Pollinating insects (Pi)—Insectivorous birds (Ib)—Birds of prey (Bp)
  (2) Plants (Fp)—Frugivorous animals (Fa)—Arctic fox (Fo)
  (3) Etc

  Let us take the example of chain (1). We can see that if the biomass of pollinating insects decreases, there will be fewer insectivorous birds, which will entail fewer birds of prey; the entire situation will favour the increase of the biomass of the vegetal resource. There are many other chains of this type. Each chain mentioned has a similar structure.
- The compartment corresponding to the abiotic factors (AF) contains the subcompartments representing the biophysical parameters of the environment (EN). Here, the subcompartments interact likewise and have emergent properties. For example, soil is a complex system comprising notably a certain amount of run-off water that contains a given concentration of *mineral salts* and a definite pH. In the area unit characterised by these soil properties, the quantity of run-off water, the various concentrations and types of mineral salts and the spectrum of alkalinity/acidity correspond to subcompartments that vary and interact and thus tune and entail altogether a given plant composition of a biocoenosis. A significant variation of one of the subcompartments will thus result in differentiation in the plant *species composition and richness*, thus modifying the fauna composition depending on it.
  The subcompartments of OB and AF can mutually interact and entail the composition of LB. The soil system, for example, is a mix of subcompartments belonging to OB and AF. Indeed, soils contain organisms, i.e. saprophytic fungi and bacterial decomposers, forming a chain of agents carrying out mineralisation. These interactions occurring between the two compartments of EN (in the sense of the first definition) determine the existence and efficiency of a given LB. Thus, the compartments of the AF system owe their existence to the biocoenoses.

• Figure 6.2c describes the functioning of an ecosystem *sensu* the second definition, with the organisms gathering according to their *mode of nutrition*. The autotrophs, heterotrophs and decomposers/mineralisers are collected in three distinct compartments. The elements, belonging to an AF compartment (the soil system), like lithospheric elements [e.g. *sulphur* (S), *phosporus* (P), *calcium*

(Ca)] and atmospheric elements [e.g. *oxygen* (O), *hydrogen* (H), *nitrogen* (N), *carbon* (C)] penetrate TB via the compartment of the autotrophs. All these chemicals will then travel inside TB from the autotroph to the heterotroph compartments and inside the heterotroph compartment across the primary, secondary, tertiary consumers subcompartments successively. These two interactions correspond respectively to the biosynthesis and transfer of organic matter within the ecosystem.

Decomposition and mineralisation relate the TB subcompartments to the EN in the opposite direction, hence closing the cycle. These interactions start from the dead organic matter of corpses and excreta. The mineral and atmospheric chemical elements, once part of the organic matter of living beings when they were alive, are eventually released in their most simple, mineral forms (elements, simple compounds) in the environment (EN), at the disposal of the producers.

6.4 Mathematical Modelling of Compartmental Systems

We will now resume all the notions seen above more formally.

Axiom 6.1 Ecosystems and their parts are identified to compartments in the sense of multicompartment model theory.

Compartments are system structures, i.e. sets whose primitives interact with each other, a condition for the whole to exist as an entity. Generally, sets are not system structures (e.g. sets of numbers).

Axiom 6.2 Compartments are intrinsic sets.

They have an existence *per se*, i.e. are contingent, auto-replicate and auto-regulating and thus have a history. On the contrary, classical sets are extrinsic, with primitives (numbers, lines, angles, etc) being tools constructed for the sake of mathematical processing.

Axiom 6.3 Compartments are complementary or nested but never intersect.

Axiom 6.4 Compartments are recursive sets: their primitives are similarly structured.

Primitives of a given compartment C are themselves compartments that interac to elaborate emergent properties characterising C, and this pattern is repeated. On the other hand, primitives of classical sets are unique and do not interact with each other.

Definition 6.1 A compartment C is said to be subsumed under another compartment C' if C is a primitive of C'.

Definition 6.2 If compartment C subsumes compartment C' or conversely, C and C' are said to be linked by a subsumption relationship.

For example, if a biocoenosis (compartment C) comprises n populations, the latter are compartments subsumed under C.

Definition 6.3 A compartment C subsumed under another compartment C' is called a subcompartment of C'.

Axiom 6.5 Ecosystem compartments subsume analogous structures (or analogues), complementarily delineated, individualised and intrinsic.

The populations of a given biocoenosis are analogues, as are the individuals of a given population or any compartment subsuming that population.

Definition 6.4 Analogues sharing the same structural, functional and dynamic properties are said to be *of the same type*.

In a biocoenosis, the member heterospecific populations are analogues of the same type, as are the member individuals of these populations.

Theorem 6.1 *Let* A_1, A_2, $\ldots$, A_n *be analogues of compartment C, then these analogues are subcompartments of C.*

Proof The A_i's being analogues, they are intrinsic and recursive sets, and hence are compartments. Since they are subsumed under compartment C, according to Definition 6.4, the A_i's are subcompartments of C. □

Axiom 6.6 Any compartment is subsumed under itself [2] (identity property).

Axiom 6.7 If a compartment C subsumes a compartment C' which in turn subsumes compartment C", then C also subsumes C" (transitivity property).

Definition 6.5 Compartment C' is strictly subsumed under C if the former is not equal to the latter.

Cases can be found where a compartment is subsumed but not strictly subsumed under another one: (1) In many reef coral species, the colony (population), and not the 'individual' polyp, is the physiological individual. The 'individual' polyp is just part of a whole, which is the actual biological unit. The limit between individual and population is thus blurred in this instance. (2) In unicellular organisms, the compartments 'cell' and 'individual' are equal, though clearly the latter subsumes the former (see Fig. 6.3).

Definition 6.6 Compartment C directly subsumes compartment C' if there is no compartment C" strictly subsumed in C that strictly subsumes C'.

Axiom 6.8 Analogues of the same type, C_1, C_2, $\ldots C_k$, induce the same emergent properties to the compartment C directly subsuming them; C is called their emergent compartment.

Definition 6.7 An entire ecological system *sensu lato* (EESSL) (Fig. 6.3) is a network showing the emergence relationships between the compartments within the biosphere compartment.

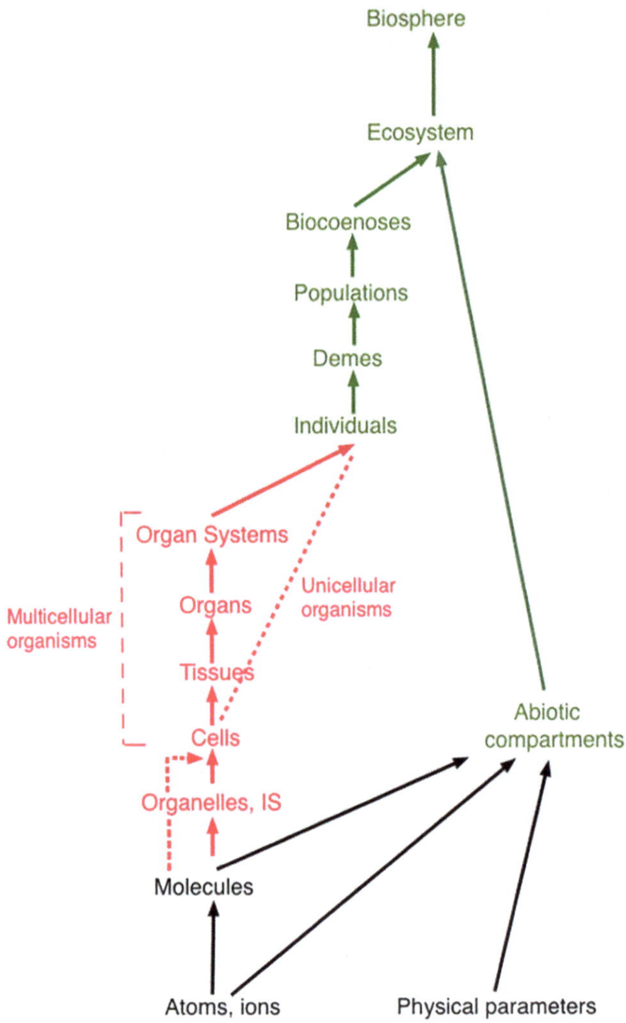

Fig. 6.3 The entire ecological system *sensu lato* (EESSL). *Solid arrows* = emergence relationships between subsumed and subsuming (emergent) compartments; *dashed arrow* = symbol showing that in prokaryotes, the molecule compartment is directly subsumed under the cell compartment; *dotted line* = equivalence between cells and individuals in unicellular organisms

Such a network comprises three subnetworks corresponding to the physico-chemical, organic-physiological and ecological domains.

Definition 6.8 Interactions between two compartments involved in a direct subsumption relationship are called direct interactions. The other interactions are called indirect.

For example, in Fig. 6.2, the interactions between Pl_1 and Ha is direct, whereas those between Pl_1 and Ly is indirect.

Definition 6.9 Let $\mathcal{J}_n = (C_1, C_2, \ldots, C_n)$ be a sequence of n compartments such that C_i is directly subsumed under C_{i+1}, then $\mathcal{J}_n$ is called an interaction chain.

Axiom 6.9 $\mathcal{J}_n^{(k)}$ will mean 'the *k*th interaction under study linking *n* compartments'.

Figure 6.2b exemplifies three interaction chains :

- $\mathcal{J}_3^{(1)}$ = [Plants_1 (Pl_1), Hare population (Ha), Lynx population (Ly)],
- $\mathcal{J}_4^{(2)}$ = [Plants_2 (Pl_2), Pollinating insect population (Pi), Insectivorous bird population (Ib), Bird of prey population (Bp)],
- $\mathcal{J}_3^{(3)}$ = [Plants_3 (Pl_3), Frugivorous animal population (Fa), Fox population (Fo)].

Extra information about Fig. 6.3: Demes are conspecific populations sharing the habitat under study; IS = intracellular structures (e.g. ribosomes). The colours designate the compartment networks corresponding to the physicochemical (black), organic-physiological (red) and ecological (green) levels.

Definition 6.10 Let $\mathcal{J}_n = (C_1, C_2, \cdots, C_i, \cdots C_n)$ be an interaction chain, then the index *i* of compartment C_i is defined as the level of integration (LI) of C_i.

Corollarily, the smaller the IL of a compartment, the most subsumed the latter is. For example,in the interaction chain $\mathcal{J}_4^{(2)}$ mentioned just above, the IL of Pl_2 is 1; that of Ib, 3; etc.

Axiom 6.10 Interaction chains are interlinked and constitute an interaction network.

In such a network, an IL has meaning only within a single compartment chain since these differ in the nature and number of their component compartments. It is impossible to compare the integration level between two compartments unrelated by subsumption, like "abiotic compartments" and "Demes" in Fig. 6.3.

Definition 6.11 Let $\mathcal{J}_n = (C_1, C_2, \cdots, C_i, \cdots C_n)$ be an interaction chain within which directly subsumed compartments are equal *k* times, then the length of $\mathcal{J}_n$ is $n - k - 1$.

For example, in Fig. 6.3, the interaction chain of the unicellular organisms have their parameters *n* and *k*, respectively, equal to 10 and 1; its length is thus $l = 8$.

Figure 6.3 represents the theoretical framework of an entire biosphere, that of Earth, which subsumes all the compartments existing on our planet. Each compartment corresponds to a structured set of observed information or hypotheses supported by solid models that are also parsimonious and falsifiable in the sense of Popper [18]. This supermodel represents the entire life system, which is closed since transferring matter or energy to another biosphere seems impossible. Either there is no other planet in the universe sheltering life, or other ones exist[2] but are so distant from ours that any exchange of energy or matter, hence any interaction with them, is virtually impossible. However, a life system is necessarily open, which is paradoxical if we accept the idea that the life system is closed. One possible correction of this apparent paradox resides in the acceptance that the life system is open and that the transfer of energy and matter, unable to leave the system, reverses inside by creating everlasting dynamics, ensuring a perpetual thermodynamical disequilibrium at each instant. We will now focus on an essential aspect of the compartmental systems involved in the biological and ecological processes, i.e. *complexity*.

6.5 Complexity of Compartmental Systems

Ecosystems are assimilated to very complex compartmental models; i.e. there are many compartments and compared to their number, the quantity of potential interactions between them is still considerably larger. Furthermore, another factor drastically increases the complexity of the model—the continuous variation of ecosystems across time, which gives a dynamic to the compartmental model describing it.

Theorem 6.2 *Let n be the number of compartments. The largest number of direct interactions at time t is $\nu = \dfrac{n(n-1)}{2}$.*

Proof The largest number of direct interactions is equal to the number of combinations of all the unordered pairs of compartments among the n compartments of the EESSL. Hence

$$\nu = \binom{n}{2} = \frac{n!}{2! \times (n-2)!} = \frac{n\,(n-1)\,\cancel{(n-2)}\cancel{(n-3)}\cdots}{2\,\cancel{(n-2)}\cancel{(n-3)}\cdots}$$

□

Theorem 6.3 *Let n be the number of compartments. The largest number of interactions at time t is $\nu = \dfrac{n(n-1)}{2} + \nu'$ where ν' is the largest number of indirect interactions, impact on the subsuming compartment $\mathcal{B}$.*

Proof If indirect interactions are added to the direct ones, we trivially infer the theorem. □

[2] Namely in telluric exoplanets or perhaps Mars, though no evidence of life has been found until the present.

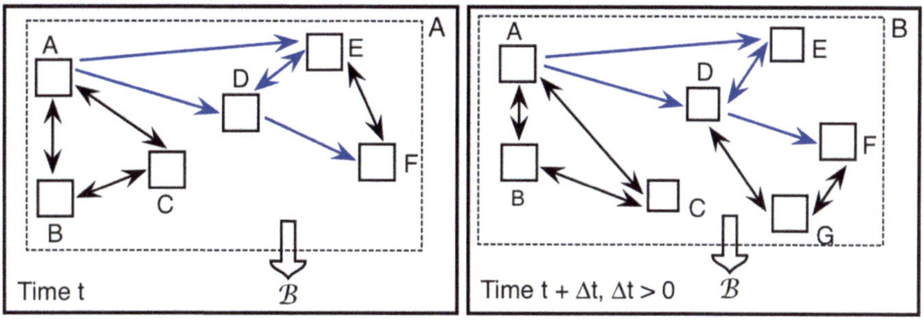

Fig. 6.4 Evolution of a compartmental system across time: a view on a level of integration, interactions between analogues. The state of each analogue represented by a *square* (their characteristics and numbers) as well as the interactions between the analogues vary across time. *Arrows* = interactions between two compartments whose direction corresponds to the flux of energy and/or matter. One-headed = trophic relationship, the arrow pointing to the consumer; double-headed = the other relationships (coevolution, competition), neutral about the matter and energy pathway. The blue arrows illustrate the example in the text. The wide hollow arrow represents the subsumption arrow between the set of analogues and the emergent compartment $\mathcal{B}$. The interactions between the analogues at time t result in a general state of the emergent level $\mathcal{B}$. The state of $\mathcal{B}$ varies parallelly

Here, we will discuss the concept of model complexity and the choice of an appropriate index from the literature [32] or that we have proposed.

A level of integration (LI) defined by 30 compartments is home to at least $\frac{30\times29}{2} = 465$ direct interactions, to which lots of indirect interactions must be added. All these data contribute to assessing the complexity of the system.

Figure 6.4 illustrates the notions of complexity and dynamics of an entire ecological system (EESSL). Let us consider that in the two subfigures A and B, the analogues represented by the squares are heterospecific populations linked by interactions and that the emergent compartment $\mathcal{B}$ is the biocoenosis subsuming them. We can see that each modification of an analogue infers the modifications of the other through direct or indirect interactions at every instant t, which instantaneously elicits emergent properties on $\mathcal{B}$. We will exemplify it as follows. Let A be a prey population, and D and E different predator populations preying on A. We will thus have a *predator-prey relationship* between A and D, a predator-prey relationship between A and E, and a competition relationship between D and E. When an individual of D consumes a member of A, it deprives E of a prey individual belonging to A; the whole process, when repeated, strengthens D and weakens E, and the role of D and E is symmetrical in this process. If D is a more efficient predator than E and that A, the only prey compartment of both, E disappears in a relatively long term. If, for example, D and E are as efficient, a dynamic equilibrium between these compartments will take place. We can see that between times t and $t + \Delta t$, the states of the components of $\mathcal{B}$, and also the emergent state of $\mathcal{B}$, changes. Several directed interactions can disappear; a compartment, e.g. G, can suddenly appear within $\mathcal{B}$ by migration.

Figure 6.4a and b also illustrate direct and indirect interactions (see the blue arrows). Let us now suppose that compartment F corresponds to a population of apex predators feeding on members of compartment D. The trophic relationship between A and F will be indirect. The trophic pressure on D by F will reduce the death rate of A due to the predation of D. So one can easily understand the equally crucial importance of both types of interactions, the direct and the indirect ones, in the dynamic of the EESSL. We will have a chain reaction where a change in one compartment will modify other network compartments via direct and indirect interactions.

Within a given compartment C, the number of subcompartments, as well as the amount of direct or indirect pairwise interactions, vary across time, either as the result of the internal dynamic just mentioned above, or as a response to a changing external environment of C at a level of integration below that of the biosphere. Between two distinct instants (e.g. t and $t + \Delta t$), the nature of an interaction between the same compartment can alter dramatically; analogues can be added because of migration, or disappear because of local extinction; a compartment that weakens first, but then strengthens. However, notwithstanding all these variations, the emerging compartment must remain stable due to the dynamic equilibrium of its member subcompartments.

6.6 The Biosphere

Within the EESSL, a particular reflection should be made about the biosphere, the compartment subsuming all the ecosystems. The early evolution of the concept is described in the frame of a chapter within the part devoted to the history of ecology (Part II, Chap. 8, p. 175). In sum:

- We owe the premise of the concept to Suess [29]: the biosphere is defined as a thin layer (thin compared to the bulk of our planet) composed of three components: (1) a solid portion comprising all the emerged land and seabeds, called the lithosphere,[3] (2) a liquid mass composed of all the oceans and seas, and brackish and freshwater volumes, called the hydrosphere, and (3) a gaseous cover encompassing gaseous elements around the Earth, called the atmosphere. For Suess, the biosphere is the location of life, the abiotic factors, and the matter cycle needed for the organisms. It is a structural definition.
- Vernadski refined the concept of the biosphere, giving it a functional character:

  1. The biosphere is the seat of converters changing cosmic rays into active terrestrial energy.
  2. The biosphere must be considered an organic entity—an alteration of one part of it impacts the whole.
  3. The organisms, the essential elements of the biosphere, are the relays of cycling material (chemical elements and compounds) whose origin is the bedrock.

[3] From the greek roots λιϑος=stone, and σφαιρα=globe; thus the part of the globe made of stone.

A classical view on the subject has been expounded in Ramade [20, ch. 1, p. 5], who defines the biosphere as the earth's fine layer that sustains macro-organic life—the biocoenoses of plants, animals, and fungi. This definition excludes the polar icecaps and the highest mountanous strata from the biosphere, as these area would not support plant life, and thus could not sustain lasting biocoenoses. According to this book, these areas only sheltering rare microorganismal spores do not truly belong to the biosphere for this reason and are thus termed parabiospheric. The author then proposes the existence of an "ecosphere", ie "the external gaseous and internal solid envelopes of the biosphere".[4] The rest of the chapter, interestingly, is devoted to exobiological considerations (i.e. the search of such "ecospheres" in exoplanets) and the history of the terrestrial "ecosphere".

Large portions of the terrestrial crust, though, considered previously unable to sustain lasting or even any life, have relatively recently been proven to compose very rich ecosystems of extremophile organisms, not only in the icecaps and high altitudes evoked by the author but also deep in sedimentary rocks, hydrothermal vents located in abyssal oceanic depths and extremely acid environments [8, 19, 26].[5]

Ramade first defines the biosphere and explains in detail the cosmogonic conditions of its existence and a scenario of its genesis. Then, after a physical description of the biosphere, he boils it down to the ecosystems. I will propose an alternate point of view on the subject, hypothesising that the biosphere varies over time and should be considered a dynamic concept resulting from the interactions between the ecosystems, explained by the concept of subsumption just explained above. In my opinion, the biosphere is the system subsuming all the ecosystems as defined above—the terrestrial and the marine ecosystems.[6] The present biosphere is the last stage of the succession of the past biospheres. Initially, it would have been restricted to limited locations on Earth and then have expanded to the entire Earth as we know it now.

6.7 The Different Disciplines Dealing with Parts of the Entire Ecological System *sensu lato* (EESSL)

Like ecology, many disciplines apply compartmental modelling to their objects within the frame of the EESSL. Table 6.1 presents various strata of the EESSL and the corresponding field dealing with it. First of all, we notice that different disciplines deal with the same analogues. We can gather the sets of analogues in two main stages:

---

[4] Quotation from the original text translated by me from French.

[5] In addition, these ecosystems are not only composed of prokaryotes (bacteria and archaea) but also of several very specialised Eukaryotes, even multicellular ones, depending on them.

[6] We must keep in mind that the interactions between the terrestrial and marine ecosystems is of the utmost importance for the overall biological equilibrium.

Table 6.1 Analogues, their level of integration (LI) and the discipline studying them

Compartment	IL[a]	Disciplines
Atoms, ions	0	Biophysics
Molecules, crystals	1	Biophysics, Biochemistry, Molecular Biology, Genetics, Genomics
Organelles, Intra-cellular structures	2	Cell Biology, Molecular Biology, Microbiology, Embryology
Cells	3	Id.
Tissues	4	Histology, Embryology
Organs	5	Anatomy, Evolutionary Biology
Organ systems	6	Anatomy, Physiology, , Evolutionary Biology
Individuals	7	Evolutionary biology, Genetics, Ecology
Populations	8	Evolutionary biology, Genetics, Biogeography, Ecology
Species	9	Evolutionary biology, Genetics, Biogeography
Biocoenoses (communities)	10	Ecology
Ecosystems	11	Ecology
Biosphere	12	Ecology

[a] level of integration (LI) assessed by an arbitrary rank

The infraindividual stage:
The concerned analogues are at the atomic, molecular, subcellular, and cellular integration levels (ILs); and for the multicellular organisms, the analogues are also at the tissular and organic ILs.

The supraindividual stage:
It concerns analogues that are at the individual, population, biocoenosis, ecosystem and biosphere ILs.

We will now briefly survey the different domains of biology involved here (cf. Table 6.1), all of which are related to ecology in some way.

- The atomic integration level (IL) interests *structural biophysics*, a discipline dealing with purely physical rules found within organisms. Living beings and their collectivities are very sophisticated physical systems where the rules of optics, mechanics, etc., apply also. However, the singular properties of living beings entail very specific physical phenomena contributing to the complex interactions between the compartments of the ecosystems.
- The molecular IL falls into the domains of *biochemistry*, *molecular biology*, *genetics* and *genomics*.
- Subcellular IL concerns the functional intracellular structures (not enclosed within membranes in prokaryotes, limited by membranes in eukaryotes). It is notably home to cellular respiration (mitochondria) or of the phenomenon of translation (ribosomes); therefore, this IL is of interest for molecular biology. But it is also the domain of cell biology, microbiology and embryology.

- The tissular IL pertains to histology and embryology in relation to the study of organs and comparative anatomy. The study of organ systems is of prime importance for physiology and comparative anatomy when taxa of higher ranks are compared.
- Last but not least, the compartments corresponding to a level of integration above 7 (see Table 6.1) belong to the realm of ecology, e.g. the individual, population, biocoenosis, ecosystem and biosphere ILs.

Other disciplines are transversal through all these subfields of biology, e.g. evolutionary biology, biogeography, and medical biology.

Remark 6.1 (1) All the disciplines mentioned above can cooperate and contribute to the understanding of ecology *sensu lato*, the ecology machinery depending on the underlying ILs below 7 in the sense of Table 6.1. (2) The compartmental model can be applied to each of the disciplines mentioned above for their own sake, without any link to ecology proper.

The relationships between these disciplines are intricate. Indeed, any population depends on the individuals that compose it and upon which natural selection acts—an environmental factor selecting indirectly the genotypes. Hence, in sexually mating species, the features conserved by natural selection are transmitted from ancestors to descendants. Any individual of a population is tributary to molecular, cellular and tissular processes that need to be understood to grasp the flow of matter and energy fuelling the whole ecological system, hence the need to involve ethology, reproductive biology, genetics, biophysics, thermodynamics and biochemistry in order to describe the structure and function of ecosystems.

Population lies at a fundamental level of ecology in all parts of ecology, namely when one deals with biological abundance, how communities work, or speciation mechanisms at the origin of the emergence of the different species—the constituents of the communities. Other disciplines than ecology are also interested in many ecological aspects occurring at the population level, e.g. *biogeography* [4] and the part of *evolutionary biology* dealing with speciation from natural populations. A number of authors integrate these disciplines within ecology *senso latu* for the following reasons:

1. The difference between geographical and habitat distributions is a matter of scale. The distinction between these two concepts is rather historical than truly objective; in fact, geographical distributions follow ecological gradients of variables like temperature, hygrometry, seasonal heat variation or other climatic parameters.
2. Evolutionary biology makes the "bricks" of ecology, i.e. the constituent elements of communities, appear. The taxa evolve, entailing the evolution of the communities they constitute.

6.8 Ecosystem Specificity Explained with Compartmental Models

Ecosystems are very specific compartmental systems; at each integration level (IL) (population, community, biome, biosphere), they associate one or more biological systems with a non-biological physicochemical system. As a reminder, a biocoenosis system comprises several interacting heterospecific populations in given time and space intervals. One can look at the composition and dynamics of populations from three points of view.

The energetic point of view:
Any individual of a population in a given species is assimilated to a relay of the energy flow through the ecosystems, whose source and distribution will be studied further. Energy is transferred between the species of any ecosystem along the food chains of the trophic network woven in it, with a certain amount of heat loss at each step of the process.

The material point of view:
Populations are composed of individuals whose matter (called organic matter[7]) must reconstitute itself continuously. This is why the individuals must continuously incorporate atoms of different elements, e.g. *carbon*, *oxygen*, *nitrogen* and *hydrogen*, but also *phosphorus*, *sulphur*, *magnesium*, *iron* and *zinc*. These elements enter the ecosystems and leave them globally in an equimolar manner, a phenomenon called matter recycling.

The biosociological point of view:
By this neologism,[8] I mean the biocoenoses viewed in terms of biological agents. A population of a given species can be defined as a system constituted of interacting individuals or social units like pairs, groups of n ($\geq$ 2) individuals of one or both sexes, theoretical "individuals" (mathematical vectors composed of average characteristics) or demes [i.e. groups of interacting individuals that either belong to the same or to different operational taxonomic units (OTUs) [23, 28]].

In addition to these three ways of tackling the ecosystems, one can view them in their different spatiotemporal scales. The size of a biocoenosis spans enormously spatially, from microscopic (e.g. the interacting microorganism populations in a water droplet) to gigantic (e.g. the biosphere examined as a total biocoenosis), but temporally as well, e.g. from the generation time of the bacterial constituents of the droplet just mentioned to the time needed for the setup of the biosphere. The two spans just mentioned, and all the intermediate ones, are intricately related.

---

[7] This matter is constituted of macromolecules like sugars, proteins, lipids, growth factors, vitamins, hormones, etc.

[8] This is not to be confused with the term *sociobiology* referring to a theory explaining the observed behaviours in animal societies, including human societies, as mainly innate and largely, if not exclusively, explained by the genes. The chief proponents of this theory were Edward O. Wilson (1929–2021), William D. Hamilton (1936–2000) and Richard Dawkins (1941–).

There are two types of existence conditions for the communities of organisms, i.e. (1) a local and short-termed one involving the proximal processes, the sequence of events controlling the maintenance of the living elements of the biocoenosis, and (2) a global and long-termed one involving the distal processes, the setup of the architecture of the whole—the species and abiotic factors supporting them forming the core of a biome or the biosphere, which needed a span of 3.8 billion years to come into existence.

Before we discuss these three approaches in detail, we need a historical overview of the discipline. To explain the notions used in ecology, we must understand their background and philosophical content, so we must jump back in time.

References

1. Allee WC et al (1949) Principles of animal ecology. Saunders Company, Philadelphia (PA) and London.
2. Cassirer E (1951) The philosophy of the enlightenment. Princeton University Press, Princeton.
3. Clapham AR (1966) What is experimental ecology? Folia Geobot Phytotax 1:88–92.
4. Cox CB, Moore PD (2000) Biogeography: an ecological and evolutionary approach. Blackwell Science, Oxford.
5. Elton (1927) Animal Ecology. Macmillan, New York
6. Gause GF (1934) The struggle for existence. Williams & Wilkins, Baltimore.
7. Grant PR (2001) Reconstructing the evolution of birds on islands: 100 years of research. Oikos 92:385–403, https://doi.org/10.1034/j.1600-0706.2001.920301.x
8. Griffith HL et al (2021) Breaking all the rules: The first recorded hard substrate sessile benthic community far beneath an Antarctic Ice shelf. Front Mar Sci 8:642040, https://doi.org/10.3389/fmars.2021.642040
9. Grinnell J (1917) The niche-relationships of the California thrasher. Auk 34:427–433.
10. Haffer J, Prance GT J (2001) Climatic forcing of evolution in Amazonia during the Cenozoic: on the refuge theory of biotic differentiation. Amazonia 16:579–607.
11. Hutchinson GE (1957) A treatise on limnology. John Wiley & Sons.
12. Huxley JS (1938) Clines: an auxiliary taxonomic principle. Nature 30:219–220.
13. Huxley JS (1939) Clines: an auxiliary method in taxonomy. Bijdr Dierk 27:491–520.
14. Jørgensen SE (2009) Ecosystem Ecology. Elsevier.
15. Mayr E, Vaurie C (1948) Evolution in the family Dicruridae (birds). Evolution 2:238–265.
16. Mayr E (1963) Animal species and evolution. Belknap Press, Harvard Univ Press
17. Odum EP, Barrett GW (2005) Fundamentals of ecology, 5th edn., Thomson-Brooks/Cole.
18. Popper K (1963) Conjectures and refutations. Routhedge and Kegan, London.
19. Quatrini E, Johnson DB (2018) Microbiomes in extremely acidic environments: functionalities and interactions that allow survival and growth of prokaryotes at low pH. Curr Opin Microbiol 43:139–147, https://doi.org/10.1016/j.mib.2018.01.011
20. Ramade F (2009) Eléments d'écologie: Ecologie fondamentale, 4 edn. Dunod, Paris.
21. Rohde K (1992) Latitudinal gradients in species diversity: the search for the primary cause. Oikos 65:514–527, https://doi.org/10.2307/3545569
22. Row JS (1961) The level-of-integration concept and ecology. Ecology 42:420–427, https://doi.org/10.2307/1932098
23. Salomon M, Dassy B (2016) Partitional classification: A complement to phylogeny. Evol Bioinform 12:149–156, https://doi.org/10.4137/EBO.S38288
24. Scholander PF (1955) Evolution of climatic adaptation in homeotherms. Evolution 9:15–26.

25. Senterre B et al (2021) Ecosystemology: A new approach toward a taxonomy of ecosystems. Ecol Complex 47:100945, https://doi.org/10.1016/j.ecocom.2021.100945
26. Smith DC, D'Hondt S (2015) Exploration of life in deep subseafloor sediments. Oceanography 19(4):58–70, https://doi.org/10.5670/oceanog.2006.05
27. Snow DW (1954) Trends in geographical variation in palearctic members of the genus *Parus*. Evolution 8:19–28.
28. Sokal RR, Sneath PH (1963) Principal of numerical taxonomy. W.H.Freeman, San Francisco & London.
29. Suess E (1892) Das Antlitz der Erde, 2nd edn, vol. 1 (In German). Tempsky & Freytag, Prague, Vienna & Leipzig.
30. Tangley L (1985) A new plan to conserve the Earth's biota. BioScience 35:334–341.
31. Whittaker RJ, Willis K et al (2001) Scale and species richness: towards a general, hierarchical theory of species diversity. J. Biogeogr. 28:453–470, https://doi.org/10.1046/j.1365-2699.2001.00563.x
32. Wood RE (1986) Task complexity: Definition of the construct. Organ Behav Hum Decis Processes 37:60–82.
33. Wright S (1942) Statistical genetics and evolution. Bull Amer Math Soc 48:223–246.
34. Yule GU (1925) A mathematical theory of evolution based on the conclusions of Dr JC Willis, FRS. Phil Trans R Soc Lond, ser B, 213:21–87.
35. Wooton T, Pfister CA (1998) The motivation for and context of experiments in ecology. In: Resetaris WJ Jr, Bernardo J (eds) Experimental Ecology, Issues and Perspectives, ch 18, 350–369, Oxford Univ Press.
36. Wu L et al (2023) Limited song-mixing without genomic gene flow in a contact zone between two song bird species. Mol Biol Evol 40:msad053, https://doi.org/10.1093/molbev/msad053

Part II
History of Ecology

Part II is over the history of ecology.

The matter making up this discipline has been elaborated after a long process of philosophical and scientific maturation, which depended on the historical evolution of humankind. In a way, the emergence of ecology as a scientific field is a response (1) to the degradation of our environment as a result of our scientific and technological development and (2) to a philosophical background reflecting a number of prejudices and fears within a society which has increasingly become more urban and less directly related to the natural world. The development of ecology has also been tributary to the state of scientific knowledge. Ecology could not tell the same story before and after the discovery of the role of DNA or before and after the awareness of the tremendous role of microorganisms in the functioning of ecosystems.

Part II is divided into two chapters. Chapter 7 is about a very long period, from the prehistorical times through the ancient Greeks and Middle Ages until the beginning of the Modern Ages when great scientific explorers discovered the wealth of nature across the entire planet. This was a time when much empirical knowledge about the relationships between animals, sometimes between plants or fungi, and their environment has been collected. This period ended with the Darwinian revolution of thought. Chapter 8 deals with the development of modern ecology, arising with the advent of evolutionary thinking, genetics, microbiology, modern statistics, populational thinking, and systemics.[1] It was during that period that ecology became a scientific discipline.

[1] Not to be confused with systematics. Systemics is science of systems, wheras systematics deals with the scientific classification of the living beings.

Chapter 7
Ecology Before Darwin

7.1 The Pathway to Ecology

Environmental problems have been cared about by *Homo sapiens* since prehistorical times. Indeed, humans soon realised that bad hunting or gathering management, or—later after the advent of agriculture—bad agricultural practice, could lead to starvation and many other tragedies impairing their survival [9]. However, it was not before the advent of the present techno-scientific society, coupled with the globalisation of the Western type of industrial and economic society, that the need for establishing scientific ecology and, corollarily, ecological practice was badly and urgently felt—as a response to the often devastating pressure of the present anthropic activities on the human ecological environment, incommensurably stronger than the anthropic pressure on environment prevailing before the Industrial Revolution. Ecology came as a reaction to spectacular damages caused by the *industrial revolution*, notably the loss of habitat and biodiversity and the pollution of habitats. Three seminal books written by the French ecologist François Terrasson (1939–2006) brought a breakthrough in understanding the profound incentive of this global destruction. The latter can be assigned to a philosophical background in the mainstream of the culture of an alienated humankind, which became more and more urban and severed from the natural world, reflecting a number of prejudices and fears against many aspects of nature,[1] in particular the forces that cannot be tamed or controlled by humans (not only outside but also inside us) [21–23].

For many scientists, nature-friendly people and persons alarmed about the future of public health or the availability of vital resources, it appeared urgent to drastically

Supplementary Information The online version contains supplementary material available at https://doi.org/10.1007/978-3-031-90656-5_7.

[1] These prejudices and fears already preexisted in several preindustrial societies—and in many Judeo-Christian societies—who strived to dominate nature. These elements can be notably found in biblical writings, e.g. Genesis.

M. Salomon, *Integrative Ecology*, https://doi.org/10.1007/978-3-031-90656-5_7

lessen the negative impact of this phenomenon and, for this, to understand the natural ecological systems on which all the human social systems are also based. Therefore, from the beginning, ecology was conceived, so to speak, as a two-faced discipline, i.e.

- a fundamental scientific domain (*fundamental ecology*) which is about the structure, function and dynamics of the ecosystems, i.e. the conditions for the existence of life in all its forms, independently of the goods and bads for human beings, and
- as a consequence of the latter, an applied version (*applied ecology*) that sees the impact of anthropic activities on the ecosystems and proposes solutions, from that knowledge, to render them compatible with the healthy functioning of a wealthy biosphere.

The first "face" of the discipline is a scientific venture with the sole purpose of broadening knowledge. In contrast, the second pertains more to a rational philosophical, ethical, and political commitment.

Modern ecology is the final stage of several things:

- the direct observation and a good amount of empiricism, but above all,
- the vast quantity of information collected by the explorers and scientific travellers around the Earth during the Modern Age (sixteenth–nineteenth century),
- the concurring scientific disciplines that emerged and developed between the end of the eighteenth century and the present, and last but not least,
- the awareness of the loss of biodiversity and environmental degradation since the beginning of the industrial era.

We owe the term "ecology" to the Prussian naturalist Haeckel [8], but the idea of studying animals in relation to their habitat dates back to Ancient Times. Already in prehistorical times, humans collected knowledge on the habits and life conditions of animals and plants in their environments. This information allowed them to determine their location in order to hunt or gather them and, much later, to domesticate them. The first writings in the Western world dealing with observations that had an ecological character are attributed to the ancient Greek philosophers. Herodotus (464–425 BC), for example, described a case of *mutualism* between a Nile crocodile (*Crocodylus niloticus*) and a stint—the bird ridding the reptile of leeches infesting its oral cavity [4]. We can also mention Aristotle (384–322 BC) and his disciple Theophrastus (371–281 BC), who made many observations on the habitus of animals and plants, on bird migration and developed a concept quite close to our present notion of *ecological niche* [10, 11].

These data constituted only a set of punctual observations—limited case studies from which these ancient authors straightforwardly drew generalisations. The Greek philosophers had an essentialist vision[2] contrary to our modern scientific approach. Nevertheless, they were the first to have a prescientific way of conceptualising the phenomena, based on observations and some experimentation. In the Western world, this way of apprehending the phenomena ended abruptly, partly due to the vitalist philosophy of Aristotle [18, p. 194] [16, p. 5]. This "freezing" lasted for 1800 years, a period that included the late Ancient Times from the second century BC and onwards through the entire Middle Ages until the Renaissance due to mediaeval scholastics. During this period, the cultural approach to animals and plants was strictly utilitarian (according to their economic usefulness or consumption value) or literary, reinforced mainly by the Judeo-Christian dogma assigning the animals and plants to be "objects" subjected to human needs. Throughout this long period, the portion of the Earth known to the Europeans was Europe and the Mediterranean basin; outside this area, Europeans had a very meagre knowledge of central Asia and the Silk Road far into China, and some parts of the Eastern coast of sub-Saharan Africa (mainly due to the contact with their Islamic neighbours, *e.g. via* crusades, trade competition, diverse commercial exchanges during peaceful periods, etc).

The Renaissance was, for the Europeans, a period of upheaval characterised by

1. the rediscovery of the knowledge of the Ancient Times,
2. the progressive and painful abandonment of the Christian dogma as the sole explanation of the world,
3. great scientific and technical advancement, but above all,
4. the discovery of vast stretches of land outside Europe (large portions of sub-Saharan Africa, Asia and the Americas) and of all the riches these lands contained.

It was in that epoch that the first long journeys across the Planet started to take place, revealing the great diversity of the living world. From that time until the beginning of the twentieth century, thanks to the accumulation of knowledge about the diversity and richness of the existing forms of life and their habitats in the diverse prospected areas worldwide, the basic rules of ecology were progressively unravelled.

7.2 The Great Scientific Travellers

At the beginning of the Renaissance, only a small proportion of the living beings dwelling in Europe, the surrounding Mediterranean Sea, and the closest areas of Africa and Asia were known to Europeans. Besides, about these species, confusion was manifold owing to religious prejudice and the lack of the tools and instruments

[2] Essentialism is a philosophical vision where the observed variations of objects, the phenomena, are only the reflection of their reality, the latter being their essence expressed as the words naming them. Plato explained this vision with his famous metaphor "The Allegory of the Cave", comparing the variations of objects as their shadows projected onto the walls of a cave and the objects proper as the essence within the words that correspond to them. On the contrary, in the modern scientific outlook, the observed facts are the only things considered real.

necessary to infer the correct concepts for objective descriptions. No non-utilitarian categorisation of organisms existed in that epoch, though a number of concepts deemed relevant nowadays were glimpsed in the Ancient Times (see above), and some of them were known at that time. It was the daring and unprecedented explorations starting then that opened the path leading afterwards, between the seventeenth and the nineteenth century, to the development of an intense naturalistic activity. These activities consisted of the harvesting of (1) animal, plant and ore specimens, and (2) geological, geographical and anthropological information from Europe and all the other continents as well. While discovering remote lands, Europeans looked for their natural resources, including living species.

A good number of European long-term travellers made descriptions of the organisms they discovered and whose specimens were brought to their respective countries and deposited in diverse collections (e.g. within curiosity displays or the first musea, which were all private then). Table 7.1 displays a sample of

Table 7.1 Table displaying some long-term travellers of the Modern Age who contributed to the increase of knowledge on fauna, flora, geology, cartography and anthropology [14]

Traveller's names	Identities	Visited areas
Tournefort (1656–1708)	French botanist	France, Spain, Portugal Greece, Turkey
Feuillée (1660–1732)	French naturalist	Canaries, Middle-East, Antilles, Venezuela
Linnaeus (1707–1778)	Swedish naturalist	Lapland
Saussure (1740–1799)	Swiss botanist and geologist	Swiss Alps
Pallas (1741–1811)	Prussian zoologist	Siberia
Adanson (1727–1806)	French botanist	Western Africa
L'Empereur (1660–1742)	French botanist and anthropologist	India
d'Incarville (1706–1757)	French botanist	China
Jussieu (1704–1779), Condamine (1701–1774)	French naturalists	Coasts of Brazil and the Guyanas, Amazon river
Mutis (1732–1808)	Spanish botanist	Colombia, Venezuela, southern Central America
Humboldt (1769–1859), Bonpland (1773–1858)	Prussian and French naturalists	South America
Brown (1773–1858)	British botanist	Australia
Leschenault de la Tour (1773–1826)	French ornithologist	Australia, Indonesia
Commerson (1727–1773)	French naturalist	Round-the-world trip
Cook (1728–1779), Banks (1743-1820), Solander (1733–1782)	British cartographer, and British and Swedish botanists	Pacific ocean, Oceania, Antartic Ocean
La Pérouse (1741–1788)	French naval officer and leader of scientific expeditions	Pacific Ocean
La Billardière (1755–1834)	French botanist	Mainly Indonesia

these discoverers. Some interesting extra information can be added about some of these explorers. Peter Simon Pallas (1741–1811) journeyed across Russia in an expedition sponsored by Catherine II of Russia. We owe Robert Brown (1773–1858) the separation between Gymnospermata and Angiospermata in vascular plants and the random motions of pollens in the air, which were subsequently called *Brownian* and generalised later in physics to all kinds of fluttering particles. Jean-Baptiste Leschenault de la Tour (1773–1826) took part with nine other botanists and zoologists in the expedition under the French naval officer and cartographer Nicolas Baudin (1754–1803).

Philibert Commerson (1727–1773) circumnavigated the world in the expedition under the naval officer Louis-Antoine de Bougainville (1729–1811) accompanied by the first female traveller disguised as a man, Jeanne Barré (1740–1807), at a time when women were forbidden to be aboard ships. Commerson discovered the bougainvillaea, a tree named so in honour of his expedition captain.

James Cook (1728–1779) was a pioneering European long-term traveller, becoming the first to circumnavigate Antarctica. His life ended tragically in Hawaii during a conflict with the native population. Jean-François de la Pérouse (1741–1788) and his expedition met a similar fate, disappearing in the Solomon archipelago. In 1791, the French navigator Antoine Bruny d'Entrecasteaux (1737–1793) was sent by the revolutionaries to find La Pérouse, but the mission was unsuccessful. Entrecasteaux died, and the ships were captured by the British; but Jacques-Julien Houtou de La Billardière (1755–1834), a member of the team, managed to survive, was captured by the British, and eventually returned to France with his specimens.

Thanks to the endeavours and great courage of these naturalistic travellers, the information related to faunistic, floristic and geological facts increased exponentially from what was known at the start of the Renaissance and led the biologists, by the mid-1700s, to conceive a fundamental and non-utilitarian classificatory unit of the organisms, the *species*. This category was initially conceived as a group of similar living beings endowed with quasi-identical body organisations, notwithstanding minor variations. The then-novel concept made it possible to name and describe collections of living beings according to a set of diagnostic features derived from comparative morphology and anatomy and, by doing so, to put some order in the plethoric biodiversity continuously discovered at that time. Carl Linnaeus (1707–1778) was, among the scientific travellers evoked in Table 7.1, the inventor of a revolutionary species concept embedded in a consistent nomenclatural system still in use nowadays. The species *sensu* Linnaeus (or Linnean species) has become the standard taxonomical unit for the nomenclature of organisms based on the description of specimens deposited in natural history museum collections. Modern ecology would have been inconceivable without the rigorous notion of the species fostered by Linnaeus.

7.3 Linnaeus (1708–1778)

Linnaeus is the inventor of the *binominal denomination* (or Linnean name), i.e. the modern scientific designation of living beings [15]. The binominal denomination of an organism is universally composed of two Greek, Latin or Latinised terms, referring either to a salient specificity of this organism or to the Latinised name of a person that the describer wishes to honour (giving thus to this person a form of dedication).

- The first term is a nominative noun,[3] always starting with a capital letter and referring to a genus comprising organisms sharing similar body organisations. Within a genus, one often recognises subsets containing organisms whose body organisations resemble one another more than two randomly sampled genus members. These subsets are called species. In sum, a genus[4] is a discretised species assembly.
- The second term is an epithet of the first term, either a genitive (see Footnote 3) noun or an adjective, always starting with a small letter and referring to a species within the genus defined by the first term.

The binominal nomenclature is a practical tool for species identification. I will take, as an example, a taxonomic group that I have studied for most of my career: the leaf warblers that are tiny insectivorous greenish Passerine species very subtly distinguishable by their morphology and habitus. They form *sibling species*, i.e. species having very similar morphologies (cf. Fig. 7.1), but emitting vocalisations that differ extremely from one another—the phenomenon about this group that at first and for long has attracted the interest of ornithologists. The observation and study of these rather elusive birds need some skills. Many evolutionary studies[5] have been carried out because they constitute a nice example of a current evolutionary process occurring presently, before our eyes, delivering much insight on speciation, the mechanism that splits a species into two or more ones. For example, the species presented in the photograph depicted in Fig. 7.1 is denominated

$$\underbrace{Phylloscopus}_{\text{Name of the genus}} \quad \underbrace{ibericus}_{\text{Name of the species}} \tag{7.1}$$

The genus *Phylloscopus* comprises 45 species [e.g., in Western Europe, the five most common species are (1) the common chiffchaff (*Phylloscopus collybita*), (2)

[3] Remark for non-latinists: in Latin or all the other languages practising the declension of nouns (like ancient Greek, German, Russian), (1) the nominative is the case used to mark the subject of a verb and (2) the genitive, the case used to indicate the noun complement expressing ownership or membership. In Latin, the declension markers are suffixes added to the noun radical.

[4] The plural of genus is genera.

[5] These studies needed a combination of several disciplines, e.g., notably, evolutionary ecology, biogeography, bioacoustics, behaviour ecology, population genetics, morphology, DNA sequence analysis, multivariate statistics.

Fig. 7.1 Photograph of a leaf warbler. This bird is an Iberian Chiffchaff (*Phylloscopus ibericus*)

the Iberian chiffchaff (*P. ibericus*[6]), (3) the willow warbler (*P. trochilus*), (4) the western Bonelli's warbler (*P. bonelli*) and (5) the wood warbler (*P. sibilatrix*)]

Leaf warblers, thus forming the genus *Phylloscopus* (literally meaning leaf inspector), have their breeding areas distributed in a large portion of the Old World [13, 19], namely:

- in the Ethiopian region (faunistic region formed by Sub-Saharan Africa and the southern tip of Arabia) (with six species),
- in the Palearctic region, from the Canary Archipelago across Europe, Northern Africa and the Middle East to boreal and temperate Asia, with an extension to tropical Eastern Asia (with 39 species).

One of the breakthroughs of the Linnean thought is that no species is distributed randomly; a species is necessarily associated with a distribution area. For example, in the genus *Phylloscopus*, the distribution area of each species is localised and distinct; it is the fruit of its history and ecological niche. The distribution of each

[6] *P.* standing for *Phylloscopus* if the complete denomination is mentioned for the first species of the same genus in a list. The reader will note that the terms of the Linnean denominations are italicised.

warbler of the genus is thus mappable. A distribution area is, therefore, an important descriptive trait in the identity of a species.

The meaning of the binominal definition for one given species can be exemplified with the common chiffchaff (*Phylloscopus collybita*); this phrase means the "leaf inspector" (the genus name) "changing the money" (the species name, which is a reference to its specific song evoking the metallic sound of handled coins). The scientific community has agreed that the description is the most relevant for the species in question.

This practice is generalisable to all the organisms except the viruses. Linnaeus did not base his system on an evolutionary explanation, as we do now. Like nearly all of his contemporaries, he was a fixist and a creationist,[7] but he defined the modern rules for the denomination and description of species that run as follows:

- Each species, designated by its binomial denomination, is described by the morphology of a specimen, i.e. the skin or any body parts of an individual of that species deposited in a referenced collection (e.g. in a natural history museum). A label is attached to the specimen with its Linnean name and sometimes some valuable diagnostic information about behavioural or other relevant traits that are not measurable. The specimen constituting the standard of the species is called the type. Only species of which the type is deposited are scientifically acknowledged.
- Any acknowledged species mentioned in a scientific publication must be associated with the specifications of its type (see just above), the place that shelters it, the name of the descriptor(s) and the year of the description.

Generally, the description of species, or any significant taxa, is published by descriptors who also become, by doing so, scientific authors. Hence, for example, the denomination of the bird represented by the photograph of Fig. 7.1 is

$$\underbrace{\textit{Phylloscopus}}_{\text{Genus name}}\ \underbrace{\textit{ibericus}}_{\text{Species name}}\ \text{Ticehurst, 1938}$$

The descriptor is the British ornithologist Claud B. Ticehurst (1881–1941), who described the type in 1938 [24] The type is deposited in the collections of the Natural History Museum of London.[8]

The Linnean species concept, also called typological, is still in use nowadays but is always completed by more modern species concepts involving the evolutionary approach of biology (since Darwin, see below) and all the scientific advances developed during the last two centuries.

[7] *Fixism* is the belief that the species have never changed since the start of life, and *creationism* is the belief that a mighty supernatural principle—the great architect of the universe—literally created these species.

[8] For more information related to the denomination of this species, see [20].

Thus, Linnaeus elaborated a fundamental tool for ecology, the first objective species concept. He also promoted an experimental approach in a number of what we can call "proto-ecological" studies, for instance, his writing on the pearl oyster and the environmental conditions fostering good pearl production.

7.4 The Predarwinian Forerunners of Ecology

7.4.1 Alexander von Humboldt (1769–1859)

Alexander von Humboldt, a long-term traveller and botanist (see Table 7.1), and his collaborator and friend Aimé Bonpland (1773–1858), were the first modern naturalists who established, in their seminal book written in French "Essai sur la géographie des plantes",[9] the basis of phytogeography, i.e. that plants did not grow randomly but formed associations very specific of different "climates", i.e. portions of the Earth defined by their latitudes and longitudes (see Sect. 3.2.1, p. 28). This novel research resulted from an expedition between 1799 and 1804, where these botanists collected approximately 60,000 plants representing 6000 new species. The cited authors remarked that some plants were asocial (individuals growing isolated from one another), while others were social (constituting societies). Thus, the plants do not distribute randomly at the species level. The social plants would group and form associations (or forms) characterised by key genera. They gave many examples of asocial and social plants[10] [12, p. 15] and a list of 15 associations with their allegedly key genera among similar associations that would cover the entire planet.

7.4.1.1 Asocial and Social Plants

Among asocial plants, the authors mention:

In Europe: the bittersweet nightshade *Solanum dulcamara*, the red campion *Lychnis dioica*, the snakeweed *Polygonum bistorta*, the Saint Bernard's lily *Anthericum liliago*, the common whitebeam *Crataegus aria*, the stubble-moss *Weissia paludosa*, the bristly haircap *Polytrichum piliferum*, the brown alga *Fucus saccharinus*, the giant club fungus *Clavaria pistillaris*[11] and the parasol mushroom *Agaricus procerus*.

---

[9] Essay on plant geography.

[10] I use the terminology of the cited authors, their taxonomy reflecting the state of the art of their epoch. That is why I included the cited brown algae and fungi within the "plants" in the text but indexed them in their proper modern taxon.

[11] Modern denomination: *Clavariadelphus pistillaris*.

In the tropics: the Guayabo de Indio *Theophrasta americana*, the Jamaican fuchsia *Lysianthus longifolius*, the quinine and alliae *Cinchona* and the rubber tree and alliae *Hevea*.

And among the social plants: The strawberry *Fragaria vesca*, the European blueberry *Vaccinium myrtillus*, the common knotgrass *Polygonum aviculare*, the brown flatsedge *Cyperus fuscus*, the grey hair-gràss *Aira canescens*, the Scots pine *Pinus sylvestris*, the sea purslane *Sesuvium portulacastrum*, the red mangrove *Rhizophora mangle*, the silver july croton *Croton argenteum*, the bayhops whole *Convolvulus brasiliensis*, the breadfruit tree [6] *Brathys juniperina*, the escalonia *Escallonia myrtilloides*, the Plumier's bromelia *Bromelia karatas*, the blunt-leaved bogmoss *Sphagnum palustre*, the common haircap *Polytrichum commune*, the brown alga *Fucus natans*, the dead man's fingers *Sphaeria digitata*, the bloodstain lichen *Lichen haematoma*, a reindeermoss *Cladonia pascalis* and the false turkey tail *Thelephora hirsuta*.

7.4.1.2 The 15 Plant Associations

1. The *Scitaminae* represented by the banana trees *M*usa, lobster claw plants *Heliconia* and crane flowers *Strelitzia*;
2. the palm trees;
3. the tree ferns (ca 600 extant species);
4. the form Arum-Pothos-Dracontium represented by the eponymous genera;
5. the Firs associated with the conifers, namely the yews *taxus* and the pines *pinus*;
6. the *Folia acerosa*, i.e.plants whose leaves are hard, always green, narrow and sharp, as defined by De Candolle in his book entitled "Théorie élémentaire de la botanique" [3].
7. the tamarinds and alliae, a form typical of the Americas, comprising notably the genera *Mimosa*[12] (600 genera in the Americas, Eastern Africa and South-East Asia), *Gleditsia* (the honey locusts) and *Porlieria*;
8. the *Malvaceae* associated to sickle bushes *Stercularia*, hibiscuses (the genus *Hibiscus*), the balsa tree (the sole species of genus *Ochroma*) and *Cavanillesia*;
9. the lianae (a mix of vines of the genus *vitis* and of lianae of the genus *Paullinia*);
10. the Orchids, namely the genera *Epidendrum* and *Serapias*;
11. The prickly pear and allied represented by all the cacti;
12. The *Casuarines* comprising the horsetails *Equisetum*
13. The Grasses;
14. The Mosses;
15. The Lichens.

[12] Not to be confused with the flowers commonly called "mimosas" in flower shops.

7.4.1.3 The Main Assumptions They Drew from Their Field Research, Weaknesses and Strengths Seen with Our Modern Viewpoint

(1) Humboldt and Bonpland point out that most botanists of his time were only interested in discovering new species and their taxonomic descriptions. (2) The authors launched the study of the geography of the plants, which was new research at that epoch. (3) They point out features of the geography of the plants, namely the local associations under the different climates, which does not mean that, according to the authors, the climate conditions were necessarily the main factor explaining the distribution, see hereunder.

- The plants differentiate not only according to the physical conditions they meet, but are strongly influenced by their intrinsic level of sociality. Certain plants grow isolated from one another, and others live closely in societies.
- Social plants are more common in temperate than in tropical zones, where (I quote)[13]

> the vegetation is less uniform, thus more picturesque.

From their long fieldwork across vast areas of South America, the cited authors widely contributed to the advancement of the knowledge of this vast subcontinent that was then widely undiscovered and made important findings as well in the physical and geological, as in the biological (mainly botanical) and biogeographical field. They namely set up the hypothesis that individuals of plant species from the plains of the tropical regions tend to grow isolated and far apart from one another and very rarely form groups, whereas those from temperate areas (Europe, North America, mountainous areas—even in the intertropical regions) would have the tendency to aggregate more and form societies.[14] They also observed this phenomenon in cryptogams, a category now completely deprecated as a taxon but used as such at that time and gathering the “inferior plants” reproducing by spores without flowers or seeds, like ferns, mosses, algae and fungi.[15] Moreover, they contended to have encountered plants in the tropical regions expected to be found in higher latitudes.

The authors elaborate that tropical forests, with their vast expanse and rich species diversity, are habitats where each species is represented by a few specimens scattered over long distances, resulting in a great multiformity of vegetation. In stark contrast, the plants of temperate habitats are represented by a few species containing many individuals that tend to form groups. This is how Humboldt explains the prevalence of heath in northern and northwestern Europe.

---

[13] Translated by me from French.

[14] But we can see from the list above that there seem to be exceptions to this hypothesis (e.g. the sea purslane and the red mangrove).

[15] The “superior” plants or phanerogams (members of the clades gymnosperms and angiosperms) reproduce from cones or flowers, which produce seeds that develop into adult individuals.

Then, the author expresses a theory that might appear strange to our contemporaries but not improbable in ancient times when most scholars were fixists and believed in spontaneous generation. Humboldt noticed that a number of tropical areas, generally mountainous localities, behaved like northern temperate regions (like Canada or Norway). The author explained this by describing the history of the planet's geology in the frame of the theory of catastrophism. The history of the plants would follow the history of the continents and the seas;[16] Humboldt and Bonpland are more convinced by this theory than by the dispersion due to an adaptation to climatic factors. Their theory is as follows: (1) The primitive forms, i.e. the cryptogams, found everywhere, are generally social and form large societies and extensive distribution areas. They adapt quickly to all sorts of habitats and climates. For example, similar moss forms vast meadows in Columbia, where the temperature swings between 28° C and 30° C and in Norway. (2) The more complex ones (phanerogams) are less adaptable and more restricted in their distribution due—for the author—to the intrinsic degeneration of their life conditions. (3) Thus, in explaining the two last points, climate played scarcely any role.

For Humboldt, the debate is as follows. (1) Did the cryptogams appear spontaneously in many places of the Earth? (2) Were they carried away from one continent to another by oceanic currents? Their territorial expansion was very large. The phanerogams, on the other hand, would have adapted afterwards to the territories left. Due to his interpretation of the fossils, Humboldt opted for point (1). We must not forget that for Humboldt, rocks and the geological movements of the Earth seemed almost eternal. In contrast, life appeared relatively recently (with an implicit biblical interpretation, though he did not explicitly hypothesise an age to the fossils). In addition, as a believer in the caloric theory,[17] he interpreted the discovery of apparently tropical fossils of disappeared animals [18] in high latitudes as the consequence of movements of calorics on the surface appearing periodically during the past ages. He thus considered that man almost always accompanied the animals and the plants on earth and that he was as active as the winds and the birds as dispersal agents of the plants.

Biology was a very young science at that relatively remote epoch. Indeed, in 1805:

1. Explanatory factors, like physical variables (like temperature, precipitation, acidity-alcalinity gradient, etc) or geological features, were notably absent from the minds of the scholars of that time. However, relatively shortly after, these variables would be proven essential for understanding phytosociological features.

[16] In Humboldt's scenario, the continents are depicted as drifting. This was an early visualisation of the continent drift theory, which was later elaborated and considered very heterodox at that time when the prevailing belief was that continents were fixed.

[17] According to this theory, abandoned in the mid-nineteenth century, heat consisted of a weightless, self-driving fluid called caloric flowing from hotter to colder spots.

[18] Cuvier's theory of catastrophism explained that fossils were the remnants of living beings who were victims of big catastrophes—an implicit reminiscence of the Flood.

2. The true nature of species, due to the state of the art of that time, was not known. The technology and concepts were not sufficiently advanced to determine the "cryptogams".[19]
3. The authors shared the fixist vision of that epoch, close to the concepts of Buffon and Linnaeus.
4. They believed in the reality of spontaneous generation, at least among primitive "crypotogams".[20]
5. Implicitly, though without any mention of it, they shared Cuvier's theory called *catastrophism*, which explains how the fossils were the remnants of disappeared plants drowned during the Flood as explained in the Book of Genesis, the existing species being the survivors of this biblical catastrophe.
6. They draw, from points (3) and (4), the hypothesis that the cryptogams and other primitive plants would have appeared spontaneously everywhere on Earth, taking as an example a number of species they asserted to find in Europe as well as in equatorial areas, e.g. the broom moss *Dicranum scoparium*, the common haircap *Polytrichum commune*, and the wart lichens *Verrucaria sanguinea* and *V. limitata*, and hypothesised that the other plant species, namely the phanerogams, would be the result of a degeneration that rendered them constant and not flexible and fixed in restricted areas (for the authors less successful, an idea that is at light-years from evolutionary thinking!).
7. The physiognomy of the plant associations throughout the Earth proposed by Humboldt and Bonpland [12, p. 31] was admittedly based on subjective perceptions of stable species groupings called *forms*.

Nevertheless, despite all these 'youthful mistakes,' Humboldt's work was crucial and constituted the first steps towards elaborating modern ecology and biogeography.

7.4.2 Augustin de Candolle (1769–1859)

Candolle [3] highlighted the vital role of light in plants' lives, discovered the fundamental difference in the mode of nutrition between animals and plants, and prefigured the food chain's concept. This Swiss botanist proposed theories that later scientists largely confirmed.

- Candolle showed that though all plants vitally need light, they do not need the same quantity of it from one species to the other. Hence, several species need more light than others, and a given quantity of light might be harmful

[19] The cell concept was still not unravelled; chemistry was nascent. It was fifty years before the Darwinian revolution and more than a century before the advent of genetics.

[20] They equated taxa that were later proven to be taxonomically distinct, nevertheless showing ecological convergence due to adaptations to similar environmental conditions.

to some species while the same quantity is obligated to others. This explains the very specific distribution of plants on Earth, thus the differentiation of the plant communities according to altitude and/or latitude. Thus, Candolle is also pioneering what will be called later *phytogeography*.

- He expounded in this work the fundamental difference between the mode of nutrition of plants and animals, which he explained as follows:
  - Animals are active. They are sensitive and mobile and must actively search for food in solid substances. They find it in organic substances, generally in other organisms (plants or animals) they prey upon. They build their substance by ingesting them and transferring their organic matter to their stomachs. Their vessels transport the molecules inwards and are never projected. Their bodies are limited.
  - On the contrary, plants are fixed and lack motility and sensitiveness. Their nutrients are inorganic and dissolved in water; thus, they feed on liquids. Their vessels develop outwards, and their bodies expand.
    Thus, this author carries the seeds of the concept of heterotrophy and autotrophy, expressed in the language of his time. Since the atom was only a hypothesis at this epoch, and cells were still not known as elementary components of animals, the thinking could thus not be cellular-oriented.
- The author clearly expressed that inorganic elements absorbed by plants passed, via feeding, into the stomachs of the animals. This analysis was a formidable scientific breakthrough that made it possible to conceive the principle of the food chain.

In the same book, (1) the author rejected the principle of spontaneous generation contrarily to most of his contemporary colleagues; (2) he noticed that when animals and plants are compared, the simpler their organisation is, the more their inner structure resemble each other; and last but not least, (3) he pointed out a striking feature shared by both kingdoms of life, sexual reproduction. However, despite these arguments, which strongly supported evolution, Candolle refused to make the leap. He refused to adopt a stance on the subject, preferring to say that the diversification of living beings remains a mystery.

7.4.3 Rudolf Amandus Philippi (1808–1904)

A novel approach was proposed in 1836 by the German naturalist Rudolf Amandus Philippi [17] (1808–1904) by carrying out numerical methods[21] on florae of phanerogams (1) of similar size and taxonomic composition (where all species were

[21] It is no wonder since that epoch witnessed the emergence of the first true chemists (Lavoisier, Dalton and pioneering followers.) with the introduction of quantitative methods (e.g. counting, measure of weight or other physical parameters).

included), (2) at similar heights above sea level (3) in given geographic coordinates. He did not a priori pose any hypothesis on the relations between the plants in the area studied. His method consisted of comparing the percentages of plant genera between florae not far apart and in a gradient from the hottest to the coldest climate or from a temperate climate to increasing or decreasing temperature conditions.

The author compared the phanerogam flora of Sicily with those of the surrounding areas (extending his scope of study to Europe, North Africa and the Middle East). He hypothesised that climate was the main factor of the variation of the percentage. He gave to the gradual progression of the plant species from one to the neighbouring florae an image of clinal variation—with typical florae separated by broad transitional territories.

He was the first known author to have a gradualist approach to phytogeographic studies based on climatic variables.

7.4.4 August Heinrich Rudolf Grisebach

For the German botanist August Heinrich Rudolf Grisebach [7], the surface of the Earth decomposes into a number of florae separated by strict boundaries. The boundary of a given flora F prevents plants from neighbouring florae from crossing it and settling in F. The cited author called these florae *formations*, which, in fact, resemble the Humboldtian associations.

With this point of view, Grisebach expresses a strong disagreement with Philippi (aforementioned). The transition zone between two neighbouring formations is very abrupt. Formations would be exclusive and hermetic as far as their taxonomic composition is concerned and almost contiguous.

7.4.5 The Ecogeographical Rules

Before the Darwinian revolution that enabled the scientific community to understand the true nature of species and, corollarily, the correct interpretation of interindividual variation within a species, ecology remained a scientific domain in its infancy, whose findings were secondary to the discoveries in other domains of biology like anatomy, physiology or cell biology for example. However, some ecological research works before or contemporary to the writings of Darwin were very interesting and made significant contributions. They showed that the individuals of a given species tend to vary for given traits along a geographical, thus an ecological gradient. Indeed, the diverse geographical locations present some physicochemical variations due to climatic or other environmental modifications. Within the same species, individuals of different regions will tend, on average, to show variation for one or another character because they adapt their physiology to the local physicochemical conditions; their physiology will tend to diverge in parallel with ecological variation. Morphological or physiological variations, functions of ecogeographical variation, are the object of a subdomain of ecology, ecophysiology that appeared very early, well before "classical" ecology. Three pioneering ecophysiologists established well-known ecogeographical rules that were subsequently named after them, i.e. Gloger, Bergmann and Allen. These rules reflect the regular variation of a morphological character or trait (size, weight, dimension of body parts, body colour) observed along ecogeographical gradients between populations within many species of *vertebrate endotherms*.

The German ornithologist Constantin Gloger (1803–1863) stated that within a bird species, the more humid the climate is in which an individual lives, the darker and duller its plumage will tend to be [5]. This rule seems to be generally verified. Later, the same rule was also globally observed with the fur of mammals.

Carl Bergmann (1814–1865) [2] established that in vertebrate endotherms, the colder the climate is in which an animal lives, the larger its size will tend to be. The gradient will run along an increase of altitude or latitude. This German anatomist gave a physiological interpretation. Bulkier animals tend to have smaller body surfaces in proportion, thus perspire less water proportionally.

Joel Asaph Allen (1838–1921)[1] was an American naturalist and a curator of the American Natural History Museum of New York for a long time. Allen was not a Darwinist but rather a Lamarkian evolutionist and a saltationist. In this conception, evolution proceeds by jumps, new organisms appearing very quickly within populations under the influence of the environment and in response to it. One owes him his eponymous rule: within a species, the colder the climate is, the smaller the outer appendages are; this is generally observed in birds and mammals. The explanation would be physiological. In warmer regions, larger outer appendages help to evacuate toxins quicker, whereas in colder areas, small outer appendages are more efficient in preventing the loss of body heat.

References

1. Allen JA (1905) The evolution of species through climatic conditions. Science 22:661–668, https://doi.org/10.1126/science.22.569.661
2. Bergmann C (1848) Über die Verhältnisse der Wärmeökonomie der Thiere zu ihrer Grösse (In German). Vandenhoeck & Ruprecht, Göttingen.
3. de Candolle AP(1813) Théorie élémentaire de la Botanique. Déterville, Paris
4. Egerton FN (2001) A history of the ecological sciences: early greek origins. Bull Ecol Soc Amer 82:93–97.
5. Gloger CL (1833) Das Abändern der Vögel durch Einfluß des Klima's. In Commission bei August Schulz und Comp, Breslau
6. Green T (1823). The universal herbal, or botanical, medical, and agricultural dictionary. 2nd edn, Henry Fisher Caxton, London.
7. Grisebach AHR (1838) Über den Einfluß des Climas auf die Begränzung der natürlichen Floren (In German). Linnaea 12:159–200.
8. Haeckel E (1866) Generelle Morphologie der Organismen. G. Reimer, Berlin.
9. Harari YN (2014) Sapiens: a brief history of humankind. Random House, London.

10. Hughes JD (1975) Ecology in ancient Greece. Inquiry 18:115–125, https://doi.org/10.1080/00201747508601756.
11. Hughes JD (1985) Theophrastus as ecologist. Environ Hist Rev 9:296–306.
12. von Humboldt A & Bonpland A (1805) Essai sur la géographie des plantes. Levrault, Schoell et Co, Paris.
13. Irwin DE et al (2001) Cryptic species in the genus *Phylloscopus* (old world leaf warblers). Ibis 143:233–247, https://doi.org/10.1111/j.1474-919X.2001.tb04479.x
14. Laissus Y (1981) Les voyageurs naturalistes du jardin du roi et du Muséum d'Histoire Naturelle: essai de portrait-robot (In French). Rev Hist Sci 34:259–317, https://doi.org/10.3406/rhs.1981.1768
15. Linnaeus C (1758) Systema naturae. Laurentii Salvii, Stockholm.
16. Mayr E (1963) Animal species and evolution. Belknap Press of Harvard UP, Cambridge, MA.
17. Philippi RA (1836) Ueber die Flora Siciliens im Vergleiche zu den Floren anderer Länder (In German). Arch Naturgesch 2:337–366.
18. Polya G (1954) Induction and analogy in mathematics. Vol 1: Of mathematics and plausible reasoning, Princeton University Press.
19. Salomon M (1997) Quel statut taxinomique donner au pouillot véloce ibérique? (In French) Alauda 65:63–81.
20. Salomon M et al (2003) On the taxonomic status and denomination of the Iberian Chiffchaffs. Ibis 145:87–97, https://doi.org/10.1046/j.1474-919X.2003.00122.x
21. Terrasson F (1988) La peur de la nature (In French). Sang de la terre, Paris.
22. Terrasson F (1994) La civilisation anti-nature (In French). Sang de la terre, Paris.
23. Terrasson F (2002) En finir avec la nature (In French). Edition du rocher, Paris.
24. Ticehurst CB (1938) A systematic review of the genus *Phylloscopus* British Museum, London.

Chapter 8
The Emergence of Modern Ecology

8.1 Darwin and His Direct Followers

8.1.1 Charles Darwin (1809–1886)

The father of the modern theory of evolution established two principles that are the basis for modern ecology: natural selection and species evolution [8].

Natural selection: This is the set of environmental factors varying with time to which organisms of a species are subjected and that "select" individuals. The individuals "chosen" by natural selection are, by definition, organisms able to survive environmental hazards at a given time and transmit this aptitude to their descendants. As a result, several physical traits are favoured while others are eliminated. Thus, natural selection is by no means a vitalist but a materialist and Malthusian principle.[1]

Species evolution: Natural selection impels organisms to change regarding their ancestors for a couple of reasons: (1) environmental conditions change across time, (2) in parallel, the aptitudes necessary for the organisms to adapt to current environmental changes also vary. These two properties have two fundamental consequences.

- The differentiation between individuals is continuous and negligible between a generation and the following one. However, the differences accumulate after many generations, so the present organism and the ancestor at the start of the chain of ancestors are different, to the point of having become members of two

Supplementary Information The online version contains supplementary material available at https://doi.org/10.1007/978-3-031-90656-5_8.

[1] This adjective refers to the tendency of populations observed by Malthus to increase exponentially if not checked by limiting factors. Natural selection is almost equivalent to "limiting factors".

M. Salomon, *Integrative Ecology*, https://doi.org/10.1007/978-3-031-90656-5_8

distinct species. This principle would explain all the diversity of life. Darwin was thus a *gradualist*.

- Analogous ecological conditions exist between habitats of separate continents characterised by analogous climatic or phytosociological conditions. Nevertheless, because of the unique history of each species, associated with its unique geographical origin, the same species or kin species can never endemically occur in two continents that have been disjointed for a longer time than the history of the incriminated species. If this occurs, and unfortunately that has been increasingly the case, it is due to introduction—rarely a natural one (e.g. via gusts of wind, storm, natural rafts, transports on animals)[2]—but too often caused by humans. In two disjointed continents, one may find ecological equivalents, i.e. species showing a superficially similar look in response to similar environmental pressure. Indeed, in this situation, the body shapes of unrelated organisms are constrained, e.g. the pisciform aspect of vertebrates adapted to aquatic life (fishes, ichthyosaurs, whales). The forms vesting similar aspects due to ecological convergence are also called parallel forms. Other examples of such parallel forms invoked by Darwin are duikers vs brockets, respectively, African bovids and South-American cervids. Both mammal species, though unrelated, have identical looks and feeding habits.

The consequences of these principles were fundamental for modern ecology.

- The species' unique histories are interdependent in each habitat and form unique communities. A habitat is thus unique for its community composition, thus also for the interactions between the member species. Thus, the disappearance of one species may naturally entail a spate of extinctions.
- In this interdependence system, Darwin shows that kin species, because they have similar nutritional requirements, tend to compete and consequently vary their diet and foraging systems so that their vital needs are not too similar, which allows them to coexist. The concept of ecological niche was on its way. He also showed that this competition principle causes the species to take part in an unending struggle for life, e.g., an increase of prey efficiency to escape predators, who themselves are more and more efficient in hunting and seizing prey; the same type of race between plants and herbivores; rough competition between plants for light and soil water.

8.1.2 Alfred Russel Wallace (1823–1913)

This man was Darwin's unfortunate competitor for the paternity of the discovery of the evolutionary character of life diversification under the principle of natural selection. Wallace published his ideas in 1859, like Darwin, independently. They

[2] E.g., it would explain the presence of monkeys in South America.

both exposed their theories in a meeting of the Linnean Society of London, where Darwin's work was presented before Wallace's, which saved, *in extremis*, the primacy of Darwin. Despite this situation, Wallace accepted his "defeat" with fair play and became a staunch follower and friend of Darwin. Wallace's findings resulted from two succeeding expeditions, the first to Brazil, which failed tragically with the wreck of the ships transporting his specimens during the return voyage, and the second to Malaysia and Indonesia. One owes Wallace the eponymous line that separates two neighbouring but very distinctive biogeographical areas, the Indo-Malaysian and the Australasian regions. These two zones encompass very specific fauna and flora associations, witnessing their having experienced very different life histories [49, 50].

Wallace made two major contributions to modern ecology:

- the concept of biocoenosis [49];[3]
- the role of sexual selection in the evolutionary process, notably (1) the contribution of courtship in partner choices for pairing and matings, in particular assortative matings, and (2) the relationship between these sexual behaviours and reproductive isolation—a mechanism often invoked in the formation of new species.

8.1.3 Ernst Haeckel (1832–1919)

As mentioned at the beginning of Chap. 2, Haeckel was the inventor of the term "ecology", and he was the first author to establish it as a science with a modern look [17, p. 286] based on the acceptance of Darwin's view on evolution via natural selection. Indeed, Haeckel was an enthusiastic Darwinist. We owe Haeckel the famous phrase "ontogeny recapitulates phylogeny", based on the striking resemblance between the embryos at an early stage of species belonging to different classes of vertebrates. About "phylogeny", Haeckel was the inventor of this term applied to the generation process of the taxa grasped as a metaphor of a genealogical tree.

Unfortunately, his conception of the struggle for life concerning the individuals within a species was erroneous since it was based on the theory of blending inheritance. Like almost all of his contemporaries, Haeckel thought that the parents' traits were blended during the generation process and that the result of the mix was transmitted to the young. Roughly, if, for a given character, one parent had traits of good quality and crossed with the other parent, who was endowed with traits of bad quality, the product would necessarily have a mediocre quality for these traits. Guided by this principle, Haeckel became a staunch defender of a policy applied to humans that favoured the crossing between individuals of identical qualities. He

[3] The concept was Wallace's but the term is attributed to Möbius article on the ecology of Oysters [32].

fostered a movement and a school of thought called "social Darwinism," developing eugenic and racialist ideas. Evolutionary biology and ecology were then in their infancy. Haeckel did not know the work of one of his contemporaries, the Austro-Hungarian monk Gregor Mendel (1822–1884), on the segregation of characters. This very important contribution would eventually lead to the emergence of genetics one generation later, a discipline strongly related to modern ecology.

8.2 Towards a Science Called ... Ecology

Modern ecology cannot be dissociated from the modern concepts of species, notably the biological one. Species defined according to the biological concept are called "biological species" or "biospecies".[4] This concept can only be applied to eukaryotes,[5] the only organisms sexually reproduced. Nevertheless, this concept has proven extremely useful for elaborating the basis of modern ecology. As a consequence of the use of the biological concept to define a species within an ecological frame, the populations constituting a given species, say A, will be characterised by a gene pool associated with an ecological niche. It is upon the individuals that natural selection acts, favouring a number of variations in the gene pool and the associated ecological niche. But clearly, by selecting individuals, natural selection gradually favours a number of genes and eliminates others; and the same process functions on all the other species interacting with species A, imposing a constraint not only upon the genetic structure of A but also upon the genetic structure of the other species of the biocoenosis whose species A is a member. Thus, each biocoenosis is also forged by natural selection, with consequences on all its subcompartments and its environment, thus also on some of the abiotic factors supporting it.

However, this is only a part of the story that was still inconceivable within the frame of knowledge before the end of the nineteenth century. The basic concepts exposed above and many other elements of modern ecology, notably the complex dynamics of the ecosystems, needed a level of understanding only possible after a large number of scientific and technological breakthroughs that occurred only since the end of the nineteenth century, notably the advent of genetics and the revolutionary progress in physics, the discovery of the biochemical machinery, the important advances in microbiology and cell biology, the increased sophistication of mathematical and statistical tools, and last but not least the advent and progress of informatics.

[4] A biospecies is an assembly of individuals capable of interbreeding.

[5] Eukaryotes are species constituted of a cell or cells endowed with a nucleus containing different linear chromosomes. This nucleus is surrounded by a cytoplasm that contains organelles and structures (e.g. ribosomes) assuming specific tasks (e.g. transcription, translation). Two types of organelles, specific to eukaryotes, contain DNA strands (mitochondria, plastids). The cell is limited by a membrane and, in the case of autotrophs (e.g. plants), enclosed within a cellular wall.

8.3 The Emergence of Genetics and Its Impact on Ecology

The main founders of genetics were Weissmann, de Vries and Morgan.

- The German biologist August Weismann (1834–1914) carried out much experimental work on sea urchins and rats and drew from his research two important findings [52]:
  - There is an absolute separation between the cells destined to construct the body (somatic cells) and the cells involved in reproduction, i.e. gametes and their predecessors (germinal cells).
  - Correlatively, hereditary information is passed from parents to offspring exclusively via gametes.

  This finding led to the abandonment of the theory of evolution via the inheritance of acquired characters (a theory called *Lamarckism*[6]) and the adoption of the principle that natural selection is the sole force driving evolution. It was Weissmann who discovered that the principle of heredity lay in the spermatozoids and ovules, but the Mendelian transmission was unknown to him.
- The author who rediscovered the principle of character segregation without knowing Mendel's work was the Dutch biologist Hugo de Vries (1848–1935) [9] . He gave a theoretical explanation of the phenomenon, which proved revolutionary: the characters are the expression of particles that he called "pangenes", confirmed experimentally later and presently named genes. A "pangene" was thus, for de Vries, a particle causing, by an unknown mechanism, the formation of a physical trait. We were in 1889. De Vries can thus be considered the very first geneticist of history. It was by accident that he fell upon the article of Mendel on the character segregation [31] while reading a collection of scientific writings by the end of the 1890s. From then on, De Vries quoted the latter very frequently, restoring posthumously justice to him. The work of the Austrian monk would strengthen him in his theory. De Vries was the first author who popularised Mendel's concepts and terms of character dominance and recessivity. One owes De Vries the concept of mutation.
- The American biologist Thomas H. Morgan (1866–1945), one of the most brilliant and prolific scientists of the twentieth century, synthesised the works of Mendel, Weissmann and de Vries and confirmed their findings experimentally, notably (1) the Mendelian character of gene transmission, (2) the existence of

[6] The theory of evolution via the inheritance of acquired characters has been proposed by Lamarck (1744–1829); this theory stated that it was the environment that directly transformed the organs and that the descendants acquired the modifications if the latter were beneficial to them. Darwin did not exclude this Lamarckian principle but considered that natural selection was the main driving force of evolution. Recently, a stream of research on epigenetics gave some credit to a form of Lamarckism, epigenetics being the discipline studying the epigenetic traits, i.e. "stable heritable phenotypes resulting from changes in the chromosomes without alterations on the DNA sequences" [2].

mutation, (3) the fundamental difference between the somatic and germinal cells, the first being mortal and the second theoretically immortal.

In addition, he made such a fundamental contribution to biology that his findings became part of modern culture, belonging to the general background.

- He is the discoverer of the two categories of chromosomes, the autosomal and sexual ones .
- He also highlighted the uniqueness of the formation of the genotypes occurring when the gametes merge during fecundation, that one of the genes originates from the father and the other from the mother, that the chromosomes of the same type exchange their genetic material, and that a genotype is always located in a given site on the chromosome, that he named "locus".

In his opinion, evolution and coevolution are the consequences of genetic mechanisms where randomness is nearly always involved; consequently, the principle of natural selection would not be necessary. Morgan's approach to evolution was non-Darwinian. Most of his results can be found in his two seminal books [33, 34]. Morgan opened the path to *molecular genetics* on the one hand and to *population genetics*—strongly related to ecology—on the other hand.

Population genetics is mainly interested in the distribution and evolution of the genetic structure of populations and uses mathematical tools for that. Many disciplines benefited from this rather formalistic and abstruse domain and would only have their present look and efficiency with it, e.g. medical epistemology, population history reconstruction, and last but not least, modern scientific ecology. The first author who used mathematics in heredity biology was Galton, a nephew of Darwin. Unlike Morgan, Galton was a staunch Darwinist. The very first authors who applied mathematics to heredity science were, by chronological order, Galton, Morgan, Pearson, Hardy, Weinberg, Fisher, Wright and Haldane.

- The British scientist Francis Galton (1822–1911) was one of the inventors of applied mathematics. His interests were manifold, notably anthropology, psychology and heredity biology. We owe Galton the concepts of linear correlation[7] and linear regression.[8] In psychology, Galton invented the first psychometric tests. In biology, this prolific scientist was famous for his research work on twins, where he tried to disentangle the part from heredity and the part from the environment in their development and "quality", the latter being for him the target of natural selection [14].
  Galton, like Haeckel, did not know about the Mendelian transmission of characters and the concept of genes. As the promoter of biometrics, he favoured continuous magnitudes and rejected the choice of discretised characters preferred

[7] A statistical parameter assessing the concomitant variation of two Gaussian variables if the relationship between them is linear.

[8] Statistical technique consisting of fitting a straight line in a scatter plot with the least-squares method.

by Darwin and Mendel. This philosophical stance prepared him, as was the case with Haeckel, to adopt the theory of blending inheritance and the idea that the influence of hereditary factors prevailed over the influence of environmental ones upon the quality of the individuals. Like Haeckel, Galton was tempted to apply the principle of natural selection to human society and founded an ideology, eugenism, militating for the elimination of the individuals judged unfit (e.g., mentally or physically disabled or people belonging to a race assessed inferior). Despite this stance that we can judge very shocking and totally unacceptable retrospectively, Galton remains nevertheless a great scientist who paved the way for the foundation of modern statistics and population genetics.

- The British mathematician Karl Pearson (1857–1936) was a disciple and friend of Galton and a staunch supporter of eugenism. He was the inventor of what would be called later "inferential statistics", one of the fundamentals of biometry and population genetics. One owes him the first biometrical research work that revealed the discreteness of morphological variation, the application of standard deviations to establish intervals around mean values, the first practical application of correlation to biometric variables, and the invention of the chi-square test that he applied (1) to fit observations to a known distribution, and (2) to verify the independence or relatedness between two characters.

  Pearson applied all these tools to biometric variables inherited from one generation to the next and tackled the qualitative characteristics of populations, but without addressing the problem of the genes. His innumerable writings on all these topics exist in the form of articles, mostly in the journal Biometrika he founded and of which he was the chief editor [1, 26, 37, 38].
- Godfrey Hardy (1877–1947) was an English biomathematician and anthropologist. In a small note published in 1908 in the journal Science [24], this scientist constructed the basis of population genetics by combining the Mendelian concepts of character segregation and dominance/recessivity on the one hand and the concepts derived from Pearson's statistics (randomness, character independence, etc) on the other hand. He applied this method to a theoretical study about the outcome of the crossing between brachycephalic and dolichocephalic persons.[9] In this research work, brachycephalism (a rare recessive character) was designated by the letter 'a' and dolichocephalism (a common dominant character) by the letter 'A'. He then hypothesised that (1) the population was large, (2) the characters were combined and transmitted randomly, (3) there were as many men as women, and (4) the fertility rate was constant. From this hypothesis and mathematical reasoning, he established that from an initial state, an equilibrium was shortly reached where the alleles 'A' and 'a' on the one hand,

[9] Brachycephalism means having a skull proportionally broader than high; it corresponds to a higher cephalic index (CI), as defined by the ratio of breadth to height. On the contrary, dolichocephalic corresponds to skulls proportionally higher than broad, thus to a lower CI. The threshold varies according to the authors, but lies between 70 and 80.

and the genotypes 'AA', 'Aa' and 'aa' on the other hand, remained constant in proportion from one generation to the next.

Independently of Godfrey Hardy, the German physician Wilhelm Weinberg (1862–1937) published nearly exactly the same model the same year [51]. It was about the transmission of human traits, notably eye colour. Therefore, this model of gene and genotype transmission tending to a genetic equilibrium after two generations, which was revealed later to be consistent and robust, was called the *Hardy-Weinberg model*.

- Ronald A. Fisher (1890–1962) was one of the greatest British scientists in his domain, linking statistics and genetics outstandingly. He was certainly one of the foremost promoters of population genetics and quantitative ecology, two intricately related disciplines. Fisher introduced the mathematisation of numerous concepts. To study the evolution of a phenotypic trait (a visible elementary morphological characteristic), Fisher assimilated each individual to a genotype. The populations were thus identified to a "gene pool". To these genotypes, the author associated quantities and principles reflecting many Darwinian concepts [13], notably:

  – *Natural selection*: This principle is described by a number of parameters, the most important being fitness, a quantity measuring the duration capacity of a genotype.
  – *Sexual selection*: This phrase refers to the choices of sexual partners and the reproduction isolation sometimes induced by the choices.
  – *Mimicry*: This is a deception strategy, that is, for prey, a means to escape from predators, and for predators, a way to approach a prey by surprise. It includes camouflage or body imitations of other species (e.g. resemblance to an aggressive or an unpalatable species).
  – *Evolution of dominance*: This expression refers to the evolution of the property of a gene: a dominant gene can become recessive, and a recessive gene dominant with time, under the action of modifier genes. Unlike the previous ones, this concept has been controversial since its proposal.

  We owe Fisher successful mathematic interpretations of Darwinian concepts of an ecological nature, for example, (1) the unbounded exponential increase of any population with time when it is not checked by natural selection (application of the demographic model of Malthus (1766–1834) [29]; and (2) the equation and curve of the logistic function of Verhulst (1804–1849) [43]—a model where natural selection is involved in checking the increase of a population). These equations applied to populations indirectly demonstrate the validity of the concept of natural selection and other concepts he developed during his prolific professional life. Though he was a Mendelian, his Darwinian convictions and his education in Cambridge as a student and disciple of Galton led him to follow eugenism, which is a paradox since he was himself visually impaired, a potential victim of his own ideology.

Two contemporaries played a major role in the further development of population genetics, with consequences on the development of behavioural ecology.

- Sewall Wright (1889–1988) was an American geneticist and the inventor of the inbreeding coefficient [54] and founded an entire domain of biology devoted to relatedness. With an integrative approach involving population genetics, ecology and mathematics, he tackled the problems of selection, mutation, migration, and genetic drift, to understand how these factors are major agents working on evolutionary mechanisms. He was also the first author who applied game theory in biology, in the form of a landscape metaphor, to explain (1) the relations between genotype and phenotype, and (2) the maximal adaptation conditions in case of phenotype change [55].
- John BS Haldane (1892–1964) was one of the most brilliant British biologists and biometricians of his time and an original left-wing person. This scientist is famous for having developed many mathematical models in issues of population genetics like the effect of mutation and migration on allelic frequencies [19–23]. He was also a pioneer in genetic cartography [18].

In parallel with these genetic and mathematical studies, the concept of system and holistic ecology, which took place in the same period (approximately between 1875 and 1940), paved the way to the modern concept of ecosystem.

8.4 Holistic vs Individualistic Conception of Communities

A debate between two schools of thought concerning the fundamental property of biological communities, holism and individualism, took place during the first half of the twentieth century. Holism asserts that communities are organic entities and, as such, can be assimilated to organisms; conversely, individualism asserts that the only biological reality is the species and that communities are coincidental formations that are variable in space and time.

- Among the main proponents of the holistic conception, we can mention Eduard Sucss (1831–1914), Vladimir Vernadsky (1863–1945) and Frederic E. Clements (1874–1945).
- The main leaders of the individualistic conception of communities were Henry A. Gleason (1882–1945), Arthur G. Tansley (1871–1955) and Eugen P. Odum (1913–2002).

8.4.1 *The Holistic School of Thought*

- Eduard Suess was an Austrian geologist and paleontologist. He was the first author who talked about a zone of the earth where life dwells—a thin layer of the lithosphere, hydrosphere and atmosphere[10] [40]—, thus the concept of *biosphere* (later to be termed biosphere by Vernadsky) was born.

- Vladimir I. Vernadsky was a Russian geologist who coined the term biosphere corresponding to the unnamed concept of Suess. He developed a more accurate and functional conception of biosphere, within a global and systemic theory simultaneously involving geology, biology, chemistry, physics and ecology in a book [44, 45] along with other findings, notably:
  - There exist major *biogeochemical cycles* of *atmospheric elements* (*oxygen, hydrogen, carbon, nitrogen*) and *lithospheric (or mineral) elements* (*iron, copper, silicon*, etc).
  - Biosphere is the seat of converters changing cosmic rays into active terrestrial energy.
  - Biosphere must be considered an organic entity—any alteration of a part of it has an impact on the rest of it.
  - Living beings are the relays of circulating *chemical elements* whose source is the *bedrock.*

  Vernadsky is the inventor of the concept of negentropy—the fact that unlike classical physical systems, living beings depart from thermodynamical equilibrium, increasing by doing so their free energy. The branch of ecology that Vernadsky developed is called "global ecology". For the anecdote, Vernadsky is the first known scientist that assessed the impact of deforestation on climate.
- Frederic E. Clements was an American botanist and one of the first modern theorists of modern ecology. We already saw that despite having been the proponent of a theory that proved later erroneous, the reactions against his proposal were fruitful and led to the conception of ecosystem.

8.4.2 The Individualistic School of Thought

- Henri A. Gleason (1882–1975) was an American ecologist and botanist. He was a staunch supporter of the open community concept and favoured the individualistic conception of the biological communities, which he conceived as a section of an ecological succession. Gleason speaks of plant association; in this, he is an heir of the Humboldtian concept [48]. For him, a community is a fortuitous association of mutually compatible and interacting species and a common successional history. This dynamic varies spatially and temporally.
- Arthur G. Tansley (1871–1955), as already mentioned in Sect. 2.1, was the first scientist who conceived the concept of ecosystem as a true *system* in the sense of the physicists, consisting of two components, a biocoenosis (that he called biome

[10] In our planet, the lithosphere corresponds to the totality of the land masses and the hydrosphere, the totality of the water masses.

or biotic community) and abiotic factors [42]. An ecosystem and its components are, for Tansley, a system in the physical sense, thus open. However, the notion of interaction and complexity was lacking in his concept.

- The American ecologist E. P. Odum (1913–2002) popularised the concept of ecosystem and adopted Vernadsky's biosphere concept. For holists, a system is more than the sum of its components due to the interactions between the latter—a phenomenon hardly quantifiable that makes the components work to ensure the maintenance of the system. Whereas Tansley saw an ecosystem as a classical physical system, Odum looked at it as an organic supersystem where a biocoenosis interacts with its environment, this relationship composed of intricate and dynamic relations between elements of the former and elements of the latter [35, 36].

  Odum differed from his contemporary colleagues as he was the first proponent of ecology being an autonomous science and not an extension of other biological or geological disciplines. This stance led him to fund specifically ecological studies and, in particular, field ecology studies.

8.5 History of the Understanding of the Inner Composition and Dynamics of an Ecosystem

Different concepts are related to that of ecosystem, namely (1) the concepts of *biocoenosis*, *biotope* and *abiotic factor system*, (2) that of *interactions*, and (3) that of *ecological succession*. The latter concept is associated with the notion of optimal equilibrium or climax. All these concepts, older than that of ecosystem, are necessary to elaborate the latter.

- The notion of biocoenosis has been known since Wallace [49, 50] but the term was ascribed by Möbius [32] (cf. Sect. 8.1.2). The term "biotope", used by continental Europeans and synonymous to "abiotic factor system" was proposed by the German zoologist Friedrich Dahl (1856–1929) [7].
- The interactions between species of a biocoenosis were described in a modern fashion for the first time by Charles S. Elton (1900–1991) [12]. This British zoologist was mainly focused on *trophic interactions*, highlighting the strong correlation between the trophic level of an organism in a food chain and its size or population size:

  – A prey species tends to be smaller than a predator species, which on average has a lesser size than a superpredator, also called apex predators;
  – A prey species tends to be more numerous than a predator species, which in turn has a larger population size than a superpredator.

  This trend has proven valid in many studies since. Of course, there are quite a lot of exceptions to it, e. g. the case of the large herbivore mammals which adopted another survival strategy than most phytophagous species.

- Ecological successions have been well-known phenomena since the end of the eighteenth century [11, 53]. But the earliest thorough modern studies on ecological successions, based on field semi-experimental work, are far more recent and date back from the beginning of the twentieth century, the first of them being the PhD dissertation of Cowles (1869–1939), an American botanist [6] . This study was about the plant succession of sandy dunes along lake Michigan. Other famous early studies refined the concept of ecological succession, namely the works of Clements [4], Gleason [15] and Tansley [41]. Clements [3] was the inventor of very important concepts, notably primary succession and secondary succession. About thirty years later, Tansley made a distinction between autogenic successions and allogenic successions. In the same epoch, the American ecologist Cooper [5] (1884–1978) made a detailed description of an ecological succession resulting from the impact of glacier retreat on plant species associations; he established a correspondence system between ecogeographical gradient and time gradient, and thus was able to reconstruct an ecological succession in a definite biome during about three centuries. These studies and many others contributed to establish our modern concept of an ecological succession.

8.6 The Advent of Community Ecology

The conceptualisation of the notion of community (cf. Sect. 13.2) is related to that of succession, due to the fact that the same authors proposed these concepts and terms in the same publications [4, 15, 16, 42]. In the 1930s, a fierce debate opposed the proponents of the closed communities (following Clements), of a quasi-organic nature, to those of the open communities (following Gleason) formed of random associations of coadapting population of different species.

We owe Alfred J. Lotka [28] (1880–1949) and Volterra [46, 47] (1860–1940) the first studies striving to explain inter-species interactions of a community with a mathematical model. They both described prey-predation interactions via the same mathematical model found independently[11] within a very simple biocoenosis seemingly composed of two partners, a population of prey and a population of predators. It is a kind of game involving a third partner playing the role of the dummy that we can call the primary resource (e.g. plants), so to say fuelling the system. The mathematical model consisted of a couple of non-linear differential equations with two unknown functions, the number of prey individuals and the number of predator individuals, and involving coefficients with an easy biological interpretation (see Sect. 13.2.3); the solutions of this system of differential equations tend, with time, to a dynamic equilibrium state—(1) an elliptic limit-cycle when the

[11] This is what the story tells. But when we look at the publication dates, I cannot prevent myself from thinking that there was a kind of gentlemen's agreement between the two authors.

initial conditions are close to the equilibrium and (2) an ellipsoid limit-cycle when the initial conditions are less close to the equilibrium. When the initial conditions diverge too much from the dynamic equilibrium corresponding to a suboptimal set of prey and predator numbers, the solutions tend to a chaotic trajectory, entailing a strong risk of stochastic extinction for the two populations.

Lotka was an American biomathematician reputed for his studies on population dynamics, whereas Volterra was an Italian mathematician famous for his research on integral equations. As mentioned just above, both published the same model, Lotka, a couple of years before Volterra, but still the model in question is called the Lotka-Volterra model. Volterra inferred it from a study on the seemingly paradoxical decrease in the number of fish in the Adriatic Sea during the period 1914–1918 when fishery activities were almost at a standstill because of the war. The explanation was that the reduction of fishery activities enabled the shark population to increase dramatically and thus have an impact on fish prey which proved stronger than that of fishery activities before the war, still rather small-scaled in that epoch. And Volterra's model correctly explained the phenomenon.

8.7 The Advent of Evolutionary Ecology

One of the consequences of the facts revealed by evolutionary biology was that communities change with time because, impacted by natural selection, their component species evolve with time. As a consequence, these species coevolve, inducing the change of their community. After the advent of the genetic, biomathematical and conceptual revolutions of the beginning of the twentieth century, as exposed above in the previous sections of this chapter, evolutionary ecology acquired its autonomy from related biological disciplines, namely biogeography and evolutionary biology, though the three disciplines overlap considerably. One of the leading scientists who could brilliantly relate biogeography and evolutionary biology to ecology, thus literally creating evolutionary ecology in the modern sense, was the German-American biologist Ernst Mayr (1904–2005). The latter was renowned mainly for being one of the foremost leaders of the modern synthetic theory of evolution, an integrative approach linking many disciplines like genetics, systematics, statistics, mathematical modelling, population biology, ethology, biogeography and ecology to apprehend evolution and evolutionary mechanisms. Ecology plays a great role in this synthesis [30].

Mayr considered that the processing unit in evolutionary biology and ecology is population and militated for the use of statistics. He rejected the *typological (or Linnean) species* concept and promoted the *biological species* or *biospecies* concept, viewed as a set of totally interbreeding populations: biospecies are groups of living beings that are similar to one another because they are generated by similar individuals of the previous generation and are capable of generating similar individuals via sexual reproduction, with no reduction of reproductive potential across the successive generations. Such groups are reproductively isolated from the other species defined the same way. The impossibility of interbreeding is called a reproductive barrier that can be partial or total. The proponents of the modern synthesis invoked this mechanism as the main motor of evolution [10, 27, 30, 39].[12] The concept of biospecies was not completely novel, though its application was. Indeed, this concept was already in germ and implicitly expressed by Lamarck (1744–1829) [25, p. 41]:[13]

> Nature did not actually form constant classes, orders, families, genera or species but only individuals that succeeding one another, each similar to those that produced them.

Corollarily to this definition, two non-interbreeding populations do not belong to the same species. From this species conception, Mayr writes a section on evolutionary mechanisms and another on variation within a species (intraspecific variation):

- About the section of evolutionary ecology:
  (1) Mayr exemplifies studies on the behaviours leading to pair formations and the degree of assortative mating. (2) He describes the role of geographical barriers and the effect of population size, showing that small isolated subpopulations are good candidates to become new species. (3) He gives interpretations of the various biogeographical situations of neighbouring populations separated by a contact zone, according to the breadth of the zone and the degree of intergradation or introgression in it, to assess their degree of differentiation. Mayr hypothesises that most of these contact zones were formed after a geographic isolation period and called them secondary contact zones.
- About the studies on intraindividual variation:
  Mayr showed great interest in the intraspecific variation of morphological traits in response to environmental pressures, invoking the ecogeographical rules, notably Gloger's, Bergmann's and Allen's (see Sect. 7.4.5), and gives a Darwinian interpretation of them.

8.8 Ecology Today

We saw some of the key milestones of the history of ecology. Two disciplines of this science were already established by the 1950s, i.e. (1) global or holistic ecology developed by Vernadsky, Tansley and Odum, and (2) evolutionary ecology fostered by the proponents of modern synthetic theory of evolution, in particular Mayr. In addition, evolutionary biology and biogeography, two disciplines sharing many topics with ecology, added to the latter some conceptual findings, stemming from modern synthetic theory of evolution as well as phylogenetics and phylogeography.

---

[12] But this concept is only valid for eukaryotes.

[13] Translated from French by the author.

At present, ecology has considerably diversified, providing vast scientific production. A good number of subdisciplines came into being, of which I give here a non-comprehensive list:

- *Ecophysiology* studies the relations between physiological processes and environmental factors.
- Autoecology deals with the relationship between one species and the rest of its ecosystem.
- *Synecology* tells about the relationship between a biocoenosis (containing at least two species) and the rest of its ecosystem.
- *Mathematical ecology* dates back to the first works of population genetics and the model of Lotka and Volterra. This discipline unveils underlying mathematical structures in an ecological phenomenon; or constructs mathematical models describing and predicting ecological events, from a hypothesis on the structure and functions of parts or all of any ecosystem. To this discipline, descriptive or inferential statistical analyses on diverse ecological parameters are applied.
- *Behavioural ecology*, also called *eco-ethology*, studies how the individuals' behaviours of an animal population can ensure the consistency, adaptation potential and reproduction of this population. This discipline can be extended to all organisms but keeping in mind that a true behaviour presupposes the possession of a kind of nervous system coupled with some motility capacity; and that therefore, when this discipline is applied to sessile animals, plants, fungi and microorganisms, the term tropism will be used rather than behaviour.
- Microbial ecology is interested in the crucial role of microorganisms (bacteria, archaea, unicellular or minute pluricellular eukaryotes) in the functioning of the ecosystems, namely in the following processes:
  - the mineralisation of organic matter - a fundamental part of the material cycle of ecosystems;
  - the redox of certain chemical elements, essential to provide the vital energy of the individuals of the populations composing the biocoenoses.

In addition, microorganisms constitute a very important proportion of the total biomass (the dry mass of all the living beings on earth) and constitute biocoenoses interesting in their own right.

All these subdisciplines overlap and interact. Despite its variety of approaches, ecology has a remarkable unity concerning its purpose (cf. Chap. 2). In addition, as we mentioned it above, a number of other disciplines are narrowly intertwined with ecology:

- population genetics, above all the part dealing with natural selection and intraspecific variation;
- biogeography;
- evolutionary biology, mainly natural selection and all the impact of this principe and other evolutionary concepts on populations.

These three scientific fields are often integrated into ecology *senso latu*.

References

1. Beeton M, Pearson K (1901) On the inheritance of the duration of life, and on the intensity of natural selection in man. Biometrika 1:50–89, https://doi.org/10.2307/2331672
2. Berger SL et al (2009) An operational definition of epigenetics. Genes Dev 23:781–783, https://doi.org/10.1101/gad.1787609
3. Clements FE (1904) The development and structure of vegetation. In: Pound R et al (eds) Botanical survey of Nebraska, University of Nebraka, Lincoln.
4. Clements FE (1916) Plant succession: an analysis of the development of vegetation. Carnegie Institution of Washington.
5. Cooper WS (1939) A fourth expedition to Glacier Bay, Alaska. Ecology 20:130–155, https://doi.org/10.2307/1930735
6. Cowles HC (1899) The ecological relations of the vegetation on the sand dunes of lake Michigan. Part I -geographical relations of the dune floras. Bot Gaz 27:95–117.
7. Dahl F (1908) Grundsätze und Grundbegriffe der biozönologischen Forschung. Zool Anz 33:349–353.
8. Darwin C (1859) On the origin of species by means of natural selection, or the preservation of favoured races in the struggle for life. Murray, London.
9. De Vries H (1889) Intracellulare Pangenesis. Gustav Fischer, Jena.
10. Dobzhansky T (1937) Genetic nature of species differences. Am Nat 71:404– 420, https://doi.org/10.1086/280726
11. Dureau de la Malle A (1825) Mémoire sur l'alternance ou sur ce problème: la succession alternative dans la reproduction des espèces végétales vivant en société, est-elle une loi générale de la nature. Ann Sci Nat 5:353–381.
12. Elton CS (1927) Animal ecology. McMillan, New York.
13. Fisher RA (1930) The genetical theory of natural selection. Clarendon Press, Oxford.
14. Galton F (1889) Natural inheritance. Macmillan, London.
15. Gleason HA (1926) The individualistic concept of the plant association. Bull Torrey Bot Club 53:7–26
16. Gleason HA (1939) The individualistic concept of the plant association. Am Midl Nat 21:92–110
17. Haeckel E (1866) Generelle Morphologie der Organismen. G. Reimer, Berlin
18. Haldane JBS (1919) The combination of linkage values, and the calculation of distances between the loci of linked factors. J Genet 8:299–309
19. Haldane JBS (1924) A mathematical theory of natural and artificial selection. Part I. T Cambridge Phil Soc 23:19–41
20. Haldane JBS (1924) A mathematical theory of natural and artificial selection. Part II. Proc Cambridge Phil Soc - Biol Sci 1:158–163, https://doi.org/10.1111/j.1469-185X.1924.tb00546.x
21. Haldane JBS (1926) A mathematical theory of natural and artificial selection. Proc Cambridge Phil Soc 23:363–372
22. Haldane JBS (1927) A mathematical theory of natural and artificial selection. Part V: Selection and mutation. Proc Cambridge Phil Soc 23:838–844
23. Haldane JBS (1930) A mathematical theory of natural and artificial selection. Part VI: Proc Cambridge Phil Soc 26:220–230
24. Hardy GH (1908) Mendelian proportions in a mixed population. Science 28:49–50, https://doi.org/10.1126/science.28.706.49
25. Lamarck (chevalier de Monet) JBA (1809) Philosophie zoologique, ou exposition des considérations relatives à l'histoire naturelle des animaux. Dentu et the Author, Paris
26. Lee A, Pearson K (1901) Data for the problem of evolution in man - vi - a first study of the correlation of the human skull. Phil T R Soc Lond A 196:225–264, https://doi.org/10.1098/rsta.1901.0005

27. Lewontin RC (1974) The genetic basis of evolutionary change. Columbia UP, New York, London.
28. Lotka AJ (1920) Analytical note on certain rhythmic relations in organic systems. Proc Natl Acad Sci USA 6:410–415
29. Malthus TR (1798) An Essay on the Principle of Population As it affects the future improvement of society with remarks on the speculations of Mr. Godwin, M. Condorcet, and other writers. J. Johnson, London
30. Mayr E (1942) Systematics and the origin of species, from the viewpoint of a zoologist. Harvard University Press, Harvard
31. Mendel G (1866) Versuche über Pflanzenhybriden. Verhandl Naturforsch Ver Brunn 4:3–44
32. Möbius KA (1877) Die Auster und die Austernwirtschaft. Wiegandt, Hempel & Paret, Berlin
33. Morgan TH (1915) The mechanisms of Mendelian heredity. Henry Holt & Co, New York
34. Morgan TH (1935) The scientific basis of evolution. The scientific basis of evolution. 2nd Ed., Norton & Co
35. Odum EP, Barrett GW (2005) Fundamentals of ecology, 5th edn. Thomson Brooks, Cole (CA)
36. Odum EP, Odum HT (1953) Fundamentals of ecology, 1st edn. Saunders, Philadelphia and London
37. Pearson K (1893) Contributions to the mathematical theory of evolution. Proc R Soc Lond 54:329–333
38. Pearson K (1900) Mathematical contributions to the theory of evolution. VII. On the correlation of characters not quantitatively measurable. Phil T R Soc A:1–405
39. Poulton EB (1903) What is a species? Proc Entomol Soc Lond 115:77–116
40. Suess E (1875) Die Entstehung der Alpen. W. Braunmüller, Vienna
41. Tansley AG (1929) Succession: the concept and its values. In: Proc. 4th Int. Congr. Plant Sc, Ithaca, 1926, pp 677–686
42. Tansley AG (1935) The use and abuse of vegetational concepts and terms. Ecology 16:284–307, https://doi.org/10.2307/1930070
43. Verhulst PF (1845) Recherches mathématiques sur la loi d'accroissement de la population. Nouv Mém Acad R Sci B-lett Brux 18:1–40
44. Vernadsky VI (1926) Biosphera (The biosphere) (in Russian). Nauchnoe khimiko-technicheskoye izdatel'stvo (scientific technico-chemical publisher), Leningrad
45. Vernadsky VI (1929) La biosphère, translation in French of [44]. Félix Alcan, Paris
46. Volterra V (1926) Variazioni e fluttuazioni del numero d'individui in specie animali conviventi. Mem R Accad Naz Lincei 2:31–113
47. Volterra V (1928) Variations and fluctuations of the number of individuals in animal species living together. J Cons Int Explor Mer 3:3–51
48. von Humboldt A & Bonpland A (1805) Essai sur la géographie des plantes. Levrault, Schoell et Co, Paris
49. Wallace AR (1860) On the zoological geography of the malay archipelago. J Proc Linn Soc Lond Zool 4:172–184
50. Wallace AR (1863) On the physical geography of the malay archipelago. J R Geogr Soc Lond pp 217–234
51. Weinberg W (1908) Über Vererbungsgesetze beim Menschen. Z indukt Abstamm Vererbungsl 1:277–392
52. Weissmann A (1892) Das Keimplasma. Eine Theorie der Vererbung. Gustav Fischer, Jena
53. Willdenow CL (1792) Grundriss der Kräuterkunde. Haude & Spener, Berlin
54. Wright S (1922) Coefficients of inbreeding and relationship. Am Nat 56:330–338, https://doi.org/10.1086/279872
55. Wright S (1942) Statistical genetics and evolution. Bull Amer Math Soc 48:223–246

Part III Structure and Functions of Ecosystems: Three Approaches

The mechanisms ensuring the structure consistency and functioning of an ecosystem will be studied by three interlinked main approaches:

The energetical approach:

This is the topic of Chap. 9. *Energy fluxes* entering any ecosystem originate from sunlight or chemical redox reactions—first incorporated within the autotrophic and then within the successive heterotrophic components (see Sect. 2.3). At each step, this energy is changed into work or heat, ensuring (1) the body maintenance, growth and reproduction, and (2) the sustainability of the populations. At each step, the energy not incorporated dissipates in the form of heat outside the system to conform to the second law of thermodynamics. We will see why the organisms and the ecosystems have a thermodynamical property rarely found among the physicochemical systems: when they are functional (i.e. when the organisms are alive, and the ecosystems are sustainable), they are in permanent thermodynamical desequilibrium and increase their internal energy.

The material approach:

In parallel, any ecosystem (Ec) recycles its matter. The renewal period of a quantity of a given chemical body of Ec, i.e. the time between it leaves and returns to a given point of Ec, is stochastically constant. The mechanism that ensures the recycling of material elements is mineralisation. Chapter 10 will treat the circulation of the main chemical agents, (1) from their distal origin in the bedrock or the atmosphere/hydrosphere to the biogeological cycles, and (2) in the biogeochemical cycles. Chapter 11 will treat the impact of the main physical parameters (temperature, humidity, pH etc) on the ecosystems.

The biosociological approach:

The living part of an ecosystem is structured in species—entities sharing a common evolutionary destiny[1] and exploiting the same portion of energetic and

[1] The most frequently invoked "common evolutionary destiny" in sexually reproducing organisms is the potentially successful interbreeding that give products with the same capacity of reproduction.

material resources, i.e. sharing the same *ecological niche* (Chap. 12). The latter is an ecomathemathical concept including the exploitation of the same portion of the total energy and matter available: same (1) type of resources consumed, (2) energy spent for foraging, and (3) strategies to balance the energy spent and the matter consumed [1]. The species are physically the relays of energy and matter. The organisms are organised in heterospecific populations structured in communities. Populations and communities are described by models. The most simple are processed mathematically, and the more complex have a verbose description. Trophic webs playing a significant role in these structures terminates the chapter.

These approaches are the main facets of an ecosystem, delineated in a given *spatial scale* and in a given *temporal scale*. Spatial scale can range, for example, from that of a seawater droplet containing interacting planktonic species [2] to that of the entire biosphere. The temporal scale can span from (1) the time of interactions between individuals within a given population to (2) the time necessary for an entire biome or biosphere to build up (geological periods). Most case studies deal with in-between situations, i.e. biocoenoses containing several interacting species in a stretch of land or sea accessible to human observers in a part of their life span.

References

1. Hutchinson GE (1957) Concluding remarks. Cold Spring Symp Quant Biol 22:415–427
2. Van der Gast CJ et al (2008) Temporal scaling of bacterial taxa is influenced by both stochastic and deterministic ecological factors. Environ Microbiol 10:1411–1418

Chapter 9
The Energetical Approach

Section 9.1 refreshes concepts like energy, work, heat, and thermodynamic principles for readers needing to revise these notions. Those familiar with these concepts can directly proceed to Sect. 9.2, where these notions are applied to ecology.

9.1 Thermodynamic Concept of Energy

9.1.1 Systems and Their States

The scientific concept of *energy* was established at the beginning of the nineteenth century within a branch of physics, thermodynamics [32]. I will give some elements for a good comprehension of the concept of energy in biology and, thus, in ecology. All physical phenomena changing with time, namely biological/ecological processes, can be modelled in the frame of a thermodynamical *system* (cf. Definition 9.1). We will be interested only in irreversible processes, the part of thermodynamics involved in chemical, biological and ecological systems.

Definition 9.1 A system is a dynamic object evolving with time. It is generally a three-dimensional region of physical space of interest endowed with three properties: (1) It contains material points carrying out work[1] or absorbing or dissipating heat. (2) A boundary, real or virtual, limits it. (3) It is involved in a temporal succession of *states* depicted by a *state function*. The latter is a mathematical function having a definite value for each state, independently of the manner in which the state is reached [32, p. 293].

Supplementary Information The online version contains supplementary material available at https://doi.org/10.1007/978-3-031-90656-5_9.

[1] We will remind the definitions of work and energy in Sect. 9.1.2.

M. Salomon, *Integrative Ecology*, https://doi.org/10.1007/978-3-031-90656-5_9

Definition 9.2 Currently, three types of systems are distinguished according to the degree of permeability to energy and matter [33, 36]: (1) *open systems*, whose boundary is permeable inwards or outwards to energy and to matter; (2) *closed systems*, whose boundary is permeable inwards or outwards to energy and not to matter. (3) *isolated systems*, whose boundary is impermeable both to energy and matter, which means that it contains all its energy and matter for its functioning inside.

Definition 9.3 The outside of a system is called its *surroundings*. A system and its surroundings constitute the *universe* (for an illustration, see Fig. 9.4, p. 195)

Although, by definition, it is unbounded, a consensus agrees that the universe is any space large enough to enclose the thermodynamic processes under study [38]. For instance, the universe may be assimilated to a vessel in experimental sciences. The universe concept is more problematic in natural sciences and prone to debate [7, 9]. Literally, the universe is supposed to have no surroundings, which would logically make it an isolated system. Figure 9.1 is, within a Cartesian coordinate system, a graphic representation of the state changes, each corresponding to the system's evolution.

We will remind, throughout this section, several well-known physical variables: (1) *Pressure*, (2) *Volume*, (3) *Kinetic Energy*, (4) *Potential Energy*, (5) *Internal Energy*, (6) *Enthalpy*, (7) *Entropy* and (8) *Gibbs Free Energy*. These variables exemplify state functions and can thus be represented by the function φ of Fig. 9.1.

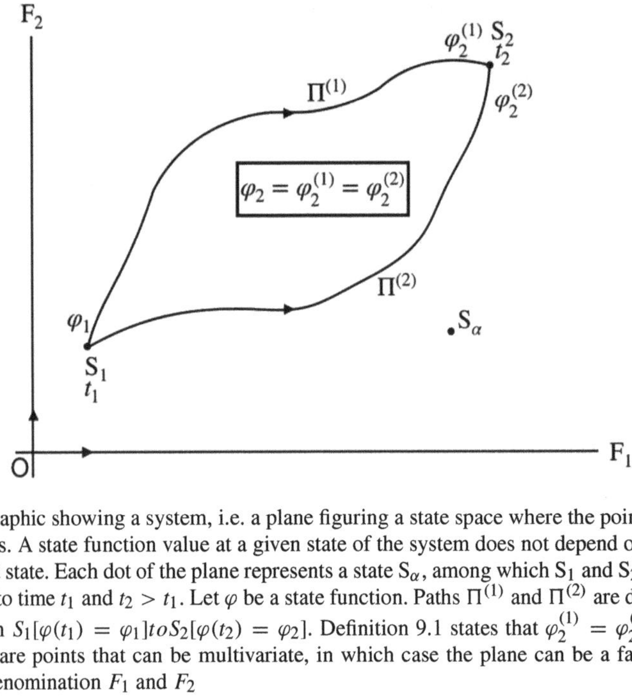

Fig. 9.1 Graphic showing a system, i.e. a plane figuring a state space where the points $\mathbf{S}_\alpha$, $\boldsymbol{\alpha} \in \mathbb{R}$ are the states. A state function value at a given state of the system does not depend on the manner to reach that state. Each dot of the plane represents a state S_α, among which S_1 and S_2 respectively correspond to time t_1 and $t_2 > t_1$. Let φ be a state function. Paths $\Pi^{(1)}$ and $\Pi^{(2)}$ are different ways to pass from $S_1[\varphi(t_1) = \varphi_1] to S_2[\varphi(t_2) = \varphi_2]$. Definition 9.1 states that $\varphi_2^{(1)} = \varphi_2^{(2)} = \varphi_2^{(\alpha)} = \phi_2, \forall \alpha$. S_α are points that can be multivariate, in which case the plane can be a factorial plane, hence the denomination F_1 and F_2

We will also extensively talk about two fundamental quantities, work and heat, that cannot be represented by function φ. Heat or work at S_2 will differ at a state which is the endpoint of path $\Pi^{(1)}$ or path $\Pi^{(2)}$ [see Appendix E (SI, link p. 189) for a demonstration by an example].

9.1.2 *Concept of Work, as Applied to One Point Object*

9.1.2.1 Work Carried Out by a Constant Force in a Space Free from Interactions

Let $\mathbf{F}_a$ be a constant force applied on a point object P of mass m in a given space free of all external interactions [27, p. 134]. P moves from an initial position A_0, at time $t = 0$, to a final position A_τ, $t = \tau$, with respective coordinates (x_0, y_0) and (x_τ, y_τ) along a trajectory (C) within an inertial reference frame. The constancy of vector $\mathbf{F}_a$ means the following:

1. The vector $\mathbf{F}_a$ is constant in magnitude, sense and direction.
2. The same is true for the vector acceleration $\boldsymbol{\gamma}$ applied to P, which impels a movement called uniformly accelerated, since by definition, $\mathbf{F} = m\boldsymbol{\gamma}$.
3. The trajectory (C) is a straight line.

The displacement (**s**), velocity (**v**), acceleration ($\boldsymbol{\gamma}$), and force (**F**), being collinear vectors, admit the same unit direction vector, **u**.

Definition 9.4 Let $\mathbf{F}_a$ be a force, not necessarily constant, applied to point object P. During a given time τ, the latter will cover a curvilinear distance $\boldsymbol{\ell} = \mathbf{s}(\boldsymbol{\tau})$. We will define the work W done by $\mathbf{F}_a$ as:

$$W = \mathbf{F}_a \cdot \boldsymbol{\ell} \tag{9.1}$$

For any vector **v**, we will use the following notation: $v = ||\mathbf{v}||$, which will apply, for example, to force, acceleration, velocity and displacement.

Theorem 9.1

$$\text{If } \mathbf{F}_a \text{ is a constant force, } W = F_a\,\ell \tag{9.2}$$

Proof left to the willing readers, see Problem 9.1 (p. 217).

Corollary 9.1 *Another way to write Eq. (9.2) is* $W = F_a \sum_{i=1}^{n} \Delta s = \sum_{i=1}^{n} F_a \Delta s$*; when* $n\Delta s = \ell$*. Or if one writes this equation in terms of vectors*

$$W = \sum_{i=1}^{n} \mathbf{F}_a \cdot \Delta\mathbf{s} \tag{9.3}$$

The point object to which constant force $\mathbf{F_a}$ is applied has a constant acceleration and uniform increase of velocity. Here, the concept of kinetic energy comes into the picture.

Definition 9.5 (Kinetic Energy) Let P be a material point of mass m endowed with velocity $\mathbf{v}$. Its kinetic energy E_k is defined as follows:

$$E_k = \frac{1}{2}mv^2 \tag{9.4}$$

Theorem 9.2 *Work W done on point object P by a constant force $\mathbf{F}_a$, along a distance l between instants 0 (set arbitrarily) and t, is equal to the variation of the kinetic energy of P between these two instants. In sum, (1) if $x(t)$ and $v(t)$ are the respective displacement and velocity of P at instant t, and (2) if the initial conditions give $x(0) = x_0$ and $v(0) = v_0$, then*

$$W = F_a\underbrace{(x(t) - x_0)}_{l} = \underbrace{\frac{1}{2}mv(t)^2 - \frac{1}{2}mv_0^2}_{\text{Change of kinetic energy}} \tag{9.5}$$

Proof left to the willing reader, cf. Problem 9.2 (p. 217)

Now, we will consider the case where the point P, of mass m or charge $q \lesseqgtr 0$, moves

- not due to the impulsion of an applied force in a portion of the universe devoid of interactions, but
- under the influence of an external action, a physical effect called potential like gravitation or electomagnetic potential.

Such a potential induces potential energy [10]. We will introduce the concept of potential energy in the case of gravitation.

Definition 9.6 (Potential Energy) Let P be a material point of mass m in position A above the Earth. Gravitation spontaneously launches a movement of material point P from A to position B, located at the Planet's surface in the Earth's vertical axis, passing through A (a fall from A to B). Its potential energy E_P is the work needed to overcome this movement, i.e. the work resulting from the minimal force needed to revert from B to A.

Let us consider an object P of mass m raised above the earth's surface (S) at height h (see Fig. 9.2). Under the attraction of the Earth, its weight $\mathbf{F}_g = -m\mathbf{g}$[2] (Newton's second law), of intensity mg, $\mathrm{g} = 9.81\, m \cdot s^{-2}$ (free-fall acceleration). The object P, if freely released, moves vertically downwards towards the centre of

[2] The force is represented by a vector pointed downwards in the Cartesian coordinate system, where by convention, the positive sense on the y-axis is given by the vector $\mathbf{j}$ pointed upwards. Hence, the negative sign of $\mathbf{F_g}$.

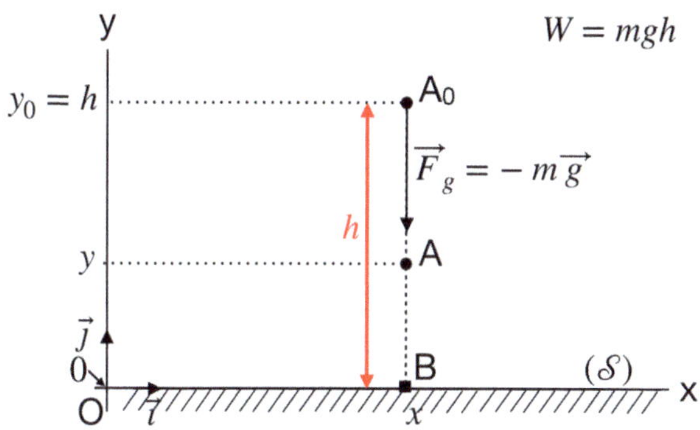

Fig. 9.2 Graphic illustration of potential energy. The situation depicted in a Cartesian system of axes $\{O, \imath, \jmath\}$, with the x-axis merged with earth's surface $\mathcal{S}$, supposed flat, and the y-axis representing the heights to $\mathcal{S}$ (the Earth's soil) of any potential objects. The nonnegative quantity h is the height of A_0, and we can easily see that $y_0 = h$

the earth. This capacity to elicit work at height h is the potential energy depicted in Definition 9.6. We will pose that at height h, at time $t = 0$, $v(0) = v_0 = 0$. Let us call W_g the work done on P by $\mathbf{F_g}$.

Let us take the data from Fig. 9.2. The physicists show that $E_P = mgh$ if we consider that the potential energy at soil level is zero, which will be our hypothesis (for more details, see [27, ch. 5]), hence our decision that B is at ordinate 0.

Theorem 9.3 *Under the assumption that P is released at* $A_0 = (x, h)$ *with initial velocity* $v_0 = v(x, h) = 0$, *if at any point* $A = (x, y)$ *between* A_0 *and B,* $E_P(x, y)$ *and* $E_K(x, y)$ *are respectively the potential and kinetic energies of falling P, and* $v = v(x, y)$ *its velocity at point A,*

$$\begin{cases} E_P(x, y) + E_K(x, y) = E_P(x, h) \\ E_P(x, y) = mgy \\ E_K(x, y) = \dfrac{1}{2}mv^2 \end{cases} \tag{9.6}$$

I leave the willing reader to check this result (cf. Problem 9.3, p. 217).

Theorem 9.4 *Initial potential energy is entirely transformed into kinetic energy at the end of the course of P.*

Proof From the equation system (9.6), we can infer the following: when P hits the soil, $y = 0 \Rightarrow E_P(0) = mg \times 0 = 0 \Rightarrow E_P(x, h) = E_K(x, 0)$. □

Definition 9.7 (Mechanical Energy) Let P be a material point in a system Σ endowed with potential energy E_P and kinetic energy E_k. The mechanical energy of P is:

$$E_{\text{Mech}} = E_P + E_k \tag{9.7}$$

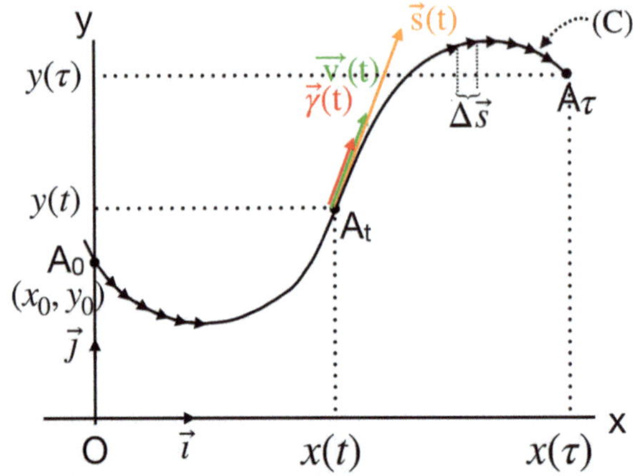

Fig. 9.3 Illustration of the work W done by an object exerted by a variable force $\mathbf{F}$. A_t is the location of the moving point object P of mass m (*black bullets*) whose coordinates are $(x(t),\ y(t))$. The *black line* (C) sprinkled with arrowheads shows the sense of the movement = trajectory of P. Coloured arrows: *yellow*=displacement, *green*=velocity, and *red*=acceleration at instant t. The work carried out by the object during time interval $[0,\ \tau]$ is given by the formula (9.5)

9.1.2.2 The Force Acting on a Single Particle Is Variable

In the most general case, force $\mathbf{F}$ is not constant; its direction varies with time. Consequently, the trajectory (C) is curvilinear in an XY-plane, where the abscissae and ordinates are, respectively, the horizontal and vertical components of displacement $\mathbf{s}$ (like the plane represented in Fig. 9.3). Let force $\mathbf{F}(t)$ be exerted on P that will move from position A_0 (at instant 0) and A_τ (at instant τ). To handle this situation, one mentally decomposes such a curve in minute segments of straight line corresponding to the minute interval of displacement $\Delta\mathbf{s}(t)$, during which $\mathbf{F}(t)$ can be considered constant (with a relative error that is extremely small). Time t being implicit, we will consider that $\Delta\mathbf{s}(t) = \Delta\mathbf{s}$ and $\mathbf{F}(t) = \mathbf{F}(x, y) = \mathbf{F}(\mathbf{s})$. Displacement $\mathbf{s}$ is a vector such that $s = ||\mathbf{s}||$ is the curvilinear distance between A_0 and $A_t (0 \leqslant t \leqslant \tau)$.

Now, we will pose the definition of work in the general case, which is, in fact, a theorem. We will see that the definition of work in the general case is an expression similar to Eq. (9.3).

Theorem 9.5 (The General Definition of *Work*) *Let* $\mathbf{F}(\mathbf{s})$ *be a force applied to point object P that circulates along trajectory (C), then*

$$W = \iint_{(C)} \mathbf{F}(\mathbf{s}) \cdot d\mathbf{s} \tag{9.8}$$

This type of integral is called a curvilinear integral. Proof left to the willing reader in Problem 9.4 (p. 217).

9.1.3 *Elements of Thermodynamics of Irreversible Systems*

Let us consider an open system Σ containing multiple body points. Each of these particles is mainly driven by electric interaction.[3] Such a system is called a composite system. From the kinetic and potential energies of the material points, we will infer the kinetic and potential energies of their member structure.

- Each particle P_i of Σ, of mass m_i and velocity v_i, is endowed with individual kinetic energy $E_k(i)$ and potential energy $E_P(i)$, thus of mechanical energy $E_{\text{Mech}}(i) = E_k(i) + E_P(i)$ (see Definition 9.7).
- The variables $E_k(i)$, $E_P(i)$ and $E_{\text{Mech}}(i)$ are additive, and their sums respectively constitute E_k, E_P and E_{Mech}, the kinetic, potential and mechanic energies of the whole system Σ. Thus, if the latter contains n particles,

$$\begin{cases} E_k = \sum_{i=1}^{n} E_k(i) = \dfrac{1}{2}\sum_{i=1}^{n} m_i v_i^2, \\ E_P = \sum_{i=1}^{n} E_P(i) \\ E_{\text{Mech}} = \sum_{i=1}^{n} E_{\text{Mech}}(i) \end{cases} \tag{9.9}$$

A composite system Σ is characterised by a global motion in the universe as a whole and, at the same time, an "internal motion of its particles forming the system relative to the centre of the mass" [28]. Its energy at a given time t can be decomposed as follows:

$$E = \frac{1}{2}MV^2 + U$$

where M is the mass of the system, V the velocity of its centre of mass, and U the internal energy, related to the behaviour of particles, as defined below (cf. Definition 9.8).

Definition 9.8 (Internal Energy) The internal energy U of a composite system Σ is the sum of the kinetic energy of the internal motion of its particles and the potential energy of their interactions [28].

Definition 9.9 (Heat) Let Σ be a system containing multiple body points and $\mathcal{E}$, its surroundings. In the case of spontaneous evolution, heat Q is the directional energy transfer between Σ and $\mathcal{E}$ caused by the temperature (cf. Definition 9.10) difference

[3] Because of their extremely light masses and significant electric charges, the particles (referring generally to atoms or molecules) are subject to electric interactions immensely stronger than the effect of gravitation.

between these two compartments, with the direction from the compartment endowed with the higher temperature to the one having the lower temperature.

Definition 9.10 (Temperature) Let Σ be a system containing n interacting particles P_i ($1 \leqslant i \leqslant n$, $n \geqslant 2$), each of them with kinetic energy $E_k(i)$ at a given time t. The temperature T of the system is the weighted arithmetic mean of the $E_k(i)$, i.e. $T = \overline{E_k(i)}$.

Definition 9.11 (Thermal Energy) Let Σ be a system containing n interacting particles P_i ($1 \leqslant i \leqslant n$, $n \geqslant 2$). The thermal energy of Σ is the sum of the mechanical energy $E_{\text{Mech}}(i)$ of the P_is [45].

Generally speaking, these body points refer to molecules. A system containing multiple particles exerts a pressure P on its boundary . It has a given volume V, two variables intimately related to temperature T (measured in K), thus to the arithmetic mean of the kinetic energies of the particles.

Definition 9.12 (Enthalpy) Let Σ be a composite, non-isolated system whose internal energy is U, exerting a pressure P on its surroundings $\mathcal{E}$ and having a volume V; at a given time t, then the enthalpy H will be defined as follows:

$$H = U + PV \tag{9.10}$$

Axiom 9.1 Sign of the quantities involved in the functioning of systems.
By convention, quantities (heat amount or work) entering a system are positive; those leaving a system are negative.

Axiom 9.2 First law of thermodynamics
It is one of the formulations of the law of conservation of energy: energy can neither be created nor destroyed and can only be transformed from one form into another. If during a period of time Δt, the variation of internal energy of a non-isolated system Σ is ΔU and its inner heat and work amounts are respectively Q and W, then

$$\Delta U = Q + W \text{ (cf. Fig. 9.4)} \tag{9.11}$$

The interpretation of Fig. 9.4 means that within a lapse of time Δt:

1. A quantity of heat Q_{in} and an amount of work W_{in} enter system Σ,
2. The variation of internal energy of Σ is ΔU,
3. A quantity of heat Q_{ex} and an amount of work W_{ex} leave Σ.

The first law of thermodynamics states that during any period Δt, energy is neither created nor lost.
$\underbrace{(Q_{\text{in}} - Q_{\text{ex}})}_{Q} + \underbrace{(W_{\text{in}} - W_{\text{ex}})}_{W}$ represents the totality of the heat and work amounts within Σ, thus the mechanical energy of the particles of Σ, which is the definition of the internal energy ΔU (cf. Definition 9.8). Hence, Axiom 9.2.

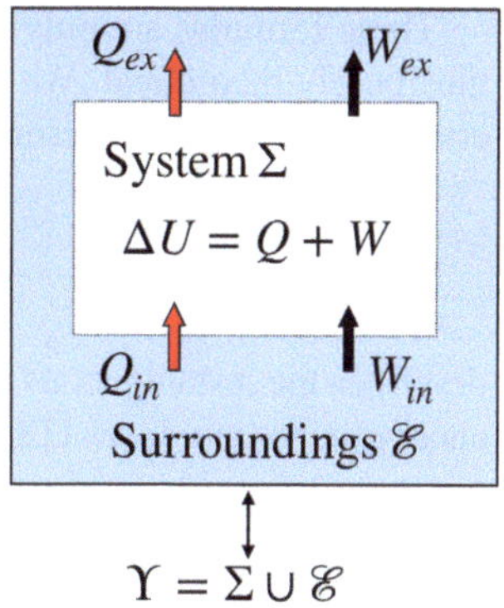

Fig. 9.4 System Σ with surroundings $\mathcal{E}$ and universe $\Upsilon = \Sigma \cup \mathcal{E}$. *White inner rectangle* = non-isolated (closed) system Σ; *blue striped area* = surroundings $\mathcal{E}$; *outer rectangle* = universe Υ (isolated system). *Red and black arrows*, respectively, energy and flux entering or leaving; ΔU = internal energy. Part of Q_{in} is transformed into work

We will keep the notations of Fig. 9.4. At the beginning of this section, we wrote that ecological systems are thermodynamically irreversible, and thus we would only care about thermodynamically irreversible systems. In irreversible systems, spontaneously, Q_{ex}, the dissipated heat increases, which entails a decrease of internal energy, thus of internal heat and work. The concept of entropy that we owe the German physicist Rudolf Clausius (1822–1888) can explain this phenomenon. Entropy is the potential of energy to be transformed into work. It is a physical concept that Clausius coined and defined mathematically in the case of reversible thermodynamical processes, i.e. processes where at a given temperature T, a given amount of heat is completely transformed into mechanical energy, thus potentially work—a situation that the German author called *equilibrium*. Clausius described this ideal situation he experimentally approached with a cycle of very slow chemical reactions and defined his famous formula in 1865 [13] as follows. At temperature T, (measured in K), the entirety of a very small amount of heat δQ is transformed into a minute quantity of mechanical energy dS by the formula

$$dS = \frac{\delta Q}{T} \tag{9.12}$$

dS being the minute variation of mechanical energy produced by δQ in a reversible process. If the system goes from state S_1 (at instant t_1) to S_2 (at instant t_2) along a cyclic trajectory (C), then the entropy variation as expressed by Clausius between states S_1 and S_2 [13, pp. 390–391] is:

$$\Delta S = \int_{(C)} \frac{dQ}{T} \tag{9.13}$$

These formulas are only valid for thermodynamically reversible systems and are thus purely theoretical. We owe Josiah W. Gibbs (1839–1903) an extension of the concept of entropy to thermodynamically irreversible systems:

$$dS > \frac{dQ}{T} \tag{9.14}$$

describes the instantaneous variation of entropy, which is the amount of mechanical energy the system loses [18].

Axiom 9.3 Second law of thermodynamics
The entropy of the universe increases with time.

The second law expressed this way means that a physicochemical, and thus a biological/ecological, system Σ spontaneously[4] loses mechanical energy during any period of time. The amount of the lost energy goes to the surroundings $\mathcal{E}$ with a part of it that will never revert to Σ, which induces a decrease of entropy in the system ($\Delta S_\Sigma \searrow$) and an increase of entropy in the surroundings ($\Delta S_\mathcal{E} \nearrow$) greater than ΔS_Σ. The gain of entropy during period Δt is $\Delta S = \Delta S_\mathcal{E} - \Delta S_\Sigma > 0$ is equivalent to the impossibility in the course of time, for an increasing part of the heat dissipated from Σ, to revert in it, thus to be converted into work in it.

Gibbs free energy G of system Σ is the potential providing the maximal quantity of work done within the internal energy, i.e. by the particles P_i, of Σ, hence the following theorem, which is a practical way to define G (proof and result in [10]). In a biological organism or ecosystem, G is called *useful energy.*

Definition 9.13 (Gibbs Free Energy G**)** Let Σ be a non-isolated irreversible system characterised by an enthalpy H, an absolute temperature T and an entropy S. Then, the Gibbs free energy G is calculated as follows:

$$G = H - TS \tag{9.15}$$

Theorem 9.6 (Another Formulation of the Second Law of Thermodynamics) *Let Σ be a non-isolated irreversible system. Then, if during a period Δt, the system Σ undergoes a variation of enthalpy ΔH, a variation of entropy ΔS and an absolute temperature T*

$$\Delta G = \Delta H - T\Delta S \tag{9.16}$$

A well-known result is the following:

Theorem 9.7

- *A chemical reaction cannot spontaneously start if $\Delta G > 0$.*
- *They will start spontaneously start if $\Delta G < 0$.*

[4] i.e. when no extra energy is added from outside to the system.

Proof Four cases will be analysed to verify when a reaction is spontaneous, hence its feasibility, without having to add energy to the system [5].

Case 1: $\mathbf{\Delta} \boldsymbol{H} > \mathbf{0}$ and $\mathbf{\Delta} \boldsymbol{S} > \mathbf{0}$: Endothermic process involving an increase in system entropy

- Subcase 1.1: $\Delta H > T\Delta S$. Temperature T is lower than a given threshold.[5] Then $\mathbf{\Delta} \boldsymbol{G} > \mathbf{0}$
- Subcase 1.2: $\Delta H < T\Delta S$. Temperature T is higher than the threshold above. Then $\mathbf{\Delta} \boldsymbol{G} < \mathbf{0}$

The thermodynamic process is spontaneous at high temperatures and non-spontaneous at low temperatures.

Case 2: $\mathbf{\Delta} \boldsymbol{H} > \mathbf{0}$ and $\mathbf{\Delta} \boldsymbol{S} < \mathbf{0}$: Endothermic process involving a decrease in system entropy
$\mathbf{\Delta} \boldsymbol{G} > \mathbf{0}$ whatever the temperature.
The thermodynamic process is non-spontaneous at all temperatures.

Case 3: $\mathbf{\Delta} \boldsymbol{H} < \mathbf{0}$ and $\mathbf{\Delta} \boldsymbol{S} > \mathbf{0}$: Exothermic process involving increase in system entropy

- Subcase 3.1: $\mathbf{\Delta} \boldsymbol{G} < \mathbf{0}$ whatever the temperature.

The thermodynamic process is spontaneous at all temperatures.

Case 4: $\mathbf{\Delta} \boldsymbol{H} < \mathbf{0}$ and $\mathbf{\Delta} \boldsymbol{S} < \mathbf{0}$: Exothermic process involving decrease in system entropy

- Subcase 4.1: $\Delta H > T\Delta S$. Temperature T is lower than a given threshold. Then $\mathbf{\Delta} \boldsymbol{G} < \mathbf{0}$
- Subcase 4.2: $\Delta H < T\Delta S$. Temperature T is higher than the threshold above. Then $\mathbf{\Delta} \boldsymbol{G} > \mathbf{0}$

The thermodynamic process is spontaneous at low temperatures and non-spontaneous at high temperatures.

□

In conclusion,

- Spontaneous reactions are only possible when $\Delta G < 0$.
- The physicochemical processes belonging to ecological systems refer all to subcase 1.2.
- We will see that during their life time, the organisms have intrinsic mechanisms that function counter to the second law.

According to the second law, chemical reactions, particularly those involved in biological or ecological processes, are open systems[6] involved in a spontaneous,

[5] Note that temperature T is measured in K, and is thus always > 0.

[6] As a reminder, this means that these systems exchange heat and matter with their surroundings, cf. Definition 9.2 (p. 187).

thus irreversible thermodynamic process. We saw that heat transfer is the energy flowing from the system to its surroundings if the temperature of the former is, on average, higher than the latter's counterpart. When in a system-surrounding pair, the average temperature of both compartments are approximately equal, hardly any exchange of energy occurs, and an equilibrium is reached, the *thermodynamic equilibrium*, which is the topic of the next section.

9.2 Thermodynamic Equilibrium and the Case of Life, thus of Ecosystems

9.2.1 Thermodynamic Equilibrium

An open system Σ is at thermodynamic equilibrium if, at macroscopical level, the flux of matter or energy entering Σ is equal to the flux of matter or energy leaving it at a given time t. This happens when its Gibbs free energy G is minimised, a situation occurring when thermal equilibrium—temperatures of the system and of its surroundings are the same—is reached, as a consequence of the second law, which entails "no further changes involving the conversion of heat into useful work" [20]. All forces are then balanced, and no significant external driving force exists. The maximum possible entropy is reached.

Biological and ecological systems, notably ecosystems, are exceptions among thermodynamic systems: they never reach thermodynamic equilibrium, i.e. their thermodynamic potentials (free energies) are never minimised. A complex pathway involving specific organic molecules prevents these living beings from achieving thermodynamic equilibrium when they are alive; the same applies to ecosystems. But when organisms are dead, they become classical physical systems prone to thermodynamic equilibrium.

9.2.2 The Flux of Energy in an Ecosystem

9.2.2.1 The Relays

- *Autotrophs* are classically defined as organisms that synthesise all their organic matter by feeding from chemical elements or simple compounds found in their environment, fuelled by an external source of energy, (1) either sunlight, (2) or redox chemical reactions. This process is called biosynthesis. Another definition, preferred by microbial ecologists, is related to the biosynthetic behaviour of an organism regarding its carbon source, the autotrophs being the organisms that take in their organic carbon atoms from carbon dioxide.[7] Both definitions are equivalent when plants and animals are analysed. However, the latter is the only valid one if we consider microorganisms, and above all, bacteria and archaea. Thus, it is sensible to adopt the second definition since the prokaryotes constitute the main actors of the ecological systems. Autotrophs divide into two categories, photoautotrophs and lithoautotrophs.

Photoautotrophs, fuelled by solar light energy for their biosynthesis via a process called photosynthesis, take their core organic elements, i.e., carbon, oxygen, hydrogen and nitrogen from atmosphere (see Footnote 7), and other essential organic elements, e.g. phosphorus, sulphur, magnesium, sodium, potassium, calcium, iron, zinc, copper, from soil interstitial water [11].
The source of the chemical species taken from the atmosphere and transformed into organic elements are the following:

- Carbon atoms are extracted from carbon dioxide.
- Oxygen atoms originate from dioxygen and water vapour.
- Hydrogen atoms stem from water vapour.
- Nitrogen atoms come from dinitrogen indirectly—nitrogen atoms are incorporated only when they are parts of a nitrate or nitrite molecule or an ammonium ion.

Photoautotrophs are typically exemplified by plants and a great number of bacteria belonging to one of the seven following phyla—Cyanobacteria, Chlorobi, Chloroflexi, Acidobacteria, Firmicutes, Gemmatimonadetes, and Proteobacteria). These organisms overwhelmingly use for their photosynthesis either two reaction centres (Plants, Cyanobacteria) or one reaction centre (the other six bacterial phyla) of a molecule governing it, *chlorophyll* or *bacteriochlorophyll* [24, 47].

Chemoautotrophs carry out their biosynthesis by using the chemical energy contained in a number of chemical species, e.g. ferrous iron, hydrogen sulfide, sulphur atoms, ammonia, molecular hydrogen, water. The only organisms carrying out this type of autotrophy are prokaryotes, either bacteria or archaea. Most of them are extremophiles, found, for example, (1) in the abyssal zone of oceans unreached by sunlight, on ocean floors close to hydrothermal vents, or (2) in volcanic springs. They are generally extremely specialised (e.g. nitrifiers, sulphur- or iron-oxyders) and can be aerobic or anaerobic [23].

- Heterotrophs are classically defined as organisms that need to feed by ingesting organic matter and digest it to obtain the chemical elements or simple compounds from which they can reconstruct their own organic material. This applies well to animals, fungi and phagocytic unicellular eukaryotes. A loser definition is

[7] The main components of atmosphere are, in volume proportion, dinitrogen (78.08%), dioxygen (20.94%), water vapour (0.01–5%), argon (0.93%), carbon dioxide (0.04%) and trace proportion of other gases, e.g. neon, helium, methane, etc.

preferred by microbial ecologists who define heterotrophs as organisms that derive all their organic carbon atoms exclusively from organic carbon, which makes it possible to include a good number of prokaryotes, namely those that absorb their carbon from organic molecules by osmotrophy [3], into the heterotrophs.

All eukaryotic and many prokaryotic heterotrophs are organotrophs, i.e. they derive a part of their digestion products to construct energy-laden molecules, containing the energy transferred from autotrophs to them. Glucose is the primary energy source of these organotrophs. This energy passes from this molecule to other energy-laden molecules that are produced in a chain of successive reactions whose primary reactant is glucose, through various chemical pathways (e.g. glycolysis, Krebs cycle, oxidative phosphorylation). There exist a group of prokaryotic photoheterotrophs, i.e. microorganisms deriving their organic carbon from organic matter but taking in their primary energy from sunlight *via* a specific molecule, rhodopsin, e.g. the α-proteobacterium *Rhodobacter* sp. [6], or the halophylic[8] archaeon *Halobacteria* [22, 42].

9.2.2.2 Analysis of the Flow of Energy

The energetic resources of the species of a given ecosystem **Ec** constitute only a tiny part of the huge total energy entering **Ec**. However, this very small portion of the energy exploited by these species is of the uttermost importance for the functioning of the ecosystem. In addition, the overwhelming part of the energy entering an ecosystem is dissipated, and globally the second law of thermodynamics holds. The *dissipation of energy* accompanying the decomposition of corpses and excrements takes place during the process of *mineralisation*; the latter is governed by a whole chain of species continuously decomposing the organic matter of the ecosystem into inorganic chemical elements or simple molecules eventually available to the autotrophs of the ecosystem. In Sect. 10.2, we will see that at this point, the elements and simple compounds released by mineralisation will again enter the ecosystem via the autotrophs in an efficient recycling system called material recycling. For that, the autotrophs need added energy, which comes from the steady energy flux we mentioned at the beginning of this section.

Using descriptors exposed in Sect. 9.4.3.2, we now have a clear image of how the energy flux works. At the first trophic level, the autotrophs (or *primary producers*) trap a part of the initial energy received by the ecosystem in the form of potential energy associated with energy-laden molecules. At the second trophic level, some heterotrophs (or first-level consumers) feed upon some of this autotrophic biomass, thus take in a portion of the potential energy. Since the passage from the first to the second trophic level is accompanied by energy dissipated in the form of heat,

[8] Adjective applied to organisms adapted in environments characterised by a high concentration of salt (sodium chloride).

less energy in the form of potential energy in energy-laden molecules is left for the organisms belonging to the third trophic level (or second-level consumers). And the process goes on recursively until the highest trophic level whose level cannot be *a priori* determined. Every species takes its part of energy and returns it , allowing the whole system to function.

Caution! Energy comes in a permanent flux; there is no such thing as energy recycling. In order to understand this concept, we can use the metaphor of water constantly squirting out of a spring.

9.3 The Originality of the Biological System

One of the consequences of the first and second laws of thermodynamics is the tendency of most physical systems towards thermodynamic equilibrium. We saw that, on the contrary, biological systems depart from thermodynamic equilibrium and consequently maintain much higher free energy than ordinary physical systems (cf. Sect. 9.2.1). In the case of an individual, thus an elemental ecological system, the deviation from thermodynamic equilibrium ceases when it dies. If the biological system is an ecosystem, the thermodynamic equilibrium is reached when it is destroyed by catastrophes (e.g. fire, hurricane, volcanic eruption, strong anthropic pressure[9]) that dismantle the communities or entail severe losses of abiotic factors.

The increase of free energy results from biosynthesis or feeding, via complex chemical pathways (the metabolism) that initially integrate chemical species initially in their mineral form (simple compounds, ions) into organic molecules, and involve, for this, energy-laden molecules and enzymes. (These molecules are also the product of the metabolism.) The relentless energy degradation that spontaneously occurs either through direct heat dissipation or accompanying metabolism (second law of thermodynamics) is slowed down by diverse mechanisms (colour patterns, reduction of external surface, chemical mechanisms, etc) [46]. The process accompanying energy intake is called *assimilation*, and the process accompanying energy dissipation is *dissimilation*.

As mentioned in Sect. 6.2, ecological systems are best described by compartmental models. We can use this property to illustrate the pathway of energy and the process by which the biological systems store free energy (at the individual level) (cf. Fig. 9.5). The compartments are, in the sense of the energy flow, the primary producers (e.g. plants), the consumers 1[10] (herbivores), the consumers 2 (first-degree predators) and the consumers 3 (second-degree predators, or apex predators), in the case—for example—of a food chain of length 3.

[9] . . . like excessive hunting or fishing, or habitat destruction or disruption.

[10] Consumers i are also called ith degree consumers; besides, consumers 1 also called primary consumers, consumers 2 secondary consumers and consumers 3 tertiary consumers.

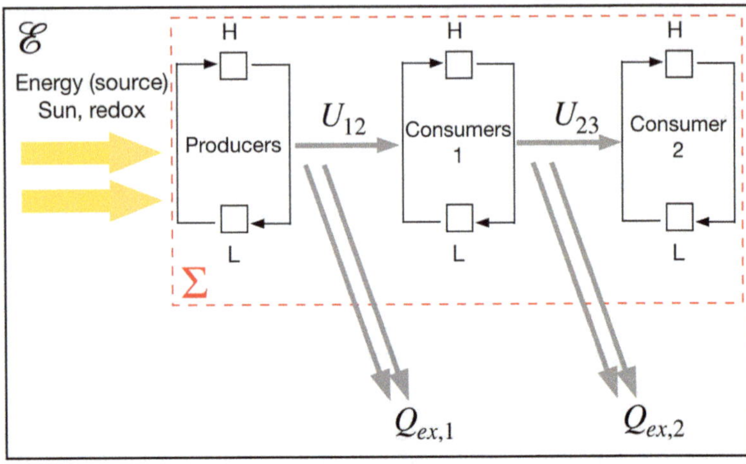

Fig. 9.5 Compartmental model of the trophics levels. *Dashed rectangle* = biocoenosis = thermodynamic system Σ; *outside the dashed rectangle*, surroundings $\mathcal{E}$. H = compound with high energy potential, L = compound with low energy potential. Energies: the *arrows* are the direction of the flow—(1) source: flow from solar or chemical energy; (2) U_{ij} part of free energy transferred from consumers i to consumers $i+1$; (3) Q_{ex} heat dissipated from system Σ

Let us describe the main mechanism that depends on sunlight or *lithochemical energy* (chemical energy originating from rocks) (see Sect. 9.4.4.2). The primary assimilatory process, *photosynthesis* or *chemosynthesis*, is ensured at the lowest trophic level. It is relayed at higher trophic levels by another assimilatory level, ingestion. Each level, able to increase its internal free energy, is a compartment. Most of the assimilated energy is degraded during dissimilation. The higher the trophic level is, the less efficient assimilation is (assimilation efficiency being defined as the rate of assimilated energy turned into internal free energy). This fact explains why food chains have finite, and generally rather small lengths.

If one considers the energy balance of the total biocoenosis of an ecosystem viewed as a single compartment (cf. Fig. 9.6), (1) the only energy assimilation process is the photosynthesis (or chemosynthesis) carried out by the primary producers altogether (the assimilation processes of the consumers being merely energy transfers between its subcompartments); and (2) the only energy dissimilation process of this single compartment is the totality of the energy dissipated by the member populations of the total biocoenosis,[11] either in the course of the mineralisation phase during the decomposition of corpses and excreta, or during metabolism when the organisms are alive.

Figure 9.6 will help us to analyse the total dissipating energy.

[11] Let us recall that an ecosystem in the sense of the second definition integrates a total biocoenosis and its environment restricted to its abiotic factor system (AFS) (cf. Definition 2.7, Sect. 2.4).

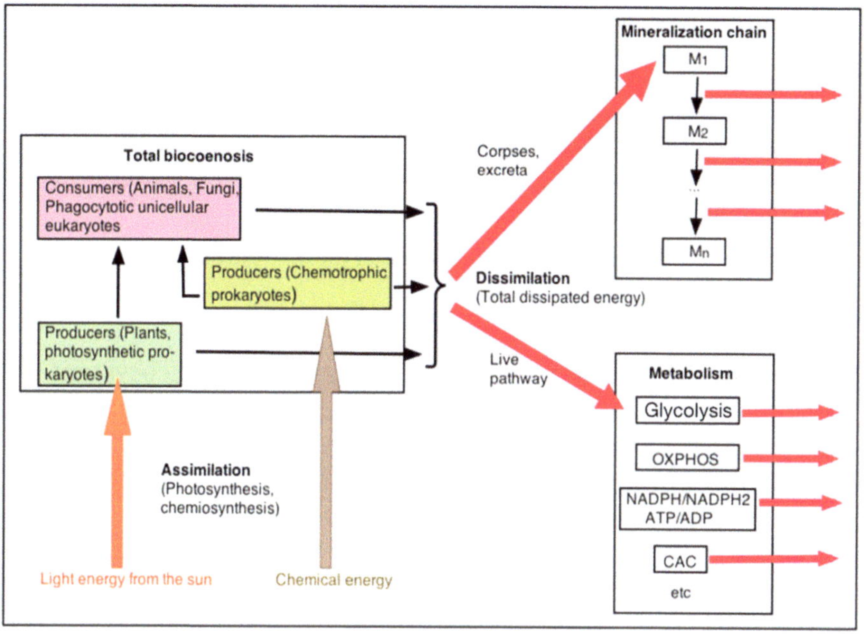

Fig. 9.6 Total biocoenosis of an ecosystem considered as a single compartment. The *arrows* represent the energy flux, from the source to the dissipated heat. *Orange and brown arrows* = energy taken in by the total biocoenosis (TB); *thin black arrows* = transfer of free energy within the TB; *red arrows* = dissipated energy at the level of different subcompartments the compartments. $M_1, \ldots M_N$ successive relays of the process of decomposition/mineralisation, *OXPHOS* oxidative phosphorilation, *CAC* citric acide cycle

- A part of the total dissipated energy is related to the formation of the simple chemical compounds needed by primary producers, the entry of matter into the total biocoenosis[12] studied.
  - During mineralisation, dissipated energy accompanies the decomposition of the organic matter of corpses and excreta into the lithospheric elements and simple compounds diluted in soil interstitial waters. This process is carried out by the successive relays $M_1, \ldots M_N$ depicted in Fig. 9.6, respectively corresponding to organisms of decreasing size, i.e. scavenging and fossorial vertebrates and invertebrates, saprophytic fungi, eukaryotic and prokaryotic microorganisms, and at the end of the chain, mineralisers (several bacteria and archaea that are able to release simple chemical compounds and ions from extremely degraded organic matter [14]).

[12] As a reminder, these simple chemical species found in the environment are almost the complete source of the nutrient intake of the primary producers.

- In the course of metabolism, the atmospheric elements in the form of gaseous molecules (e.g. dioxygen, dinitrogen, water vapour) are released in the atmosphere. These lithospheric and atmospheric elements, essential for the organic molecules, redox molecules and related factors and cofactors found in all the organisms of the total biocoenosis (and structurally very similar), are listed in Table 2.1 (p. 15), associated with some of the vital organic molecules which they are part of.

- The rest of the dissipated energy is lost out of the ecosystem.

9.4 From the Primary Source of Energy in Most Ecosystem

This section will primarily treat the mainly-occurring type of photosynthesis, i.e. oxygenic photosynthesis, where the initial electron donor is water and which releases dioxygen (O_2) as a waste product (in the atmosphere or hydrosphere). A summary overview will evoke less frequent and conspicuous forms of global energy intake, e.g. anoxygenic photosynthesis and chemosynthesis (Sect. 9.4.4, p. 215). Throughout this section, except in Sect. 9.4.4, photosynthesis will only refer to oxygenic photosynthesis.

9.4.1 *Simplified Description of Photosynthesis*

Solar energy induces the synthesis of organic molecules,[13] among which many energy-laden molecules that induce the kinetics of electrons and protons within the organisms constituting the populations of each ecosystem's total biocoenosis. These macromolecules mainly contain the four atmospheric and ca. 20 lithospheric elements depicted in Table 2.1 (p.15) (Sect. 9.3). Sunlight energy and these elements are integrated into the organic matter (i.e. the totality of the organic molecules) by a process called photosynthesis that takes place within the lithophototrophs or primary producers (e.g. plants, cyanobacteria).

Photosynthesis transforms gaseous *carbon dioxide* and water vapour extracted from the atmosphere into glucose, gaseous dioxygen, and water in its liquid form (cf. Eq. (9.17)).

- *Glucose* is a very powerful energy-laden carbohydrate that carries the primary potential energy within the body: it is the source of the kinetic energy that circulates in the organism via all the pathways and cycles of metabolism.

[13] The main classes of organic molecules, found with a remarkable uniformity among all the species, are the carbohydrates, lipids, proteins and nucleic acids (DNAs and RNAs).

- *Oxygen* is released by the phototrophs into the atmosphere in the form of gaseous dioxygen. The atom of oxygen is an electron acceptor. The electron flux throughout the metabolism pathways is launched from the glucose, the electron donor, to oxygen. It is now widely accepted that the abundance of this molecule in the present Earth's atmosphere[14] is due to this excretion that has lasted at least ca. 2.5 billion years.[15]
- *Water* in its liquid form is essential for life, for it includes all the necessary solutes and electrolytes needed to trigger the kinetic energies of metabolic activities.

The overall chemical equation of photosynthesis is, per mole of carbon:

$$6\,CO_2\,(g) + 12\,H_2O\,(g) + \underbrace{\text{sunlight energy}}_{676\text{ kCal}} \xrightarrow{\text{Chlorophylls}} \underbrace{C_6H_{12}O_6}_{\text{Glucose}}\,(s) + 6\,O_2^{\uparrow} + 6\,H_2O\,(aq) \tag{9.17}$$

This chemical reaction triggers electron transfers launching a round of other chemical reactions, which generate molecules with high potential energy, e.g. ATPs, or chemical pathways and cycles producing all the organic matter of the primary producers.

9.4.2 *The Different Phases and Steps of Photosynthesis*

In the case of plants, photosynthesis takes place in organelles specific to the phototrophs called chloroplasts that have two characteristic structures: (1) the thylakoid network where a sequence of chemical reactions take place called the light phase of photosynthesis and (2) the stroma housing another sequence of chemical reactions called the dark phase of photosynthesis (Fig. 9.7). These two phases are complementary. *Light* and *dark phase* have been well-explained, in a educational and synthetic way, within one of the seminal books of Ramade [43, p. 14], and I reproduce the explanations of this author hereunder in sum.

Light phase: occurs under the influence of light. The chlorophylls, embedded in the membrane of the thylakoids (Fig. 9.7), are excited by photons when light wavelengths lie between 670 and 680 μ; their molecules catalyse the photolysis of water, thus:

$$H_2O \longrightarrow \underbrace{2\,H^+}_{\text{Protons}} + \frac{1}{2}\,O_2^{\uparrow} + 2\,\underbrace{e^-}_{\text{Electrons}} \tag{9.18}$$

[14] Gaseous dioxygen was almost nonexistent, but carbon dioxide was abundant in early Earth's atmosphere.

[15] The first organisms to have ever carried out that excretion were very probably anaerobic photosynthetic phototrophic prokaryotes (bacteria or archaea).

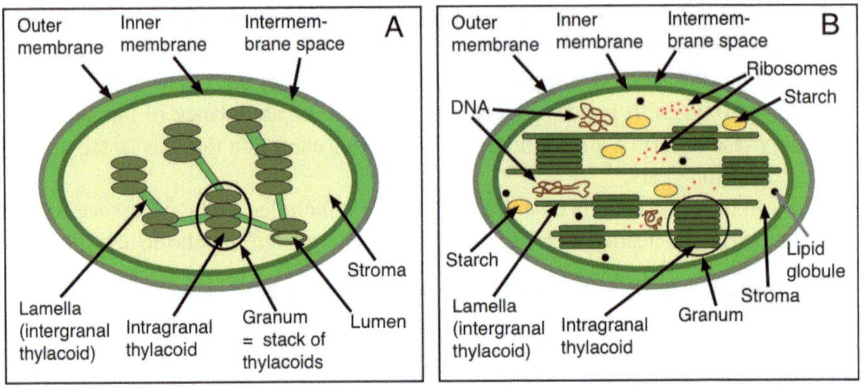

Fig. 9.7 Section of a chloroplast with the thylakoid network and the strom. (**a**) *Two different zones, the thylakoid network (in green) and the stroma, an alkaline aqueous fluid surrounding the thylakoid network (light yellow).* The thylakoids fall into two types, short ones, or intragranal thylakoids, forming a heap called granum, and long ones, or intergranal thylakoids, interconnecting the grana. Light phase takes place in the thylocoid network, and dark phase in the stroma. (**b**) *Simplified structure of a chloroplast.* This organelle is remarkable as it contains DNA and ribosomes, and is very similar to cyanobacteria. It is yet widely accepted that plants' chloroplasts have evolved from Precambrian cyanobacteria that were engulfed into eukaryotic unicellular organisms, presumably the most remote ancestors of plants

This reaction is the source of a fast-moving protons and electrons flow, causing a sequence of chemical reactions during which ATPs, i.e. molecules with a strong energetic potential, and $NADPH_2$, molecules with a strongly reducing power, are produced. Very importantly, the cycle $NADPH_2 \rightleftarrows NADP$ triggers the metabolism's chain of redox reactions. The light phase is also the phase when gaseous oxygen is released into the atmosphere.

Dark phase: takes place outside the influence of light in the stroma (Fig. 9.7). It consists of all the chemical reactions performing autotrophy, i.e. the extraction of carbon atoms from atmospheric or hydrospheric CO_2 and their incorporation into all the organic molecules elaborated during biosynthesis. The primary molecule hence constructed is glucose ($C_6H_{12}O_6$) and from it, all the other organic molecules are derived (i.e. the other carbohydrates, fat acids and amino acids, via the $NADPH_2$ and ATP molecules previously produced during the light phase). The sequence of chemical reactions whose product is glucose is obtained from a circular pathway of chemical reactions called the *Calvin cycle*.

The Calvin cycle starts with the extraction of three molecules of carbon dioxide (CO_2) from the atmosphere or hydrosphere, and their entry into the stroma. Thereafter, these molecules react with the ketone-containing 5C-sugar RuBP[16]

[16] Shortening of "Ribulose-1,5-biphosphate".

catalysed by the enzyme *Rubisco*[17] generating a series of 3-carbon intermediates. This series eventually produces six molecules of 3-PGA. A sequence of reactions continues the cycle fostered by transfers of electrons from six reactions ATP → ADP, and six others, $NADP^+ \rightarrow NADPH_2$, giving six molecules of 3-GA3P. One of these 3-GA3P is the reactant of a sequence of two reactions that escapes the cycle, towards the production of glucose ($C_6H_{12}O_6$). (From 3 moles of CO_2, $\frac{1}{2}$ mole of glucose is obtained.) The five remaining 3-GA3P are the reactants of a chain of reactions during which three molecules of ATP → ADP, and whose final products are three molecules of RuBP. After these production, three molecules of carbon dioxide (CO_2) comes into the cycle and the cycle is locked.

9.4.3 *Assessment of the Primary Energy Intake of the Ecosystems*

9.4.3.1 Ecosystems Whose Primary Producers Perform Oxygenic Photosynthesis

Solar energy is radiant energy lying in the visible wavelength spectrum of sunlight. This primary energy intake of the ecosystems occurs via *photosynthesis*, a process that we thoroughly treated in Sects. 9.4.1 and 9.4.2. As a reminder, photosynthesis produces the energy and organic matter of the primary producers and, consequently, of the consumers depending on them.

- During the *light phase* of photosynthesis, i.e. under the influence of solar radiations, the free energy of the organisms, necessary for the cellular and tissular growth, maintenance, repair and synthesis, is generated.
- During the dark phase of photosynthesis, i.e. in light-independent conditions, the organic compounds are produced in a chronological sequence—first glucose, then other simple carbohydrates, then succeedingly all the other organic compounds, e.g. the carbohydrates, lipids, amino acids, nucleic acid bases, ATP and ADP, NADP and $NADPH_2$, vitamins, growth factors; and also their polymers when they exist like, e.g. proteins and nucleic acids.

Photosynthesis absolutely needs *magnesium* (Mg) for its functioning. The primary producers find this element in the environment (e.g. as a mineral salt in the *soil system* and as a solute in the *hydrosphere*) [26]. Magnesium is a crucial atom in the composition of chlorophylls. Without this element, chlorophylls cannot be synthesised, and photosynthesis will not occur.

[17] Shortening of 'Ribulose-1,5-biphosphate carboxylase/oxygenase'; I quote: "More than 90% of the inorganic carbon that is concerted into biomass is fixed by the enzyme RubisCo ... found in all three domains of life: bacteria, archaea and eukaryotes" [21].

Photosynthesis takes in the inorganic carbon atoms from the carbon dioxide (CO_2) of the atmosphere/hydrosphere. It transfers them into the primary producers, transforming them into organic carbon atoms, the backbone of every organic compound of all the living organisms on Earth. That is done in successive steps, the first being the elaboration of glucose via the Calvin cycle (see Sect. 9.4.2), the successive others triggering all the organic molecules needed for life in numerous other pathways and cycles. This mechanism sets off diverse movements of electrons and protons, forming the chemical energy (a form of kinetic energy) that flow within the organisms.

When inorganic carbon passes as organic carbon through the Calvin cycle, carbon is said to be assimilated. Per gram of assimilated carbon, the producer (e.g. plant, cyanobacterium) gains 39 kJ of energy [41]. This quantity of energy is called "assimilated energy per gram of carbon assimilated". The second law of thermodynamics sets a balance *balance* between Gibb's free energy, and the energy loss. A part of it is transferred from one to another trophic level in each biocoenosis (e.g. from a producer to a first-order consumer). Like all physical systems, biological and ecological systems are subject to thermodynamics laws and raise their entropy in the long run despite the increase of free energy that accompanies the synthesis, growth and reproduction of the biological matter that makes the biocoenoses. We have already seen that during their lives, organisms are in thermodynamical disequilibrium at each instant, with their free energy higher than their lost energy. However, after their deaths, thermodynamic equilibrium always holds in the individuals, thus also in the ecological systems.

9.4.3.2 From Gross Solar Energy to the Useful Energy Fraction: The Case of the Plants

The total solar energy striking an ecosystem in percentage is called the *gross solar energy* (GSE), thus 100% of the energy of interest. Only 8.8% of the GSE is absorbed by the primary producers (e.g. plants). From this *absorbed energy* (AE), only 34% is changed to the useful energy fraction (UEF)—that is the energy associated with the synthesis of the carbohydrate compounds of the organism (evaluated in dry biomass). The rest of the energy is degraded into heat via a number of possible pathways (cf. Fig. 9.8). We can easily calculate that only ca 3% of the GSE is transformed into UEF.[18]

Several parameters are very useful for characterising the energetic flow through a plant during photosynthesis, among others, *photosynthetic efficiency*, *production* and *productivity*, and plant biomass [notions that will defined in sections "b-Photosynthetic Efficiency" and "a-Production, Productivity and Biomass" of this subsubsection].

[18] See Exercise 9.8, p. 218.

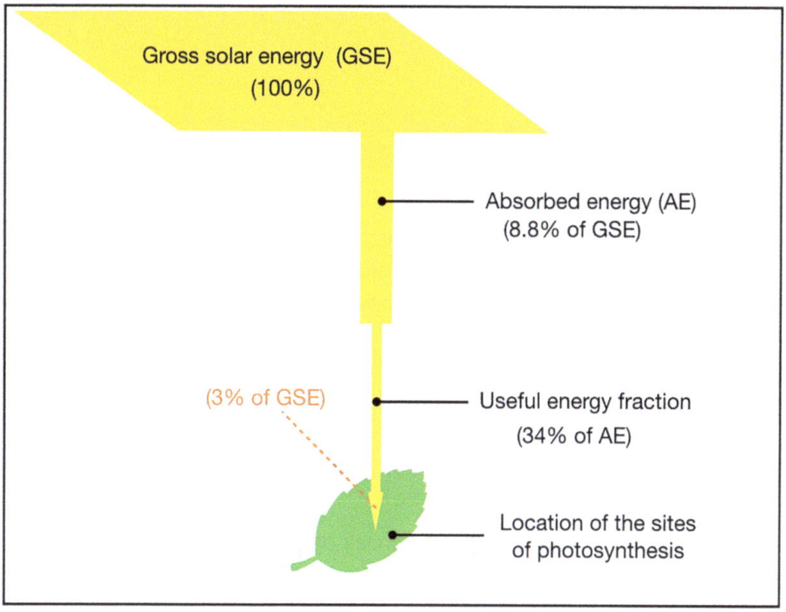

Fig. 9.8 Energy flow from the gross solar energy to the useful energy fraction. The gross solar energy (GSE) is the total energy conveyed by solar radiation that hits the surface of our planet. Of it, 8.8% is absorbed energy (AE), i.e. the part of the GSE that is absorbed by the primary producers (e.g. the plants). But only 3% of the GSE (thus only 34% of the AE) is the energy lading the biomass of the primary producers

a-Production, Productivity and Biomass

Definition 9.14 Gross production is the totality of the energy assimilated by photosynthesis.

Definition 9.15 Net production is the part of the gross production that enters in the plant biomass. It is the energy that is namely used for growth and reproduction.

Two remarks should be made here.

- Between the two types of production defined above, energy has been lost as heat via respiration, excretion and other metabolic events.
- The net production of the primary producers (plants, phytoplankton, cyanobacteria) is the energy available to higher trophic levels (animals, fungi, decomposers, saprophytes). We can easily say that the net production of plants contains the gross production of the organisms that feed on the plants. The same relationship exists recursively between the primary and secondary consumers, the secondary and tertiary consumers, ..., the consumers $n-1$ and n (for a trophic chain of length n).

Definition 9.16 Primary production is the totality of the energy assimilated by the primary producers of a given ecosystem. It is thus equivalent to the production of the primary producers.

In terrestrial ecosystems, primary production is equivalent to plant production. In oceanic ecosystems, primary production is almost equivalent to phytoplanktonic and cyanobacterial production.

Definition 9.17 *Biomass* is classically defined as the total mass of the living organisms in a given area (or population, or ecosystem) in a given time [43, p. 17]. It is usually expressed as a dry weight [44, p. 802].

b-Photosynthetic Efficiency

Definition 9.18 Photosynthetic efficiency is the percentage of light energy assimilated by plants [44, p. 819].

- If it is based on gross production, we speak of gross photosynthetic efficiency;
- In case it is based on net production, it is termed net photosynthetic efficiency.

It can be shown that the photosynthetic efficiency of a species is equal to the rate of inorganic carbon assimilated by this species during photosynthesis. Photosynthetic efficiency strictly depends on the intensity of infrared light absorbed during the production of carbohydrates and their derivatives.

1. When light intensity is low (less than $\frac{1}{4}$ of the intensity of full solar light), photosynthetic efficiency is proportional to light intensity.
2. When light intensity, on the contrary, is high, photosyntheticefficiency asymptotically decreases with the increase of light intensity. The higher the light intensity is, the stronger this phenomenon is: when light intensity is very high, photosynthetic efficiency is null. Plants saturated by light are not able to modify atmospheric CO_2.

Photosynthetic efficiency at a given light intensity is determined by two constants, compensation point and saturation point [44, p. 194].

Definition 9.19 Compensation point is the light intensity where assimilated CO_2 equals lost CO_2 (e.g. via respiration).

Definition 9.20 Saturation point is the light intensity where an increase of light intensity is not accompanied by an increase of biosynthesis.

These points are measured in $W \cdot m^{-2}$. For comparison:

1. Full solar light has an intensity of $500\ W \cdot m^{-2}$.
2. The compensation points of most plant species lie between 1 and $2\ W \cdot m^{-2}$.
3. The saturation points of most plat species are situated between 30 and $40\ W{\cdot}m^{-2}$.

Let us remind the reader of two important properties of photosynthetic efficiency verified experimentally.

- The photosynthetic efficiency of a species is the rate of inorganic carbon assimilated by this species during photosynthesis.
- This parameter strictly depends on temperature. Hence, the following definition.

Definition 9.21 The temperature defined as optimal for a species is the temperature that maximises its photosynthesis efficiency.

As a general rule, the plant species of the temperate zones have optimal temperatures averaging 16° C, whereas the optimal temperatures of tropical plant species lie about 38° C. In several species, the optimal temperature varies as a function of light intensity, e.g. the Alpine heath species of the genus *Loiseleuria* [31].

The energy corresponding to the maximal photosynthetic efficiency of the producers is estimated in the order of 1–2% of the total solar energy. Between 25 and 75% of the latter is reflected by the leaves and other plant surfaces during evapotranspiration [39, p. 32]. The rest of the energy is absorbed by the plants but a large portion through other pigments than the chlorophylls. At every step of the process, energy is degraded as heat that leaves the biocoenoses. Only a tiny portion becomes actually free energy. Let us take the example of an oak forest. The incident radiation is $1.9\,\mathrm{kJ}\cdot\mathrm{m}^{-2}$; 56% of this energy is absorbed by the leaves and passes on principally to metabolism. To this energy mostly lost, one can add the energy lost to evapotranspiration, which makes a total of $1.1\,\mathrm{kJ}\cdot\mathrm{m}^{-2}$. The rest of the energy, which is a tiny proportion, is transformed into biomass.

Biomass is the mass of the species' energy-laden molecules that governs all their biological activities. But this notion is quite polysemic. Different parameters have been applied according to the type of studies and the traditions of the authors: wet biomass, dry biomass, potentially digestible dry biomass, and the mass of organically bound carbon.

- The wet biomass of a group of organisms in a given spatiotemporal unit (area, population, community or ecosystem) is defined as their total mass *in situ*. This parameter is not widely used in fundamental ecology; but it is common practise in applied ecology, e.g. in aquacultural or fisheries studies [1, 4, 25], where the water content of organisms is important[19] or in industrial applications [12, 30, 40].
- The dry biomass of a group of organisms in a given spatiotemporal unit (area, population, species, community or ecosystem) is the sum of the dry masses of their members, i.e. their masses minus their water content. Once the dry mass of a given amount of biological material (for instance, 1 g) is estimated, the dry biomass of the spatiotemporal unit is extrapolated to the estimated number of organisms of the spatiotemporal unit. This parameter is the most commonly used one to measure biomass in fundamental ecology, based on the viewpoint that the

[19] In these studies the nutritional value of the studied organism, substantially depending on water solutes, counts and constitutes a commercial argument.

sheer biological matter lies in the organic molecules that pass on from one trophic level to the next one of a higher degree[20] (1) In some cases, this information is obtained directly by heating it at 105° C at standard atmospheric pressure until all its water content is lost. (2) In general, dry mass is obtained by indirect biomathematical methods, e.g. using a variable proportional to leaf thickness and leaf density (in the case of plants), the last parameters often being estimated with optical instruments [16, 34, 48]. In some cases, the extruded or egested parts (cellulose, bones, shells) are not counted in the biomass and, therefore, excluded from the dry mass. (3) Another commonly used way of calculating a biological material's dry mass is to weigh only its organically bound carbon.

Gross primary production (GPP) is the gross production of the primary producers, and *net primary production* is the net production of the primary producers (NPP) needed for the primary producers' maintenance, growth and reproduction. Part of the biomass of the primary producers constructed by the NPP is ingested by the species of the trophic level just above, forming the *primary consumers*, also called *first-level consumers* (the herbivores or the fungi directly dependent on the plants). The part of the ingested energy that is not egested (like that related to skeletal parts, pellets, cellulose, etc) and that accompanies digestion is at their disposal for the synthesis of their own biomass, and form the *gross secondary production* (GSP). After respiration, the energy left for the construction of their biomass is the *net secondary production* (NSP). The biomass of the primary consumers is hence made up. The energy lading of this biomass is available to the *secondary consumers* (or *second level consumers*) that are the species feeding on the herbivores and fungi feeding on the fungi parasitising plants. Part of this energy passes on to the secondary consumers that ingest it and egest a portion of the ingested energy, constituting the gross secondary production (GSP). The respiratory processes of the secondary consumers extract the net tertiary production (NSP) to form the biomass of the secondary consumers. At each step, energy is degraded into heat out of the system. The flows of the productions and degraded heat and the resulting biomasses, at each main step, is shown in Fig. 9.9.[21] The process goes on likewise with the third level, fourth level, ..., *nth level of consumers*. At each step, a portion of the then-available energy is degraded into heat out of the system via different channels, i.e. chlorophyll action, chemical reactions during metabolism, respiration, ingestion and egestion. To all these energy losses, another type of energy loss must be added, i.e. the degraded energy accompanying the process of mineralisation (with the decomposition of the corpses and feces of the biocoenoses). Figure 9.10 illustrates in sum the process at the level of the consumers of a given level *n* in a biocoenosis.

[20] The dry mass of a given material is its mass less the mass of its water content. Though this definition seems to be simple at first glance, the obtention of this information is far from simple.

[21] Figure 9.9 was inspired by Fig 3–15 [35, p. 106] and Fig. 9.10 by [43, p. 74].

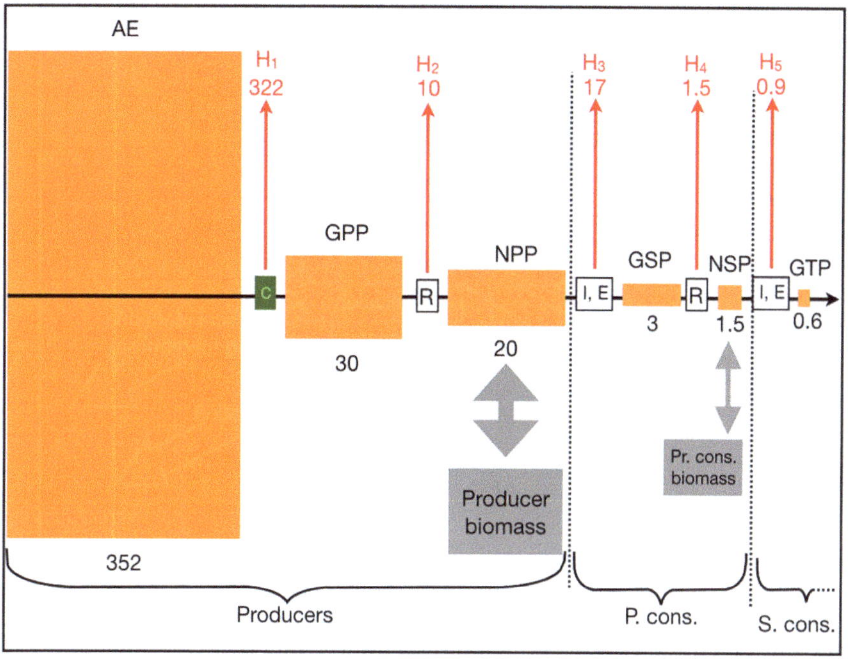

Fig. 9.9 Illustration of the different concepts of production and biomass. In the energy pathway of Fig. 9.8, steps in arrowed axis. Rectangles: *orange*=energy inputs (in kcal · m^{-2} · d^{-1}); *green*=actions of chlorophylls; *white*=action of respiration, ingestion, egestion; *upward red arrows*=energy losses. Energy input: growth solar energy (GSE) (see Fig. 9.8, not represented) = 4000; energy absorbed by primary producers (e.g. plants) (AE)=352; gross primary production (GPP)=30; net primary production (NPP)=20; gross secondary production (GSP)=3; net secondary production (NSP)=1.5; gross tertiary production (GTP)=0.6; net tertiary production, and productions of consumers of higher degrees not represented; H_1, flow of energy degraded into heat resulting from activities of photosynthesis, 3; H_2 and H_4, ibid from respiration; H_3 and H_5 ibid from ingestion and egestion. *P. cons.* primary consumers, *S. cons.* secondary consumers

To measure the primary net production of a given biocoenosis, several variables are used that are equivalent:

1. the mass in grams of the assimilated carbon atoms,
2. the mass in grams of plant tissue,
3. the energetical equivalents of these measures.

These three magnitudes are strongly correlated.

Definition 9.22 The productivity of a given ecological system E (a population, a biocoenosis, an ecosystem, a habitat of any form, a biome, or even the entire biosphere) is defined as the rate of generation of its biomass, i.e. its biomass generated per unit surface (or volume) per unit time.

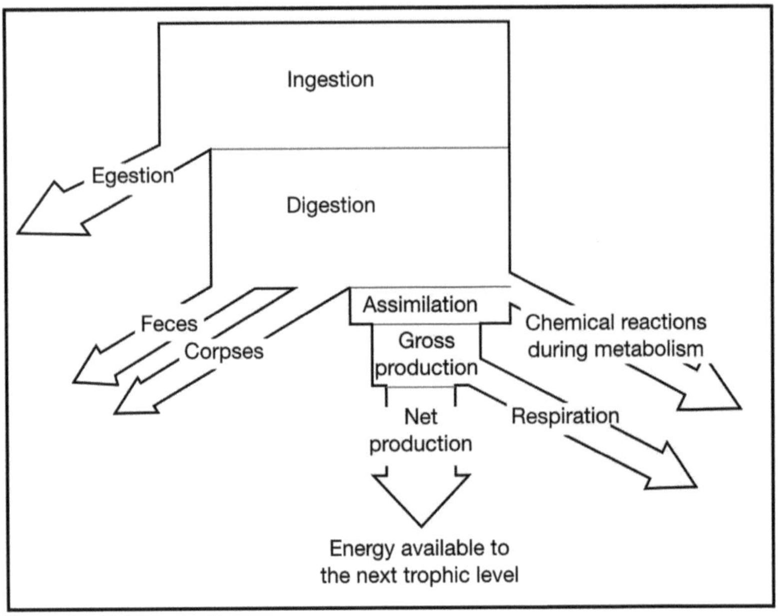

Fig. 9.10 Energy flow through a group of *n*th level consumers in a given ecosystem

Definition 9.23

- The primary productivity of E is defined as the productivity of its primary producers.
- The secondary productivity of E is defined as the productivity of its consumers.

Definition 9.24

- The gross productivity of E is defined as the rate of generation of its wet biomass.
- The net productivity of E is defined as the rate of generation of its dry biomass.

Theorem 9.8 *The net primary productivity of E is the net productivity of its primary producers (plants in terrestrial ecosystems); and the net secondary productivity of E is the net productivity of its consumers (animals and fungi in terrestrial ecosystems).*

It is common to characterise E with its net primary productivity since the net secondary productivity is a portion of the latter. To estimate the net productivity of E, (1) the dry mass of a sample, taken from a plant of E, is first calculated by means of a calorimeter or s spectrometer, (2) then the dry mass of the primary producers of E is estimated by extrapolation. This estimation is then divided by the time of the observation period and by the area of E.

Dry biomass can be evaluated by means of a *calorimeter*. The latter is a small device that measures the heat released by the total combustion of a small amount of plant material. It encloses a chamber that contains the sample to be burnt. High pressure oxygen is injected in this chamber so as to obtain the complete combustion of the sample. A socket filled with water encompasses the chamber. The liquid absorbs the heat produced, which raises the temperature of a given amount of the socket water. The temperature increase measured by a thermometer dipped into the socket water corresponds to a direct measure of the heat absorbed (via an equation, an annexe device, etc). The results obtained on 1 g of plant material with such an equipment show values comparable to the well-known theoretical conversion result of 1 g assimilated carbon, i.e. 39 kJ).

Dry biomass can also be measured with a *spectrometer*, which is based on a totally different principle. It functions with light absorption. The amount of carbon assimilated by a plant is directly related to the absorption of infrared light by its tissues at a given spectral range. First, the amount of assimilated carbon is estimated with this technique on small plant samples. Then, this value is extrapolated to all the plants of the studied habitat by means of mathematical modelling and aritheoremetical operations.

9.4.4 Brief Summary of Other Types of Global Energy Intake

9.4.4.1 Anoxygenic Photosynthesis

Anoxygenic photosynthesis is carried out when solar energy triggers, as initial electron donor, another chemical species than water, generally a sulphur compound like hydrogen sulphide (H_2S) or thiosulfate ($S_2O_3^{2-}$). As a result, the chemical processes launched by this type of photosynthesis will give a final waste product which will not be dioxygen, very frequently sulphur (S). Anoxygenic phototrophs occur in few habitats; most of them are sulphur bacteria that "form dense accumulations only in certain lacustrine environments and in intertidal sandflats" [39, p. 32]. Some of these are found in geothermal sulphur springs. Other prokaryotes practising anoxygenic photosynthesis belong to other groups, either purple non-sulphur bacteria or archaea like *Halobacteria*. These prokaryotes are very diverse, scattered among ten different bacterial and one archaeal phyla [24].

The ecosystems where anoxygenic photosynthesis occur would count for 4% of the global primary production. It has been assessed that the amount of CO_2 fixed by anoxygenic photosynthesis would count for less than 1% of the global primary production. Two functional principles are used in their respective photosynthesis by the anoxygenic phototrophs [39, p. 33]: (1) an apparatus containing bacteriochlorophylls—pigments very similar to chlorophylls but differing from the latter, from the fact that contrary to the latter, they do not release oxygen; (2) a system without photosynthetic pigments called the retinal-based archaeorhodopsin found only among the Halobacteria. In this group, the solar energy that launches this specific type of anoxygenic photosynthesis is converted into a redox gradient across the membrane, thanks to an enzyme, halorhodopsin, that capture light with the aid of dipterpenes, a type of hydrocarbon specific to the group.

9.4.4.2 Chemosynthesis

It is now widely proven that non-negligible portions of the biosphere do not depend on sunlight energy, namely the ecosystems thriving around *hydrothermal vents* in the abyss [15, 19]. These ecosystems primarily take in their energy from redox reactions of chemical compounds endowed with high redox potential,[22] near fault lines on the ocean floor; in addition, it seems now very probable that the first biocoenoses of our planet arose from this energy intake ca 4 billions years ago [8, 29].

This source of energy was discovered relatively recently. Indeed, a marine research team, headed by the American geologist Robert D. Ballard (1942–), to their astonishment, discovered in 1977 a habitat teaming with life along the Galápagos rift, at a depth of 2500 m below sea level, in total darkness. It was an immense discovery, because at that time, it was unconceivable that life could simply exist without solar energy as a primary energy source. In deep ocean, light thus solar energy is absent, so life was then wholly unexpected. Since then, more than 600 abyssal places have been discovered where such oases of life were unveiled [15].

Thus, as we mentioned two paragraphs before, due to their location at sea floor at great depths on hydrothermal vents or *cold-seep sites*, the abyssal ecosystems use as energy source different energy-laden chemical species that have a strong redox potential like dissolved dihydrogen (H_2), hydrogen sulphide (H_2S), ammonia or ferruginous compounds. This energy source is practically unending, as these energy-laden molecules flow permanently from the vents within highly-pressurised upward water streams or smokers (see Footnote 22). We will exemplify this with hydrogen sulphide as the primary energy source of a given abyssal ecosystem. All the chemoautotrophs found in the abyss are barophylic prokaryotes, either bacteria or archaea. They extract their carbon atoms from the carbon dioxide dissolved in water. The chemical equation of the chemosynthesis is:

$$18\,H_2S + 6\,CO_2 + 3\,O_2 \rightarrow \underbrace{C_6H_{12}O_6}_{\text{Glucose}} + 12\,H_2O + 18\,S^{\uparrow} \tag{9.19}$$

Here, such a reaction produces sulphur as a waste product. In addition to the hydrothermal vents, entire ecosystems composed of archaea and bacteria have been discovered still later (around 2000), deep in sediments below seabeds; some of these microorganisms organisms practise oxic and other anoxic chemosynthesis [2, 37].

[22] These chemical compounds are dissolved in very hot water mixed with water vapour issued from chimneys, forming a hot current called smokers that flow across the deep waters close to the ocean floor.

Quizzes and Problems

9.1 Give a short proof of Theorem 9.1: $\mathbf{F_a}$ being constant and ℓ, the distant covered, $W = F\,\ell$.

9.2 Prove the Theorem 9.2: $\mathbf{W_a}$ carried out between two states is equal to the difference of kinetic energy between them.

9.3 Prove that in the case of a point body released at height h above the earth with initial velocity $\mathbf{0}$, based on Fig. 9.2 (p. 191), initial potential energy is entirely transformed into kinetic level at soil level.

9.4 From Corollary 9.1 (p. 189), infer that $W = \int_{(C)} \mathbf{F(s)} \cdot d\mathbf{s}$.

9.5 Let us have a look at Fig. 9.3 (p. 192).
A particle moves in a plane along a trajectory (C), under the influence of a variable force $\mathbf{F}$ from position A_0 to position A_τ, the indices being real numbers representing time instants ($0 \leqslant t \leqslant \tau$). At any instant t, we represent the vector $\mathbf{s}$ describing the displacement from A_0 to A_t. For the sake of simplification, we will admit that the arc $\overset{\frown}{A_0A_t}$ can be decomposed in two complementary line segments, (1) an arc of circle $\overset{\frown}{A_0M}$ [with M ∈ (C) and $A_0M \parallel$ x-axis] of radius r and delineated by a sector of angle θ, determined by its depth h and width c; (2) a straight line segment MA_t (see figure). The coordinates of A_t, whatever $0 \leqslant t \leqslant \tau$, are (x_t, y_t).

(a) Draw a figure where the arcs $\overset{\frown}{A_0M}$ and $\overset{\frown}{A_0A_t}$, the chord A_0M of length c, and the $\overset{\frown}{A_0A_t}$ are represented in a orthogonal coordinate system. Represent also the three parameters determining $\overset{\frown}{A_0M}$. In this frame, sketch the circle C of radius r to which the arc of circle belongs, and draw the origin of C, the two straight line segments OA_0 and OM, and the mediator OK of the triangle OA_0M joining O to H, a point belonging to the chord A_0M. The angle delineating arc of circle $\overset{\frown}{A_0M}$ is θ in radians. The intersection of the mediator mentioned above and the arc of circle $\overset{\frown}{A_0M}$ is K.
(b) The numeric data determining the arc of circle $\overset{\frown}{A_0M}$ are c and r. If A_t is close enough to M to assimilated the arc MA_t to a straight line, calculate ℓ_{0t} the norm of the vector displacement $\mathbf{s}$
(c) Numerical application: $A_0M = 10.50$ cm, HK $-$ 2 cm, $x_0 = 2$ cm and $x_t =$ 15 cm. Calculate the norm of $\mathbf{s}(t)$

9.6 Prove that if univariate state functions $u_1, u_2, \cdots, u_n$ are combined into a multivariate function F, the latter is also a state function.

9.7 An expanding gas enclosed in a rectangularly parallepipedal chamber carries out W against a piston exerting force F. The gas expands distance Δl, and the surface of the plate that is rectangular and forms a side of the chamber is A. The expanding gas sweeps a volume ΔV.

(a) Draw the diagram illustrationg these data
(b) Demonstrate that $W = F(t)\,\Delta V$.
(c) The plate has sides of lengths 5 and 3 cm, and the length covered by the piston is 10 cm . The force is 0.7 N. What is the work carried out by the expanding gas?

9.8 In plants, from the data of Fig. 9.8, p. 209, prove that only ca 3% of the gross solar energy hitting an ecosystem is transformed int useful energy fraction?

References

1. Altieri AH et al (2012) A trophic cascade triggers collapse of a salt-marsh ecosystem with intensive recreational fishing. Ecology 93:1402–1410, https://doi.org/10.1890/11-1314.1
2. Amend JP, LaRowe DE (2016) Ocean sediments – an enormous but underappreciated mcrobial habitat. Microbe 11:427–432.
3. Andréani M, Ménez B (2019) New perspectives on abiotic organic synthesis and processing during hydrothermal alteration of the oceanic lithosphere. In: Orcutt BN et al (eds) Deep carbon: past to present, p. 447–479, Cambridge Univ Press. https://doi.org/10.1017/9781108677950.015
4. Beauchamp DA (1995) Riverine predation on sockeye salmon fry migrating to Lake Washington. N Am J Fisheries Management 15:358–365, https://doi.org/10.1577/1548-8675(1995)015<0358:RPOSSF>2.3.CO;2
5. Blaber M, Anonym (2023) ΔG° and K as functions of temperature. https://chem.libretexts.org/@go/page/24320
6. Bryant DA, Frigaard NU (2006) Prokaryotic photosynthesis and phototrophy illuminated. Trends Microbiol 14:488–496, https://doi.org/10.1016/j.tim.2006.09.001
7. Buranovich BU (2021) Natural sciences and a scientific look at the universe. Int J Innov Eng Res Technol 8:159–162.
8. Canfield DE (1998) A new model for Proterozoic ocean chemistry. Nature 396:450–453, https://doi.org/10.1038/24839
9. Chaisson E (2014) The nature science underlying big history. Sci World J 2014:384912, https://doi.org/10.1155/2014/384912
10. Chen LQ (2019) Chemical potential and Gibbs free energy. MRS Bull 44:520–523, https://doi.org/10.1557/mrs.2019.162
11. Cherif M (2012) Biological stoichiometry: the elements at the heart of biological interactions. In: Innocenti A (ed) Stoichiometry and research - the importance of quantity in medicine, pp. 357–376, IntechOpen, https://doi.org/10.5772/34528
12. Ciuta S et al (2017) Gasification of waste materials. Elsevier.
13. Clausius R (1865) Ueber verschiedene für die Anwendung bequeme Formen der Hauptgleichungen der mechanischen Wärmetheorie. Ann Phys Chem 125: 353–400
14. Cosmidis J, Benzerara K (2022) Why do microbes make minerals? C R Geosci 354:1–39, https://doi.org/10.5802/crgeos.107
15. Dasgupta S et al (2021) Interaction between microbes, minerals, and fluids in deep-sea hydrothermal systems. Minerals 11:1324, https://doi.org/10.3390/min11121324.
16. Davies H, Deeley E (1956) An integrator for measuring the "dry mass" of cells and isolated components. Exp Cell Res 11:169–185, https://doi.org/10.1016/0014-4827(56)90201-4
17. DeVoe (2022) Statements of the second law. University of Maryland. https://chem.libretexts.org/@go/page/20662
18. DeVoe (2024) The system, surroundings, and boundary. University of Maryland. https://chem.libretexts.org/@go/page/20455

19. Dick GJ (2019) The microbiomes of deep-sea hydrothermal vents: distributed globally, shaped locally. Nat Rev 17:271–283, https://doi.org/10.1038/s41579-019-0160-2
20. Drake GWF (2024) "Thermodynamics". In: Encyclopedia Britannica, https://www.britannica.com/science/thermodynamics.
21. Erb T, Zarzycki J (2018) A short history of RubisCO: the rise and fall (?) of Nature's predominant CO_2 fixing enzyme. Cur Opin Biotechnol 49:100–107, https://doi.org/10.1016/j.copbio.2017.07.017.
22. Falkowski PJ, Godfrey LV (2008) Electrons, life and the evolution of earth's oxygen cycle. Phil Trans R Soc B 363:2705–2716, https://doi.org/10.1098/rstb.2008.0054
23. Friedrich B (1987) Genetics of aerobic lithoautotrophs. In: Van Verseveld HW, et al (eds) Microbial Growth on C1 Compounds: Proceedings of the 5th International Symposium, Springer Netherlands, Dordrecht, pp 230–237
24. George DM et al (2020) An overview of anoxygenic phototrophic bacteria and their applications in environmental biotechnology for sustainable Resource recovery. Biotechnol Rep 28:e00563, https://doi.org/10.1016/j.btre.2020.e00563
25. Guidetti P et al (2003) Distribution patterns of sea urchins and barrens in shallow mediterranean rocky reefs impacted by the illegal fishery of the rock-boring mollusc lithophaga lithophaga. Mar Biol 143:1135–1142, https://doi.org/10.1007/s00227-003-1163-z
26. Kamennaya NA et al (2012) Cyanobacteria as biocatalysts for carbonate mineralization. Minerals 2:338–364, https://doi.org/10.3390/min2040338
27. Kittel C et al (1965) Mechanics. Berkeley physics course. Macgraw-hill, New York
28. Landau LD et al (1967) Particle mechanics In: General physics - Mechanics and molecular physics, pp. 1–43, Pergamon Press
29. Lenton TM, Daines SJ (2017) Biogeogeochemical transformations in the history of ocean. Ann Rev Mar Sci 9:31–58, https://doi.org/10.1146/annurev-marine-010816-060521.
30. Libra JA et al (2011) Hydrothermal carbonization of biomass residuals: a comparative review of the chemistry, processes and applications of wet and dry pyrolysis. Biofuels 2:89–124
31. Löffler J, Pape R (2020) Thermal niche predictors of alpine plant species. Ecology 101:e02891, https://doi.org/10.1002/ecy.2891
32. Mahan BH (1970) University Chemistry. Addison-Wesley, Reading (Mass)
33. Münster A (1970) Classical thermodynamics. Transl. from German, Halberstadt ES. Wiley-Interscience Series, Wiley.
34. Niinemets U (2001) Global-scale climatic controls of leaf dry mass per area, density, and thickness in trees and shrubs. Ecology 82:453–469, https://doi.org/10.1890/0012-9658(2001)082[0453:gsccol]2.0.co;2
35. Odum EP, Barrett GW (2005) Fundamentals of ecology, 5th edn. Thomson Brooks, Cole (CA)
36. Oliveira LE (2019) Is there any example of an isolated system in nature? What is the applicability of the second law of thermodynamics? RE&PQJ 17(3):441–446, https://doi.org/10.24084/repqj17.338
37. Orsi WD (2018) Ecology and evolution of seafloor and subseafloor microbial communities. Nat Rev Microbioles 16:671–683, https://doi.org/10.1038/s41579-018-0046-8
38. Ott JB, Boerio-Goates J (2001) Chemical thermodynamics: Principles and Applications. Academic Press, San Diego
39. Overmann J, Garcia-Pichel F (2006) The phototrophic way of life. In: Dworkin M et al (eds) The prokaryotes, vol. 2: ecophysiology and biochemistry, 3rd edn, pp. 32–85, https://doi.org/10.1007/0-387-30742-7_3; Springer, New York.
40. Penninger JM et al (2007) Compressed hydrogen-rich fuel gas (CHFG) from wet biomass by reforming in supercritical water. Int J Hydrog Energy 32:1472–1476, https://doi.org/10.1016/j.ijhydene.2006.10.026
41. Platt T (1986) Primary production of the ocean water column as a function of surface light intensity: algorithms for remote sensing. Deep Sea Res 33:149–163
42. Rajagopalan R, Altekar W (1994) Characterisation and purification of ribulose bisphosphate carboxylase from heterotrophically grown halophilic archaebacterium, Haloferax mediterranei. Eur J Biochem 221:863–869.

43. Ramade F (1982) Elément d'écologie, écologie appliquée: action de l'homme sur la biosphère. Edition Mc Graw-Hill, Paris
44. Ricklefs RE (1990) Ecology. Freeman & Co, New York.
45. Rossel J (1963) Physique générale. Editions du Griffon, Neuchâtel
46. Trenchard H, Perc M (2016) Energy saving mechanisms, collective behavior and the variation range hypothesis in biological systems: a review. Biosystems 147:40–66, https://doi.org/10.1016/j.biosystems.2016.05.010
47. Ward L et al (2018) Evolution of phototrophy in the Chloroflexi phylum driven by horizontal gene transfer. Front Microbiol 9:260, https://doi.org/10.3389/fmicb.2018.00260
48. Xiong D et al (2016) Leaf density explains variation in leaf mass per area in rice between cultivars and nitrogen treatments. Ann Bot 117:963–971, https://doi.org/10.1093/aob/mcw022

Chapter 10
The Material Approach-I: The Biogeochemical Cycles

10.1 Some Necessary Basics of Chemistry

10.1.1 The Basis of Redox Mechanism, and Energy

Any chemical reaction consists of the exchange of electrons between atoms during bonds that tie or untie between them, while these atoms organise themselves in distinct molecules. Atoms are the smallest discrete material particles taking part in chemical processes. They fall into at most 137 different chemical species called elements [15] characterised by a number of chemical properties, i.e. *atomic weight*, electron configuration. One hundred and eighteen elements are known and arranged in a table called the periodic table or Mendeleev's table (cf Fig. 10.1). In the viewpoint of a nuclear physicist, atoms are, of course, not the smallest material units and can be subdivided into much more little particles, but these are of no interest to the chemist and thus to the biologist and ecologist, except for nucleons (neutrons and protons) and electrons that are handled as properties of atoms. From Planck and Bohr's models, still accepted as the basis of chemistry nowadays,[1] each atom possesses a nucleus composed of n_P protons and n_N neutrons. Protons are particles bearing positive electric charges, whereas neutrons are electrically neutral. They are both approximatively the same mass [22]—ca 1.67×10^{-24} g. Basically, in the atomic nucleus of a given element, $n_N = n_P$. However, variants exist within the atomic nucleus of many elements in rare but non-negligible proportions. The basic form of an element and its variants are called isotopes and will be defined by the values of n_N respective to n_P ($n_N = n_p + 1$, $n_N = n_p + 2$, or even $n_N = n_p + 3$).

Supplementary Information The online version contains supplementary material available at https://doi.org/10.1007/978-3-031-90656-5_10.

[1] ...though the mathematical explanation of the fine atomic processes is based upon quantum physics that will not be approached here.

M. Salomon, *Integrative Ecology*, https://doi.org/10.1007/978-3-031-90656-5_10

1 **H** 1.00																	2 **He** 4.00
3 **Li** 6.94	4 **Be** 9.01											5 **Bo** 10.8	6 **C** 12.0	7 **N** 14.0	8 **O** 16.0	9 **F** 19.0	10 **Ne** 20.2
11 **Na** 23.0	12 **Mg** 24.3											13 **Al** 27.0	14 **Si** 28.1	15 **P** 31.0	16 **S** 32.1	17 **Cl** 35.5	18 **Ar** 40.0
19 **K** 39.1	20 **Ca** 43.1	21 **Sc** 45.0	22 **Ti** 47.9	23 **V** 50.9	24 **Cr** 52.0	25 **Mn** 54.9	26 **Fe** 55.8	27 **Co** 58.9	28 **Ni** 58.7	29 **Cu** 63.5	30 **Zn** 65.4	31 **Ga** 69.7	32 **Ge** 72.6	33 **As** 74.9	34 **Se** 79.0	35 **Br** 79.9	36 **Kr** 83.8
37 **Rb** 85.5	38 **Sr** 87.6	39 **Y** 88.9	40 **Zr** 91.2	41 **Nb** 92.9	42 **Mo** 96	43 **Tc** 98	44 **Ru** 101	45 **Rh** 103	46 **Pd** 106	47 **Ag** 108	48 **Cd** 112	49 **In** 115	50 **Sn** 119	51 **Sb** 122	52 **Te** 128	53 **I** 127	54 **Xe** 131
55 **Cs** 133	56 **Ba** 137	57-71 Ser 1	72 **Hf** 178	73 **Ta** 181	74 **W** 184	75 **Re** 186	76 **Os** 186	77 **Ir** 192	78 **Pt** 195	79 **Au** 197	80 **Hg** 201	81 **Tl** 204	82 **Pb** 207	83 **Bi** 209	84 **Po** 209	85 **At** 210	86 **Rn** 222
87 **Fr** 223	88 **Ra** 226	89-103 Ser 2	104 **Rf** 267	105 **Db** 262	106 **Sg** 269	107 **Bh** 264	108 **Hs** 269	109 **Mt** 278	110 **Ds** 281	111 **Rg** 282	112 **Cn** 285	113 **Nh** 284	114 **Fl** 289	115 **Mc** 289	116 **Lv** 292	117 **Ts** 294	118 **Og** 294
		Ser 1	57 **La** 139	58 **Ce** 140	59 **Pr** 141	60 **Nd** 144	61 **Pm** 145	62 **Sm** 150	63 **Eu** 152	64 **Gd** 157	65 **Tb** 159	66 **Dy** 163	67 **Ho** 165	68 **Er** 167	69 **Tm** 169	70 **Yb** 173	71 **Lu** 175
		Ser 2	89 **Ac** 227	90 **Th** 232	91 **Pa** 231	92 **U** 238	93 **Np** 237	94 **Pu** 244	95 **Am** 243	96 **Cm** 247	97 **Bk** 247	98 **Cf** 251	99 **Es** 252	100 **Fm** 257	101 **Md** 258	102 **No** 259	103 **Lr** 262

Fig. 10.1 Mendeleev's table. In each table cell, constructed in two rows, the number to the left of the element symbol (in bold) is its atomic number Z, i.e. the number of its protons; and the number below, its atomic mass units expressed in atomic mass units (AMU). Atom symbols and their full names are displayed in Table 10.1

For example, carbon (C) is an element with three isotopes: ^{12}C (the basic form with 6 protons and 6 neutrons), ^{13}C (with 6 protons and 7 neutrons) and ^{14}C (with 6 protons and 8 neutrons); to appreciate the relative frequencies of these isotopes, ^{12}C amounts to ca 98.9 %, ^{13}C, to nearly 1.1%, and ^{14}C only to trace proportions though this isotope is far from negligible in its effect! Indeed, this very rare isotope is extremely useful in biology, notably in evolutionary biology, as a tool to date subfossils (remains of organisms that are less than 40,000 years of age). Hereunder, I present a table with the meaning of the chemical symbols (cf. Table 10.1).

10.1.2 Electrons, Orbitals

An electron is the elemental electric negative charge of an atom equal, in absolute value, to the positive electric charge of the atom, the proton. Globally, in an atom in its ground state, (1) the number of electrons (n_E) and the number of protons (n_P) of an atom are equal and (2) the electrons are "located" in the orbitals or shells of the atoms, as defined just below.

Chemists view an atomic orbital in a given chemical element as a virtual space surrounding the nucleus where a definite number of electrons are probable to be found. Orbitals can be visualised as superimposed electron shells, though in reality, this image constitutes a useful metaphor.[2] Rutherford's model metaphorically represents this model. The latter depicts an atom as concentric circles whose centre is the nucleus, concentrating all the protons, and circles the shells, numbered increasingly from 1 to n with increasing distance from the centre (cf. Fig. 10.2). The maximal number of electrons in shell n is $2n^2$. If a shell contains its maximal

[2] Nuclear physicists grasp an atomic orbital as a wave function.

Table 10.1 Elements with their atomic number, symbols and names

Z	Abb.	Name	Z	Abb.	Name	Z	Abb.	Name
1	H	Hydrogen	21	Sc	Scandium	41	Nb	Niobium
2	He	Helium	22	Ti	Titanium	42	Mo	Molybdenum
3	Li	Lithium	23	V	Vanadium	43	Tc	Technetium
4	Be	Beryllium	24	Cr	Chromium	44	Ru	Ruthenium
5	Bo	Boron	25	Mn	Manganese	45	Rh	Rhodium
6	C	Carbon	26	Fe	Iron	46	Pd	Palladium
7	N	Nitrogen	27	Co	Cobalt	47	Ag	Silver
8	O	Oxygen	28	Ni	Nickel	48	Cd	Cadmium
9	F	Fluorine	29	Cu	Copper	49	In	Indium
10	Ne	Neon	30	Zn	Zinc	50	Sn	Tin
11	Na	Sodium	31	Ga	Gallium	51	Sb	Antimony
12	Mg	Magnesium	32	Ge	Germanium	52	Te	Tellurium
13	Al	Aluminium	33	As	Arsenic	53	I	Iodine
14	Si	Silicon	34	Se	Selenium	54	Xe	Xenon
15	P	Phosphorus	35	Br	Bromine	55	Cs	Cesium
16	S	Sulphur	36	K	Krypton	56	Ba	Barium
17	Cl	Chlorine	37	Rb	Rubidium	57	La	Lanthanum
18	Ar	Argon	38	Sr	Strontium	58	Ce	Cerium
19	K	Potassium	39	Y	Yttrium	59	Pr	Praseodymium
20	Ca	Calcium	40	Zr	Zirconium	60	Nd	Neodynum

Z	Abb.	Name	Z	Abb.	Name	Z	Abb.	Name
61	Pm	Promethium	81	Tl	Thalium	101	Md	Mendelevium
62	Sm	Samarium	82	Pb	Lead	102	No	Nobelium
63	Eu	Europium	83	Bi	Bismuth	103	Lr	Lawrencium
64	Gd	Gadolinium	84	Po	Polonium	104	Rf	Rutherfordium
65	Tb	Terbium	85	At	Astatine	105	Db	Dubnium
66	Dy	Dysprosium	86	Rn	Radon	106	Sg	Seaborgium
67	Ho	Holmium	87	Fr	Francium	107	Bh	Bohrium
68	Er	Erbium	88	Ra	Radium	108	Hs	Hassium
69	Tm	Thulium	89	Ac	Actinium	109	Mt	Meitnerium
70	Yb	Ytterbium	90	Th	Thorium	110	Ds	Darmstadtium
71	Lu	Lutetium	91	Pa	Protactinium	111	Rg	Roentgenium
72	Hf	Hafnium	92	U	Uranium	112	Cn	Copernicium
73	Ta	Tantalum	93	Np	Neptunium	113	NH	Nihonium
74	W	Tungsten	94	Pu	Plutonium	114	Fl	Flerovium
75	Re	Rhenium	95	Am	Americium	115	Uup	Ununpentium
76	Os	Osmium	96	Cm	Curium	116	Lv	Livermorium
77	Ir	Iridium	97	Bk	Berkelium	117	Ts	Tenessine
78	Pt	Platinum	98	Cf	Californium	118	Og	Oganesson
79	Au	Gold	99	Es	Einsteinium			
80	Hg	Mercury	100	Fm	Fermium			

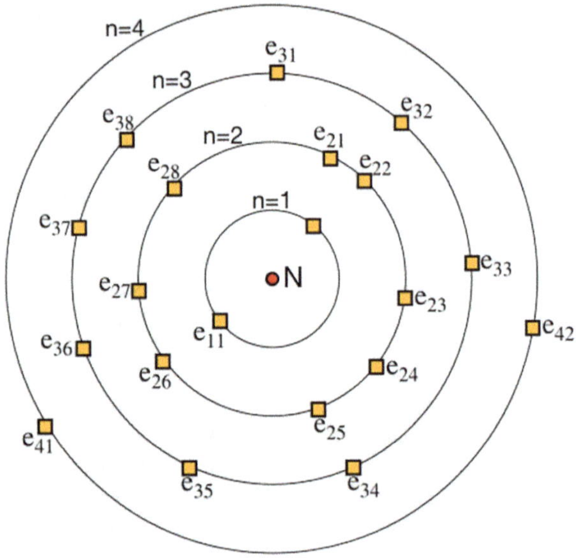

Fig. 10.2 Rutherford's metaphoric model of the organisation of an atom. Example of the calcium atom. N = nucleus (*red circle*), e_{ij} = electron i occupying shell j (*yellow squares*). Shells 1 and 2 are saturated, shells 3 ($8 < 32e^-$) and 4 (valence orbital, ($2 < 8e^-$) are not but only the valence orbital is chemically active

number of electrons, it is said to be saturated. The outer shell which plays the most active chemical role is called the valence shell or valence orbital.

The model describes the stable electron configuration of an atom in its electrically neutral form with a definite pattern: the outer shell, which is called the *valence shell*, is composed of 1-8 electrons, depending on the column location in Mendeleev's table, whereas the inner shells are determined by the row location (cf. Fig. 10.1).

The elements of the first row [hydrogen (H) and helium (He)] have no inner shell. Their electron(s) are located in their valence shell, the only shell available to these elements. The number of inner shells in an element's atom is given by its row; the elements of the n-th row have $n - 1$ inner shells. For example, an element of the 2nd row like boron (Bo) has 3 inner shell, an element of the 4th row like zinc (Zn) has 3 inner shells, an element of the 7th row like Radium (Ra) has 6 inner shells.

Let us designate the columns of the upper subtable (the first seven lines), from left to right, with numbers 1-18. Column 1 in yellow (H, Li, ..., Fr) contains the elements whose valence shell contains 1 electron. The green block comprises elements whose valence shell contains 3-8 electrons:

- 3 for the elements of column 13 (Bo, Al, ..., Uut),
- 4 for those of column 14 (C, Si, ..., Fl),
- ...
- 8 for those of column 1(Ne, Ar, ..., Og)].

The remaining elements (represented in the white cells of the table), called transition metals, are all characterised by a valence shell containing 1 or 2 electrons.

The Z number of an element represents its number of electrons when the atom is in its ground state. All the first $n - 2$ inner shells of the atoms of elements of the 2nd-7th row are saturated. The reconstruction of the $n - 1$ inner shells of the transition metals (white squares of the Mendeleev's table) is obtained with $Z - \nu$, where ν is the valence number (number of electrons of the valence orbital in the ground state).

Each orbital corresponds to a given level (or quantum) of negative potential energy. The farther the orbital is from the nucleus, the more positive (in fact, the weaker) its potential energy is, the least energy, in probability, one needs to rip off its electrons from their orbital.

10.2 Principle of Matter Recycling

The chemical species—elements, simple compounds or ions—needed by the ecosystems for their construction and energy come from two proximal sources: (1) the atmosphere or the hydrosphere on the one hand, and (2) the soil on the other hand. Their distal origins lie in the superficial layers of the bedrock. These chemical species enter the *abiotic factor system* (AFS) of the ecosystems, at disposal of the autotrophs. We already saw that there was a limited choice of chemical species available to the ecosystems for their needs and that the living beings themselves have been very active since the origin of life to select and retain them (cf. Sect. 2.5).

Crucial *chemical species* are available to the ecosystems in the atmosphere or hydrosphere as gases or solutes, notably, water (H_2O), dioxygen (O_2), dinitrogen (N_2), carbon dioxide (CO_2), dinitrogen (N_2). Others, as important, are found by the ecosystems in the soil system, generally in the form of ions or mineral salts, e.g. chemical species made up of atoms of elements like potassium (K), sulphur (S), iron (Fe), magnesium (Mg), etc. The chemical species taken from atmosphere or hydrosphere are called atmospheric, and those extracted from the soil, lithospheric.

Life and by extension biocoenoses are based on six major elements constituting the building blocks of the main organic macromolecules, namely proteins, nucleic acids, lipids and carbohydrates [13]. These six elements are (i) four atmospheric elements: hydrogen (H), carbon (C), oxygen (O) and nitrogen (N) and (2) two lithospheric elements: phosphorus (P) and sulphur (S).

We will study the cycles of water and of the elements C, O, N, P and S, in Sects. 10.5–10.9. It has been shown in recent articles that the cycles of these chemicals are intimately linked [13].

The elements constituting these chemical species enter a given ecosystem $\mathcal{E}$, transit through it, leave it, and return to $\mathcal{E}$ after an approximately periodical time. It is noteworthy that the periodicity of the atmospheric chemical species is much longer than that of the mineral chemical species. Let us follow an element E to illustrate the cycle just mentioned. Element E enters the ecosystem in its mineral form—that is, as a single atom, ion, or atom within a simple compound. Inorganic E is then taken in by the producers that incorporate it in one of the organic molecules they biosynthesise; the element takes its organic form. In this form, E transits through the food web (via the consumers) from one trophic level to another. Eventually, element E is released in its initial mineral form (or inorganic form) after decomposition and mineralisation. The cycle is completed and ready to start again. This transformation is carried out by respiration, evapotranspiration and the activities of decomposing bacteria or saprophytic fungi. Matter recycling occurs in all chemical species that fuel the ecosystems.

Such cycles ensure almost a material constancy. Little of the chemical elements and simple components that fuel the ecosystems are lost when human activities do not interfere. This feature is an important factor contributing to the dynamic equilibrium of the element that periodically reoccurs (a depleted element will eventually be replaced). Matter recycling generally preserves the elements within the ecosystems approximately equimolecularly. However, several elements like carbon (C) and calcium (Ca) seem to be exceptions. Indeed, massive amounts of these elements were trapped and incorporated into dead organic matter deposits or rocky sediments. However, compared to the total quantity of these elements on Earth, these quantities account for a small proportion of them [32, p. 47]).

Coal and oil were lengthily elaborated in the sediments of lakes and shallow seas during the Carboniferous (between −358.9 and −298.9 Ma), when surges of oxygen depletion had occurred, causing an increase of anaerobic conditions. These events slowed down the activities of the decomposers. As a result, huge masses of undecomposed organic material stemming from plants accumulated beneath the Earth's crust. It was this material that eventually turned into coal or oil. The carbon atoms trapped in that organic material could not be released in its mineral form into the abiotic factor system. They could not be recycled, thus causing them to be lost to the ecosystems.

A large quantity of carbon and calcium originated from the tests and shells of Mesozoic marine organisms. After their deaths, these chemical elements precipitated into calcium carbonate, which formed layers in the sea beds. That was how lime deposits formed at the bottom of the *Mesozoic* seas. Some of the sea beds that contained these lime deposits were subjected to the mountainous uplifts of the Cenozoic and were raised well above sea level, thus on dry land and, in a number of cases, at high altitudes close to mountain summits. We will see below the details of the carbon cycles responsible for the hydrocarbon deposits and limestone sedimentation, as well as the calcium cycle related to lime formation.

It was already mentioned just above that when there is no human interference, matter is roughly recycled. However, nowadays, the human impact on the environment, is very intense. In particular, vast deforestation, crude oil and coal exploitation as the main energy supply of humankind, linear extraction of material for industrial activities and intensive modes of cultivation entail a direct depletion of chemical elements (some becoming very rare) and wide-scale erosion and leaching that sweep away soil decaying organic matter and mineral elements. As a result, nowadays, matter is hardly or very poorly recycled in many parts of the world. However, organic matter loss has been negligible in "normal" ecological conditions, as it used to be everywhere on Earth before the industrial era. In rare cases, when this loss is significant (again in "normal" ecological conditions), for example as in the case of carbon or calcium as seen above, the depletion has taken place extremely slowly and continuously during a couple of hundred million years; which explains the colossal mass of hydrocarbon and limestone existing at present.

We will explain with more details, in Sect. 10.4 and onwards, the cycles of different chemical species playing a great role in the biogeochemical cycles. To understand their dynamics, it is useful to have an overview of the distal processes, that is the origin of the chemical elements before their circulations in their cycles in the ecosystems. As we already said it in the beginning of this section, they have their origin in the bedrock.

10.3 The Distal Origin of All the Telluric Chemical Species Used by the Ecosystems

10.3.1 The Bedrock

Bedrock forms the *Earth's crust*, constituted of solid rocks generally located under a very thin layer of rock debris and stony dust clouds called *regolith* [26]. The latter arose from various weathering (biological, physical, chemical) or volcanic activities. An interesting part of regolith is *pedolith*, the material forming the soil, which comprises (1) soil granulations (gravel, sand, silt or clay particles) forming the framework of the soils and (2) most mineral chemical species (see Sect. 11.3.2). Bedrock is a set of lithified rocks, i.e. rocks whose components have been compacted under pressure of expelled connate fluids, i.e. liquids trapped in the pores of sedimentary rocks.[3]

Figure 10.3 represents a section of the Earth showing that its peripheral crust, gathering nearly all the rocks known directly to science, is a very thin film (compared with the radius of the Earth, i.e. 6378 km). The radial thickness of the Earth's crust has a mean value of ca 35 km [25], ranging from 0 km to a maximum of 100 km but generally between 5 and 75 km [31].[4]

Our meagre knowledge of the layers below the Earth's crust (the mantle—asthenosphere plus lower mantle—and the core) can only be inferred from the interpretation of seismographic studies, lava upsurge, and modeling based on physical theories and *in vitro* experiments [8]. The detailed geology of the core is very poorly known.

Table 10.2 displays the estimated radial thicknesses and volumes of the four main layers depicted in Fig. 10.3, relative to Earth's radius and volume, respectively. (The

[3] The different rock types, among which the sedimentary rocks are defined in Sect. 10.3.3.

[4] The width of the Earth's crust is highly variable, being much thinner in seabeds (5–10 km) than in emerged land masses (30–50 km)—being the thickest in mountainous areas (up to 75 km and in rare cases up to 100 km).

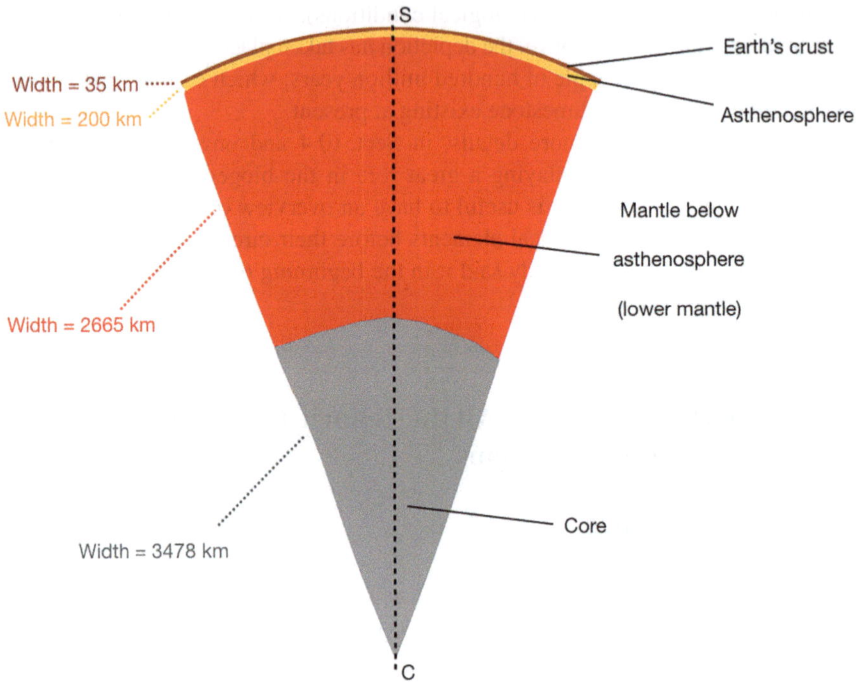

Fig. 10.3 Wedge-shaped schematic cross-section of the Earth. (1) *Brown circumference* = *Earth's crust* (bedrock); (2) *golden beige arc* underneath = *asthenosphere* = ductile "roof" of the mantle, partly composed of molten rock, on which mantle convection (i.e. the movement of continental plates in the course of geological time) takes place; (3) *red layer* = *the lower mantle* = solid and very dense block of silicate rock; (4) *grey wedge* = *core* = block made of iron and nickel, liquid above and solid below. There is a gradient of temperature and pressure from surface (S) (ca. 10° C) to centre (O) (ca. 5000° C)

Table 10.2 Relative radial thicknesses and volumes of the four main layers of the Earth

Earth layers	Relative thickness in %	Relative volume in %
Earth's crust	0.55	1.64
Asthenosphere	3.14	9.01
Lower mantle	41.78	73.13
Core	54.53	16.22

Earth has been assimilated to a sphere, the three outer layers to spheric shells, and the core to a sphere—all these layers sharing the same centre O).

These figures are values at first approximation, which need to be slightly refined (the values have been taken from different sources and are not always very homogeneous). Still, we can have a fair picture. The Earth's crust is indeed very thin, and its relative volume tiny; the relative thickness and relative volume of the bedrock account for around 1%. We will now consider the composition of the bedrock, the source of the chemical species of ecosystems.

10.3.2 The Nature of Bedrock

Bedrock is a set of rocks—solid masses composed of minerals (cf. Definition 10.1), mineraloids (cf. Definition 10.2), and/or aggregates of smaller rocky components .[5] The latter has in turn the same structuration, and this can go on recursively until the smallest rocky component, eventually composed of minerals and/or mineraloids.[6]

10.3.2.1 First, Let Us See What Is a Mineral

Definition 10.1 A mineral is a *chemical compound* with four main properties:

(1) its chemical formula is known,
(2) it is inorganic,
(3) it is found *in natura* and, most importantly,
(4) it possesses a crystalline structure, and thus is a crystal (cf. Definition 10.3 in Sect. 10.3.2.3).

Applying Definitions 10.1 and 10.3 (cf. Sect. 10.3.2.3), a mineral is always a crystal, but a crystal not necessarily a mineral. For example:

- Proteins, nucleic acids or sugars are crystals but not minerals because they are organic molecules. They do not meet property (2).
- Medicines like synthetic drugs are also crystals that are not minerals. Indeed, they are manufactured and thus do not occur in the natural world; they do not meet property (3).

10.3.2.2 Next, What Is a Mineraloid?

Definition 10.2 A mineraloid is, according to the internet platform 'mindat.org',[7]

> a mineral-like naturally occurring geological material which is non-crystalline or not definite enough in chemical composition or physical properties to be considered a mineral.

Referring to Definition 10.1, a mineraloid is thus partially a mineral, i.e. sharing property (3) but not one of the other properties, i.e. (1), (2) or (4).[8] We will give a number of examples of mineraloids in Sect. 10.3.2.4.

---

[5] The distinction between mineral and mineraloid is debated. Many mineralogists tend to disregard the concept of 'mineraloid', but most geologists find this concept useful.

[6] This structuration is reminiscent of the concept of subsumption (cf. Sect. 6.4).

[7] Run by the Hudson Institute of Mineralogy.

[8] Solid substances that are neither minerals nor mineraloids exist because they are not crystalline and are manufactured, thus not existing in the natural world, e.g. glass and plastic.

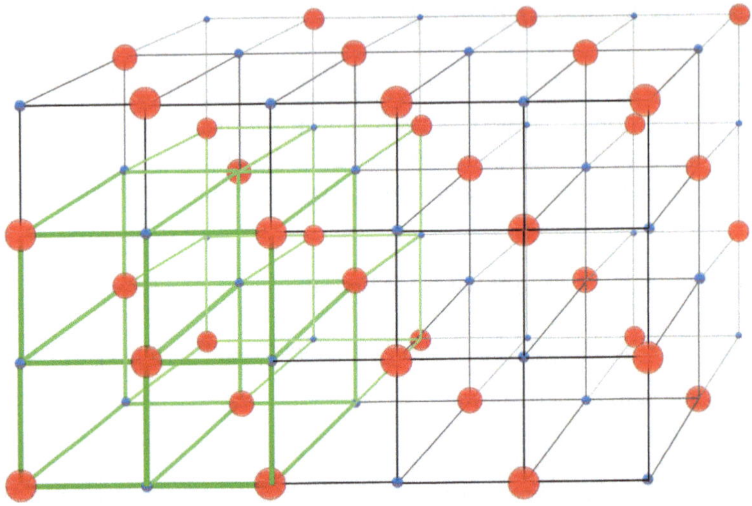

Fig. 10.4 Microstructure of a crystal: the quicklime CaO. The figure is a 3D representation of a crystal. CaO is an ionic crystal, i.e. the elementary constituents are the ions Ca^{2+} (the *large red spheres*) and O^{2-} (the *small blue spheres*) (calcium's atom is about three times bulkier than oxygen's; see Table 10.1. Hence, the representations of the two types of spheres. An example of a cell unit is the cube wrapped by the *green highlighted* links.)

10.3.2.3 What Is a Crystal

Definition 10.3 Crystal can be defined twofold:

- Microscopically, a crystal is a physical structure characterised by the periodic arrangement of its constituents, the unit cells, that are regularly repeated in a three-dimensional lattice. The unit cells are made up of elementary components, i.e. atoms, ions or molecules, and constitute the smallest crystal structure that is repeated identically and periodically along the lattice (see Fig. 10.4). The unit cells of a crystal may be composed of one or several chemical species.
- Macroscopically, crystals are the basic structure of true solids, i.e. materials that are rigid, incompressible, endowed with a characteristic geometry (as well externally as internally) and, importantly, anisotropic. Crystals have definite shapes, notably flat faces whose lead angles are characteristic of the species to which they belong. For a more detailed account, see Mahan [24, ch. 3].

Crystals are generated from a single unit cell, the seed. In the beginning, the seed makes a copy and forms with it a link. Then, the seed and the copy generate new copies on their own and build similar links. The unit cells thus form a network moulded into a lattice whose shape is specific to the crystal species.

10.3.2.4 The Minerals and Mineraloids Composing Rocks

Now, we will stress the importance of the difference between the concepts of 'mineral' and 'mineraloid', which is a main issue in understanding the different types of rocks. Solids *senso latu* are rigid and incompressible due to atoms held together with forces of very great magnitude [24]. On the opposite, gases and liquids are fluid and compressible, with atoms that are loose and not orderly distributed in a given volume[9] because of the weak bonds between their constitutive atoms. However, among solids, one can separate the true solids from the amorphous ones, the former possessing and the latter being devoid of a crystalline structure. The atoms of amorphous solids do not fit in a lattice and share with the fluids (1) an unorderly arrangement and (2) spatially a relatively large degree of freedom of their atoms.

- Thus, minerals are true solids as they are crystalline. A mineral of one type of crystal—thus generated from one single seed—is said to be *monocrystalline*. On the contrary, a mineral produced from different seeds is called *polycrystalline*. Rocks composed only of minerals are called *crystalline rocks*. Tables 10.3 and 10.4 display a list of the most common minerals in solid rocks, that is, in the Earth's crust.
- On the other hand, mineraloids are amorphous or crystalline solids that are organic. The first type of mineraloid lacks a crystalline structure, e.g. obsidian or opal. The second type of mineraloid can be exemplified by perl or petrolium. Rocks, composed partly or totally of mineraloids, are called amorphous rocks. Mineraloids have variable and more complex chemical compositions than minerals, another feature that separates them from minerals. Table 10.5 presents a list of several mineraloids.

10.3.2.5 The Most Frequent Elements in the Bedrock

Table 10.6, borrowed from a website article [31], displays the chemical elements most frequently found in the bedrock's minerals or mineraloids.

Table 10.6 explains why most of the chemical elements represented in it are called mineral elements. The oxygen (paradoxically the most frequent mineral element in the Earth's crust) and hydrogen present in the rocks do not participate in the material recycling of the ecosystems; they are trapped in the rock's composition.[10] Nevertheless, most of the other mineral elements play a crucial role

[9] In gases, the atoms are distributed randomly and shifted by Brownian motion. In liquids, they glide on one another disorderly if not set into motion by an external force.

[10] The oxygen entering the biogeochemical cycle is biogenic; it has been derived from the activities of photosynthetic organisms, mainly cyanobacteria and plants, since about 2.5 billion years ago. The hydrogen needed by the organisms, thus the ecosystems, comes from water (see Sect. 10.4).

Table 10.3 The most common mineral classes on the earth's crust-(1) silicate minerals [31]. Under each class, its % relative to all the minerals of the earth's crust

Classes of minerals	Species	Chemical formula	Characterisation
Plagioclase (Feldspar) 39%	Albite	$Na(AlSi_3O_8)$	Sodium- or calcium-rich feldspar
	Anorthite	$Ca(Al_2Si_2O_8)$	
	Labradorite	$(Ca, Na)[Al(Al, Si)Si_2O_8]$	
K-feldspar (Feldspar) 12%	Microcline	$K(AlSi_3O_8)$	Potassium-rich feldspar
	Orthoclase	Same formula	
	Sanidine	Same formula	
Quartz 12%	Quartz	SiO_2	Hard, resists chemical weathering
Pyroxene 11%	Diopside	$CaMgSi_2O_6$	Share chains of silicate tetrahedra
	Enstatite	$Mg_2Si_2O_6$	
	Ferrosilite	$FeSiO_3$	
	Hedenbergite	$CaFe^{2+}Si_2O_6$	
Mica 5%	Annite	$KFe_3^{2+}(AlSi_3O_{10})(OH)_2$	Id., but with Al, K 2 and (OH)
	Muscovite	$KAl_2(AlSi_3O_{10})(OH)_2$	
	Phlogopite	$KMg_3(AlSi_3O_{10})(OH)_2$	
Amphiboles 5%	Glaucophane	$Na_2Mg_3Al_2Si_8O_{22}(OH_2)$	With a double chain of silicate tetrahedra
	Grünerite	$Fe_7Si_8O_{22}(OH_2)$	
	Tremolite	$Ca_2Mg_5Si_8O_{22}(OH_2)$	
Clay minerals <= 5%	Kaolinite	$Al_2(OH_4)Si_2O_5$	Hydrous aluminium phyllosilicates
	Pyrophyllite	$Al_4Si_8O_{20}(OH_4)$	
	Talc	$Mg_3Si_4O_{10}(OH_4)$	
Other silecate minerals 3%	Fayalite	Fe_2SiO_4	Olivine, brought up in lava
	Forsterite	Mg_2SiO_4	

in the organisms, namely silicon, iron, calcium, sodium, potassium, magnesium, phosphorus, manganese (as we can see in Table 10.6), and other rarer but still very important ones like sulphur or zinc. Some of these are well represented in the rocks (like the series silicon-magnesium); others are less frequent (the other elements mentioned in Table 10.6 for example). But all exist in the organisms only in very small quantities compared to atmospheric elements like carbon, oxygen, nitrogen and hydrogen. They are as vital as these for the organisms and indispensable in proteins, nucleic acids, cell machinery, and metabolic and physiological pathways. Remarkably, carbon, the most central chemical element of living beings, is absent in Table 10.6. Indeed, this element as an inorganic structure of the bedrock is extremely rare (it accounts for ca 0.02% of all the elements of the Earth's crust). We will see that all the carbon necessary for life stems from atmospheric carbon dioxide, which distally originates from the magma via volcanic eruptions.

Table 10.4 The most common mineral classes on the earth's crust-(2) non-silicate minerals [14, 27, 31]. Accounts for 8% relative of the minerals present in the earth's crust

Classes of minerals	Species	Chemical formula	Characterisation of the classes
Carbonate minerals	Calcite	$CaCO_3$	Minerals that contain carbonate ions—source of mineral carbon
	Magnesite	$MgCO_3$	
	Dolomite	$CaMg(CO_3)_2$	
	Azurite	$Ca_3(CO_3)_2(OH)_2$	
	Monohydrocalcite	$CaCO_3.H_2O$	
Oxide minerals	Cuprite	Cu_2O	Oxide anions bonded to one or more metal alloys
	Ice	H_2O	
	Hematite	Fe_2O_3	
	Chromite	$FeCr_2O_4$	
	Brucite	$Mg(OH)_2$	
Sulphide minerals	Pyrite	FeS_2	Minerals containing sulphide or persulphide
	Pyrrhotite	$Fe_{1-x}S,\ x \in [0, 0.2]$	
	Astenopyrite	$FeAsS$	
	Proustite	Ag_3AsS_3	

Table 10.5 List of several mineraloids with their characterics

Mineraloid species	Crystalline	Organic	Type of material
Amber	No	Yes	Fossilised tree resin
Chlorophaeite	No	No	Hydrous amorphous pitch-like material of deuteric origin [29]
Jet	No	Yes	Very compact coal. Fossiled wood.
Lechatelierite	No	No	Amorphous glass, obtained from meltting of quartz during lightning strikes
Limonite	No	No	Iron ore of varying chemical composition. Unorderly mixture of hydrated minerals.
Obsidian	No	No	Volcanic glass
Opal	No	No	Amorphous glassy non-crystalline silicon dioxide
Pelagonite	No	No	Light orange dust stemming from the interaction of water with volcanic glass or basalt melt
Pearl	Yes	Yes	Calcium carbonate
Petrolium	No (It is a liquid)	Yes	Hydrocarbons from fossilised plants

Data from the wikipedia article "Mineraloid" and derived articles (web site: 'en.wikipedia.org/wiki/Mineraloid'); under licence CC BY-SA 3.0

The relative quantities of the mineral elements, depending on the rock species, are not identical from one to another ecosystem's soil. Combined with other physical parameters (e.g. temperature, humidity, altitude, latitude), they vary with a lot of consequences on the microbial, fungal, plant and animal populations they contribute to sustain. The rock species, according to their formation and composition, are

Table 10.6 The most common elements of the Earth's crust

Element	Symbol	Abundance (%)
Oxygen	O	46
Silicon	Si	27.7
Aluminium	Al	8.2
Iron	Fe	4.1
Calcium	Ca	3.6
Sodium	Na	2.8
Potassium	K	2.5
Magnesium	Mg	2.0
Titanium	Ti	0.4
Hydrogen	H	0.1
Phosphorus	P	0.1
Manganese	Mn	0.1

distributed in three main groups. The site's ecology will considerably differ due to the dominance of one or another of these main types.

10.3.3 The Three Main Rock Groups of the Earth's Crust

The rock species forming the bedrock are distributed into three main groups, mainly according to the way they were formed: (1) the igneous (or magmatic) rocks, (2) the sedimentary rocks, and (3) the metamorphic rocks.

10.3.3.1 The Igneous Rocks

Igneous rock forms when molten rocks (called magma) located in the mantle and mainly composed of a blend of silicates move into the Earth's crust, where temperatures are much lower, either by convection or via volcanic eruptions. There, these rocks cool and crystallise. They become crystalline rocks. Granite and basalt are typical igneous rocks. Both constitute the majority of the rocks on Earth; granite is mostly found in the continental bedrock, and basalt lies mainly in the oceanic bedrock. Granite can be found either under sedimentary rock layers (see point below) or as large areas of outcrops. Basalt, on the other hand, is the main rock formation of seafloor bedrock. Outcrops of basalt can be found on continents in former oceanic grounds that came into place by seafloor uplift. Columnar basalts are typical basalt formations found either along some coastlines or in mountainous areas of volcanic origin.

10.3.3.2 The Sedimentary Rocks

Sedimentary rocks are mostly formed on the floor of water bodies (sea or fresh waters) from dissolved material of rocky origin [minerals, mineraloids, or regolith (see Sect. 10.3.1)] and sometimes from solid organic material [e.g. fossil remains, dead organic material, isotopes bearing a biogenic signature [28] like depletion of carbon 13 (^{13}C) for example]. Initially, the dissolved rocky material precipitates into solid particles that stick together to form a surface on which other similar particles deposit and stick. These particles remain stuck and build up solid rocks thanks to natural cement, e.g. *silica* (SiO_2), calcite ($CaCO_3$) or hematite (Fe_2O_3). Some of the solid organic material mentioned above can be incorporated into the forming sedimentary rock, testifying to the latter's aqueous and often marine origin. A large part of the sedimentary rocks generated during the geologic periods have been uplifted above sea level and incorporated into continental land masses, sometimes at very high altitudes (as these rocks are commonly observed close to the summit of young mountains like the Himalayas or the Alps for example).

Sedimentary rock can be classified, according to the way they were generated, into four groups, i.e. (1) clastic sedimentary rocks, (2) biogenic sedimentary rocks, (3) chemical sedimentary rocks, and (4) a heterogeneous group of rocks of diverse origins, e.g. stemming from volcanism or impact from extraterrestrial bodies (meteorites, comets, etc)[11]

Clastic sedimentary rocks: These rocks are composed of different types of minerals, notably quartz, feldspar, mica or clay minerals, cemented by silicate minerals (see Table 10.3). Each rock species of this category has its constitutive minerals specifically clumped to form elementary sediment particles, called grains, that build them up. The volume of the grains of a given clastic sedimentary rock has a considerable influence on the fertility of the soils generated from the pedolith of this rock, hence on the biodiversity[12] of the ecosystem stemming from this pedolith. According to the Udden-Wentworth grain size scale [44, 47], the grains are distributed into four types: (1) gravel grains ($ø > 2$ mm),[13] (2) sand grains ($0.0625 \text{ mm} < ø < 2$ mm), (3) silt grains ($0.0625 \text{ mm} < ø < 2$ mm), and (4) clay grains ($ø < 0.0039$ mm).

In Sect. 11.3.2, we will see why soil fertility decreases with grain size and understand why clay grains retain far more water molecules than gravel or sand grains per volume unit. Among clastic sedimentary rocks, we can mention unconsolidated sediments like *gravel*, *sand* or *mud* and consolidated ones like *sandstone*, *claystone*, *mudstone* or *shale*.

---

[11] From the wikipedia article "Sedimentary rock" (https://en.wikipedia.org/wiki/Sedimentary_rock); under licence CC BY-SA 3.0.

[12] Species richness, diversity, and frequency distribution are some biodiversity parameters; see Chap. 12.

[13] $ø$ = diameter; $0.0625 \simeq 1/16$; $0.0039 \simeq 1/256$.

Biogenic sedimentary rocks: These rocks are overwhelmingly formed by chemical species involved in the biogeochemical cycles, i.e. that once transited through the living beings of the ecosystems and thereafter were mineralised and released to the environment (atmosphere/hydrosphere, soil) via a process described in Sect. 10.2. The chemical elements involved here function, so to speak, in isolation; they are very rare in the Earth's crust. For example, carbon, calcium and magnesium are common in biogenic sedimentary rocks and very rare in the bedrock, where they respectively represent 0.1, 3.6 and 2% of the total; on the other hand, silicon, aluminium or iron are well represented in the Earth's crust (respectively 27.7, 8.2 and 4.1%) but exist in minute proportions in the organisms, thus also in the biogenic sedimentary rocks (cf. Table 10.6). Some rocks (limestones, dolostones) are the main proximal source of CO_2 and of the ionic calcium and magnesium involved in the biogeochemical cycles. We will see in detail how these elements circulate through the ecosystems. Three properties of organic carbon, the most important element of life, are noteworthy:

1. Its distal source is the mantle via volcanic eruptions.
2. The quasi-totality of this element on the surface of the Earth is either in the organic form or in the inorganic form in biogenic sedimentary rocks or in the atmosphere/hydrosphere.
3. This element is extremely rare elsewhere in the bedrock.

We can mention three types of biogenic sedimentary rocks:

Carbonate sedimentary rocks:

They form in the bottom of aqueous, generally marine environments, frequently engulfing fossilised remains of aqueous organisms [tests of foraminiferans (a group of marine protozoa), mollusc shells, arthropods or vertebrate skeletons]. A part of this material (calcium compounds, for example) is precipitated and forms the sediments that make up these biogenic sedimentary rocks (cf. Sect. 10.5.4). Among them, we can mention

- limestone, a group of carbonate sedimentary rocks whose mineral constituent is crystal carbonate ($CaCO_3$), either in the form of calcite, aragonite or vaterite,
- chalk, a form of soft and porous limestone formed by accumulating the calcareous tests of a group of photosynthetic eukaryotic microorganisms, the coccolithophores.
- marl, a type of soft fossiliferous sedimentary rock. This rock is often a type of limestone. Some soft mudstones fall also into this group. Marls share in common their soft consistency and their great content of silt or clay.
- dolostones essentially composed of mineral dolomite ($CaMg(CO_3)_2$),

Coal:

This sedimentary rock is chiefly composed of biogenic carbon stemming from fossilised plant matters. This dead organic matter, which originated from diverse plants dating back to the Carboniferous, is the very essence of these coal seams. The atoms of this biogenic coal are organised in strata that are mineraloids because of their organic origin.

Chert:

This is a third group of biogenic sedimentary rocks composed of silica (SiO_2) originating from the skeletons of microorganisms (diatoms, silicoflagellates, radiolaria). It can occur in irregular nodules in limestone, dolostone, chalk or marl. When chert occurs in chalk or marl, it is called flint.

10.3.3.3 Metamorphic Rocks

These rocks result from the profound structural transformation, under the impact of intense heat and pressure, of former rocks that were once igneous or sedimentary. One consequence of these "metamorphoses" is that these rocks acquired a foliate or banded texture. Well-known representatives of this group of rocks are slate, gneiss, schist and marble.[14] E.g. slate is derived from a type of shale, gneiss is transformed from granite, schist is modified from mudstone or shale, and marble is metamorphosed from limestone or dolostone. Metamorphic rocks form either deep in the Earth's crust or more superficially at plate boundaries when tectonic plates collide. In both cases, the forming metamorphic rocks are driven to the surface of the bedrock by the geological uplift accompanying the geological convection movements. At the surface, the metamorphic rocks are exposed to weathering processes and may break down into sediments that will be used either (i) to form new sedimentary rocks, or (2) to form pedolith and contribute to construct a soil system.

10.3.4 Ecological Impact of this Geological Situation

Igneous or metamorphic rocks develop below the sedimentary rocks that, for physical reasons, tend to spread and form a shield. The bulk of igneous and metamorphic rocks thrust up and put pressure on the sedimentary shield that sometimes breaks and lets outcrops of igneous or metamorphic rocks appear at Earth's surface. That is why sedimentary rocks form larger areas than the other two types of rocks. Table 10.7 summarises the proportion of the main groups of rocks expressed as surface and volume percentage of the total bedrock.

The three types of rocks are subjected to weathering processes that constitute the distal origin of the overwhelming majority of the chemical elements in the soils at

[14] Information from a website of the Geologic Society of London: "https://www.geolsoc.org.uk/ks3/gsl/education/resources/rockcycle/page3459.html".

Table 10.7 Proportion (in %) of the bedrock occupied by each of the main groups of the Earth's crust's rocks

Main groups of rocks	Surface cover (in %)	Relative volume (in %)
Igneate	15	65
Sedimentary	73	8
Metamorphic	12	27

disposal of the producers,[15] with a determining influence on the types of ecosystems resulting from these soils. We will now describe the biogeochemical cycle of the main chemical species. Most of the data exemplifying the water and carbon cycles have been taken from Ramade

10.4 The Cycle of Water (H_2O)

10.4.1 Generalities on the Water Cycle

Water circulates through the biosphere and ecosystems.[16] It is the main constituent of living matter. To grasp how much this molecule is necessary to all organisms, for example, water accounts for 63% of the mass of a human being, 80 % of the mass of a fungus, and 98% of the mass of a Cnidarian. In addition, water is the source of all the *organic hydrogen*. However, which type of water do we refer to here?

Almost 97% of the total volume of water on Earth is formed of salt and brackish waters (seas and oceans, estuaries, brackish marshes, some parts of deep waters), the remaining 3% being freshwater. Of the latter, 95% is trapped in ice caps and glaciers, or under the Earth's crust as fossil aquifers. Of these, only a tiny proportion is sprayed in the atmosphere (or seldom in the hydrosphere) by erupting volcanoes in the form of water vapour or droplets, constituting thus the only portion stemming from the aquifers that participates to *circulating water* (see just below). The remaining 5% of the total fresh water is called the *superficial waters*, i.e. those in the freshwater courses (rivers and streams, groundwaters), in aqueous areas (lakes and ponds), in the moisture on solid surfaces (rocks, leaves, etc) due to evapotranspiration, dews, hazes and fogs, etc, and last but not least, on atmospheric condensed water and water vapour originating from the evaporation from all water surfaces (oceans and seas,[17] watercourses, lakes, etc) under the influence of solar radiation energy. Thus, the only portion of the total mass of water that enters the biosphere and is used for the functioning of all the ecosystems is the circulating water which stems, as we wrote just above, from the water ejected by volcanic

[15] A notable exception is nitrogen that is of atmospheric/hydrospheric origin (cf. Sect. 10.6).

[16] Most data has been borrowed from Ramade [32, 33].

[17] Water vapour rising from the superficial waters of oceans is devoid of salt and thus fresh.

Table 10.8 Volumes in km^3 of different water mass types

Types of water mass	Volume (in 10^6 km^3)
Ocean	1380
Waters in sedimentary rocks	210
Ice caps	16.7
Subterranean waters	0.25
Surface continental waters	0.025
Constant atmospheric water vapour	0.013
Precipitated and/or condensed waters(rain, snow, dew, etc)	0.319

Table 10.9 Heights in m of the water mass types (after the condensation of water vapour) on a smooth spheric surface having the same area as Earth

Types of water mass	Height (in m)
World ocean	2700
Ice caps	100
Subterranean waters	15
Surface continental waters	0.4
Atmospheric water vapour	0.03
Precipitated and/or condensed waters	0.65

eruptions and evaporated from the superficial waters; it accounts only for 0.03% of the total water volume.

Approximately three-quarters of the Earth's surface (363 million over 510 million km^2) is covered by water. Table 10.8 displays the volume distribution of different water mass types [20].

If the different water masses were uniformly spread on a smooth spheric surface assimilated to Earth's surface, after the condensation of gaseous H_2O, we would obtain layers of different heights characteristic of different water mass types,as displayed in Table 10.9.

The combination of atmospheric water vapour and circulating water build up a cyclic dynamics that recruits all the water needed for the photosynthesis of the living part of the ecosystems; some of the water taken in by the biocoenoses goes back to the abiotic factor system [namely to the surrounding atmosphere (or hydrosphere) or to the soil] through respiration, transpiration or evapotranspiration. We saw that circulating water originates with the evaporation of the liquid water contained in the superficial water layers of oceans, rivers, lakes and on solid surfaces of many kinds (rocks, leaves, etc). Only a tiny proportion of this water vapour remains in the atmosphere, though that proportion varies according to local ecogeographical conditions. Most atmospheric water vapour condenses (in the form of clouds, fogs, hazes or dews) and precipitates to the ground as rain, snow or water droplets. It has been computed that the height per annum of precipitated waters worldwide is 65 cm; whereas the height of water vapour, if artificially condensed, would be 2.5 cm (thus 26 times smaller).

Let us now look at Table 10.9. The only portion of water entering the biosphere is atmospheric water vapour, represented by a layer whose height is only 2.5–3 cm—

that is, about 1/10,000 of the height of the cumulated layers representing all the water on Earth. These water molecules constitute an abiotic factor directly available to the organisms constituting the biocoenoses.

At a given point, the atmospheric water vapour molecule should not be considered a static physical phenomenon. On the contrary, it is the dynamic result of a constant water flux. Indeed, while circulating, a single water vapour molecule condenses and precipitates, immediately replaced by another water vapour molecule that will undergo the same process, at constant rate. This process may explain the constant frequency of water vapour molecules in the atmosphere locally. The quantity of water vapour is highly variable from one biome to another according to its pluviometry. For example, the following data are observed annually about the precipitation height of water vapour once condensed between several latitudes: 2.5 mm at the North Pole, 10 mm at 45° N, 45 mm at the equator, 20 mm at 45° S, and 8 mm at the South Pole.

Let us now sketch a bit more accurately the material transfer and energy flux accompanying the water cycle.

- The cycle of water begins, as we already mentioned, with the evaporation of water from the superficial layers of the seas and freshwater stretches under the influence of solar light energy; the molecules of water vapour are transported within air masses and driven by air currents and winds, under the influence of the Coriolis force. They then condense as clouds and precipitate as rain, snow or hail when appropriate weather conditions are met, e.g. atmospheric depression or encounter between hot and cold air masses. Sometimes, water vapour condenses at the surface of cold bodies - which is the principle of dew.
- The cycle of water is a thermodynamic system in the sense that it can increase its free energy and chronically shift from thermodynamic equilibrium. The quasi-infinite source of energy is the same, i.e. solar light energy. A process analogous to assimilation in living beings, thus in ecosystems, is evaporation: a water molecule at the surface of an ocean or freshwater stretch passes from the liquid to the gaseous phase and enters into the composition of the atmosphere. While changing its state, water will raise the potential energy of its cycle by 2.24 kJ. Evaporation thus fulfils the same role as reduction in redox reactions. The mass of water evaporated per annum has been estimated to 378^{12} t; this water, if condensed, would fill a volume of 378 million km^3. The annual rise of potential energy at the beginning of the water cycle is

$$\underbrace{378 \times 10^{18}}_{\text{Mass of evaporated water in g/yr}} \times \underbrace{2.24}_{\text{Rise of potential energy for 1 g water molecule evaporated, in kJ/g}} = 8.5 \times 10^{20}\ \text{kJ/yr}.$$

  It has been estimated that 7/9 of the circulating water falls on the surface of oceans and seas, and only 2/9 on the surface of emerged lands. Thus, most of the cycle occurs between the ocean and the atmosphere. Consequently, only a small part of an already tiny portion of the total water devoted to the circulating water allows life in the continental and insular habitats that contain the largest biomass and most complex ecosystem structures. Once it has reached the surface of an emerged land, water undergoes one or several of the following movements—(1) infiltration and percolation, (2) evaporation, (3) evapotranspiration, and (4) run-off. The last water movement is the final step that will eventually bring back the circulating water to the aqueous surfaces from which the cycle began.

10.4.2 Infiltration and Percolation

Infiltration is the mechanism of absorption of precipitated water (rain, snow) in the soil, whereas percolation is the motion in the soil of infiltrated water. Percolation can result (1) in moisture retained by the soil, (2) in the formation of wells, or (3) in water run-off or leaching. Infiltration and percolation form a process that hydrates the soils and ensures the constitution of subterranean flowing or stagnant water reserves. Soils collect the interstitial water, i.e. water lying in the empty spaces between the soil solid structures and available to the autotrophs (plants and autotrophic bacteria). Interstitial water is the solvent of the mineral salts the autotrophs need, thus the biocoenoses' producers. It is mainly located in the superficial layers rich in humus and living biomass. Soils have to be able to retain a sufficient quantity of water to be rehydrated. A number of mechanisms carry out this capacity that varies according to the biome to which the ecosystem under study belongs. For example, mosses play an important role as a water-retaining agent in forest soils.

A portion of the infiltrated water is not retained. Percolation collects it into groundwaters and subterranean rivers that emerge as sources; this is the beginning of the run-off system that generates the river network. When the ground is very pervious, the percolated water is trapped in deep wells, forming layers of fossil water well under the run-off network. This portion of water leaves the water cycle. The water not taken over by percolation rises to the surface and leaches the soil.

10.4.3 Evaporation and Evapotranspiration

We already saw that water evaporation was the starting water movement of the water cycle (cf. Sect. 10.4.1). Water evaporation also continuously feeds the water cycle at later stages during its entire process. Water evaporation on land masses moves counter to precipitation, which is the precursor of infiltration and percolation. It operates from liquid water drops on surfaces of rock outcrops, soil and aerial parts of organisms; the process changes them to water vapour molecules that rise into the atmosphere and become part of it. Evaporation is a key factor in the hygrometry of an ecosystem, i.e. its degree of atmospheric humidity and water regime.

In an ecosystem, evapotranspiration involves only the plant biocoenosis and mainly occurs in tropical rainforests. It is a phenomen far more complex than plant transpiration.

- As a reminder, simple plant transpiration is a continuous process occurring in all plants where the internal water evaporates through the leaves during the sap ascent in the xylem. It takes place mainly in the stomates of the leaves but sometimes also through the cuticule of other aerial parts [46].
- Evapotranspiration on the other hand combines simple transpiration with another mechanism, more active, consisting of pumping out the water in excess in the soil through the root and project it into the atmosphere without incorporating it in the organism of the plant, contrary to plant transpiration where the excreted water stems from the sap. The water pumped in through the latter mechanism immensely exceeds the water received to elaborate sap in the xylem—the process followed by simple transpiration. This phenomenon explains why the overwhelming majority of the water the plants receive does not participate in the metabolism and directly escapes into the atmosphere. Only a tiny portion of the water the plants receive is hydrolysed by photosynthesis, meaning that the water requirement of the plants performing evapotranspiration hugely exceeds their need for water just for the fixation of hydrogen molecules.

Evapotranspiration is universal and can be observed in cultivated as well as wild plants. Per ha of cultivated plants, from 2000 t of water absorbed by them, only 3 t is used for water photolysis; thus photosynthesis accounts for only 0.15% of the water pumped in. The rest of the water is evapotranspirated. The more the atmosphere of an ecosystem is loaded with water, the larger the quantity of evapotranspirated water is. In Western Europe, one ha of forest evapotranspirates between 7300 and 18,250 t of water per year. This figure falls to 4000 t/year on dry soil for 1 ha of spruce in Sweden.

Evapotranspiration is highly variable in the different biomes of the world. In many rainforest areas, most rainwater necessary for the plants does not come directly from the precipitations but indirectly via evapotranspiration, which acts as a potent humidity enhancer, which explains that the humidity of these areas is saturated. The Congo basin, for example, functions as a closed ecosystem that depends nearly strictly on evapotranspiration for its water supply. In tropical deserts and several Mediterranean regions, evapotranspiration is very high and considerably exceeds rainwater; therefore, the water must be supplied from outside, either as superficial or subterranean waters. On the contrary, evapotranspiration is negligible in subantarctic areas.

10.4.4 Run-offs and Leaching

Infiltration and percolation, as mentioned above (cf. Sect. 10.4.2), result in two types of run-offs:

Contained run-off: I define this concept as the succession of two types of water circulation, i.e. (1) a subsurface flow corresponding to the water flow of groundwaters and subterranean currents emerging, as we already mentioned, from infiltration and percolation and possibly leading to a source, and (2) a secondary water surface run-off system that stretches in a canalised form on Earth's surface from this source, generally to the sea or ocean through streams and rivers, sometimes across lakes and ponds. Contained run-off waters contribute widely to the formation and maintenance of rivers, constitute complex networks of rivers with their affluents (basins), and contribute widely to the completion of the water cycle. It contributes to providing the main land ecosystem it crosses with water, but frequently forms secondary ecosystems like mangroves, fresh water ecosystems, and other types of semi-aquatic semi-terrestrial formations.

The proportion of contained surface run-off water in a river basin depends on the precipitation intensity. The higher the precipitation is, the greater the relative volume of water in a river that comes from contained run-off water. Thus, rainforest regions account for the largest relative volume of run-off water. For example, the Amazon River transports 15% of the waters returned in all the oceans of the planet and the Congo basin 33% of its total flow. The arid and semi-arid regions that represent 40% of the land mass account for only 2% of the contained run-off waters.[18]

Uncontained surface run-off or leaching: This type of surface run-off depicts uncontained water flows that flood land surfaces and take the form of a water spread on the ground surface, according to its topography.

This situation occurs in rainforest areas cleared by *logging* after a tropical downpour. Then, the mineral elements and salts are washed away into the rivers and vanish in the sea at ultimate stage. This phenomenon is responsible for the irreversible degradation of soil that ultimately disappears. Indeed, in rainforests, soil is maintained by the root system of trees. Once the trees have been felled on a vast area, soil cannot be kept in place and is easily removed. Eventually patches of bedrock appear, the surface of which oxidises and turns to laterite.

10.4.5 Summary of the Cycle of Water

In natural situations, when the water cycle is not disturbed (essentially by man), water entering the biosphere is recycled. As a reminder, most of the water needed by the ecosystems comes from the water evaporated from the superficial layers of the oceans. This mass of water condenses, precipitates, and reaches the ground of the ecosystems. A vast amount of that water transits through the organisms and soil, is infiltrated, percolated and enters the contained run-off system. The latter is a dynamic process that starts with the subterranean water currents, namely the groundwaters that feed the sources, which in turn feed the surface hydrographic

[18] Adaptation in English by the author of an excerpt from a Wikipedia article in French on run-off titled "Ruissellement" (web address = "https://fr.wikipedia.org/wiki/Ruissellement"); under licence CC BY-SA 3.0.

network. The latter transports its water to the seas and oceans, and the water cycle is completed.

A part of the water is, of course, retained by the organisms for their metabolism, growth and reproduction, and seems to be removed from the cycle, eventually, all living beings die, and their internal water returns to the soil with a given delay. This part of the water is also infiltated, percolated and run off, so that eventually, all the water taken by the ecosystem from the superficial waters of the world return to these.

10.5 The Cycles of Carbon (C)

10.5.1 Generalities on the Carbon Cycles

Carbon is the organiser of the chemical machinery of all living beings. All the carbon atoms of this machinery stem from carbon dioxide (CO_2), one of the constituent of air—the gas mixture that composes atmosphere. Carbon dioxide account for a very tiny percentage of air (c.a. 0.04%) [3, 39], almost a trace. This apparently negligible quantity of CO_2 in the atmosphere is fortunate for us, because this gas at a larger amount would be toxic for aerobic organisms like us and would impair the climate homeostasis of Earth since it is a powerful greenhouse gas.[19] As we will see below, carbon dioxide is also largely present in the hydrosphere, as a dissolved molecule in oceans and seas as well as in freshwater networks. Remarkably, such a small amount of carbon dioxide is at the origin of all the organic carbon on Earth, on land and water. The ratio between gaseous CO_2 and dissolved CO_2 is constant. In addition, the circulation of CO_2 is highly regulated.

1. Atmosphere's and hydrosphere's carbon dioxide is the source of all the carbon atoms needed for *photosynthesis* and chemosynthesis, thus all the carbon atoms needed for all organisms.
2. There is a permanent exchange of carbon dioxide between the atmosphere and the hydrosphere. It has been shown in various studies that the concentration of CO_2 in the hydrosphere is 50 times that in the atmosphere, and that this value has been remarkably constant since the beginning of the Phanerozoic (about 540 Ma ago) [35, 45].
3. Carbon also plays an important role in the formation of calcareous sediments and limestone, the basis of many ecosystems.

Hence, the three main cycles of carbon summarised in the next three subsections.

[19] We know, from a large body of scientific publications since the late 1950s, that carbon dioxide, emitted into the atmosphere by present human activities, is largely responsible for the global heating of the Earth that has exponentially developed since the beginning of the industrial age [3, 39, 49].

10.5.2 Cycle 1 Delivers to All Cells Their Carbon Machinery

This cycle conveys all the carbon atoms needed for the metabolism of all cells on Earth, thus needed for all the biocoenoses of the biosphere. It works as follows. The atoms of inorganic carbon, present in the molecules of atmospheric or hydrospheric CO_2, enter the total biocoenosis of an ecosystem E (see Definition 2.7 of Chap. 2) as follows.

1. They are absorbed by the autotrophs of the biocoenosis of E and extracted by the chemical reactions of their metabolisms, catalysed by diverse enzymes. Thereafter, these atoms are integrated with other chemical elements into the basic organic molecules, a mechanism called biosynthesis. This happens thus at the first trophic level.
2. A proportion of the carbon atoms of the first trophic level are then integrated into the organic matter of the heterotrophic biocoenoses of the second trophic level, via ingestion and digestion. Then, the same transfer will take place between the second and third trophic levels, between the third and fourth trophic levels, etc, and this recursively. We already saw that in a trophic network, the higher the trophic level is, the lower its biomass is and thus, the fewer carbon atoms there are. Because there is globally a balance of the carbon atoms, those that are not retained by the total biocoenosis necessarily leave it and enter a portion of the abiotic factor system, here the atmosphere, by three mechanisms where they are released as inorganic atoms within new carbon dioxide molecules, i.e. (1) aerobic respiration, (2) oxidative degradation and (3) fermentation.

Aerobic respiration: The aerobes, i.e. the organisms practising areobic respiration (all the eukaryotes and aerobic prokaryotes), carry out aerobic respiration that changes organic carbon into mineral carbon. Thus, via the sum of aerobic respiration of all its members, the aerobic part of a biocoenosis releases inorganic carbon in the form of carbon atoms, which eventually leave the ecosystem by returning to its initial form, i.e. inserted in molecules of carbon dioxide. These molecules return to the atmosphere or hydrosphere.

An atom of inorganic carbon that is integrated as organic carbon into an aerobic biocoenosis via photosynthesis is thus compensated by an inorganic carbon leaving this aerobic biocoenosis via aerobic respiration. Indeed, aerobic respiration is the chemical converse of photosynthesis, fully explained by the next equation reflecting this balance:

$$\mathrm{n\,CO_2 + 2n\,H_2O + n\,kCal} \underset{\text{Aerobic respiration}}{\overset{\text{Photosynthesis}}{\rightleftarrows}} \mathrm{+n\,O_2^{\uparrow} + n\,H_2O + [C_n(H_2O)_n]} \quad (10.1)$$

Respiration is the sense of the reaction from right to left.

Oxidative degradation: This mechanism is defined as the transformation, under aerobic conditions, of residual organic matter,[20] i.e. the organic matter found in plant debris, corpses and excreta. It consists of transforming all the organic elements, i.e. the elements trapped in organic matter, into inorganic elements released in the abiotic factor system (AFS). Dead residual organic matter, like physicochemical elements, belongs to the AFS. Oxidative degradation is done by a chain of decomposers/mineralisers like saprophytic fungi and bacteria that eventually transform organic carbon into mineral carbon, in the form of CO_2 molecules. Though the pathway of this transformation is more complex than aerobic respiration, its result concerning the destiny of carbon is identical.

Fermentation: This is a process similar to oxidative degradation but occurring under anaerobic conditions. A number of anaerobic decomposers, generally yeast species, perform this kind of transformation. This is another pathway used by anaerobes to return organic carbon to its inorganic form in the form carbon dioxide.

Aerobic respiration, oxidative degradation as well fermentation have the same result, a complete cycle of carbon: inorganic C in CO_2 transformed into organic C, which in turn is altered into inorganic C in CO_2 *ad libitum*).

10.5.3 Cycle 2 Is an Equilibrium of CO_2 *Between Atmosphere and Hydrosphere*

CO_2 permanently circulates between the hydrosphere and the atmosphere. The waters contain a large concentration of carbon dioxide.

The mass of CO_2 per volume unit dissolved in the hydrosphere has been estimated to be 50 times greater than the density of CO_2 in the atmosphere. Under the influence of solar light energy flux, 1/50 of the CO_2 contained in the waters goes in the atmosphere, and it is this tiny portion of carbon dioxide that provides all the carbon atoms fixed in all the terrestrial ecosystems. Importantly, statistically for one aqueous CO_2 molecule released in the atmosphere, one atmospheric CO_2 molecule in contact with water gets dissolved in the hydrosphere, which makes an equimolar balance between hydrospheric and atmospheric molecules of carbon dioxide [36].

10.5.4 Cycle 3 Is the Strict Control of Circulating Carbon Dioxide that Accompanies Limestone Sedimentation

This cycle involves several phenomena, of which we will depict two:

[20] Sometimes called dead organic matter.

Only a small percentage of Earth's carbon circulates as carbon dioxide for the needs of the organic material of the biocoenoses. The vast majority of carbon atoms is trapped in the rocks and participates to their structure and physicochemistry; some of these atoms are involved in a well known subcycle, that I will call subcycle 3a, involving carbon and calcium that explains the balance between the phenomenon of sedimentation and the formation of limestone on the one hand, and the permanent presence of calcium at disposal of the aquatic organisms on the other hand in a relative constant proportion in the hydrosphere (see just below).

Subcycle 3a has had spectacular effects. As time went by, from the Archaean until the present throughout the *geologic eras*, vast masses of *limestone deposits* accumulated on the planet [2], and sufficient *calcium ions* have been present in the waters, mainly in the oceans, to meet the needs of the living beings (namely for the *shells of molluscs*, *the tests of the foraminifers*, the bones of the fishes, and diverse other physiological needs of all the species). In aqueous habitats, some conditions favour the precipitation of calcium carbonate and launch its deposit on their beds, the main material of limestone; others, on the contrary, nurture the dissolution of calcium in the form of carbonate-containing ions or molecules. Calcium, in its ionised form, is carried along rivers and brought to the ocean. In given situations, the following chemical reactions take place that will entail the situation described above.

Dissolution of Carbon Dioxide and Successive Consequences vs Concentration

- When carbon dioxide is dissolved in water, carbonic acid is formed as follows:

$$CO_2 + H_2O \rightleftarrows \underbrace{H_2CO_3}_{\text{Carbonic acid}} \tag{10.2}$$

- One molecule of carbonic acid on the right side of Eq. (10.2), in two steps, provides one molecule of bicarbonate ion and one molecule of carbonate ion; which is described by the following system of equations:

$$\begin{cases} H_2CO_3 \rightleftarrows H^+ + \underbrace{HCO_3^-}_{\text{Bicarbonate ion}} \\ HCO_3^- \rightleftarrows H^+ + \underbrace{CO_3^{2-}}_{\text{Carbonate ion}} \end{cases} \tag{10.3}$$

Conditions for the Precipitation of Calcium Carbonate

We are now in a situation where, in water, calcium carbonate molecules (the main elementary component of limestone) are side by side with dissolved calcium, H^+ and CO_3^- ions. In a situation of equilibrium,

$$\underbrace{CaCO_3}_{\text{Carbonate calcium}} \leftrightharpoons \underbrace{Ca^{2+}}_{\text{Dissolved carbon}} + CO_3^{2-} \tag{10.4}$$

From Eq. (10.4) and water acidity-alkalinity conditions, we can foresee (1) when the ions will precipitate into carbonate calcium and (2) when calcium carbonate will dissociate into the ions.

- pH $<$ 7 corresponds to the acid condition (H^+ ions being dominant):
  The chemical reaction (10.4) proceeds from right to left, and we will see (1) a strong concentration of CO_2 and HCO_3 and thereafter, (2) from the dissolved carbon and carbonate ions precipitate into calcium carbonate. This situation favours the accumulation of calcium carbonate, thus the formation of limestone and other calcareous rocks.
- pH $>$ 7 corresponds to the alkaline condition (H^+ ions not prevailing):
  The chemical reaction (10.4) proceeds from left to right. Calcium carbonate disintegrates in chemical reactions increasing carbonate ions and dissolved calcium ions.
- pH = 7 corresponds to neutrality in the acidity-alkalinity gradient. We obtained the following chemical equilibrium:

$$CaCO_3\text{ (insoluble)} + H_2O + CO_2 \leftrightharpoons \underbrace{Ca(HCO_3)_2}_{\text{Calcium bicarbonate}} \tag{10.5}$$

Calcium carbonate is brought to the marine habitat from run-off waters (it is the continuous freshwater flow from rivers, surface run-offs, and precipitation). That mass of water's destiny globally is to end in the seas and oceans, as is also the dissolved carbon dioxide and carbonate calcium. The marine habitat is, on average, at pH = 8, thus slightly alkaline. However, pH variations can explain the dynamics generated by local acidity or alkalinity conditions.

10.5.5 Conclusion: The Supercycle of Carbon

The three cycles described above are linked and form the supercycle depicted in Fig. 10.5.

10.6 The Cycle of Nitrogen

10.6.1 Nitrogen, a Paradoxical Case

Nitrogen (N) is the prevailing element of the Earth's atmosphere, in the form of a gaseous molecule, dinitrogen (N_2): it accounts for ca 75% in mass and 78% in volume of dry air. This situation means that in the atmosphere, four atoms of

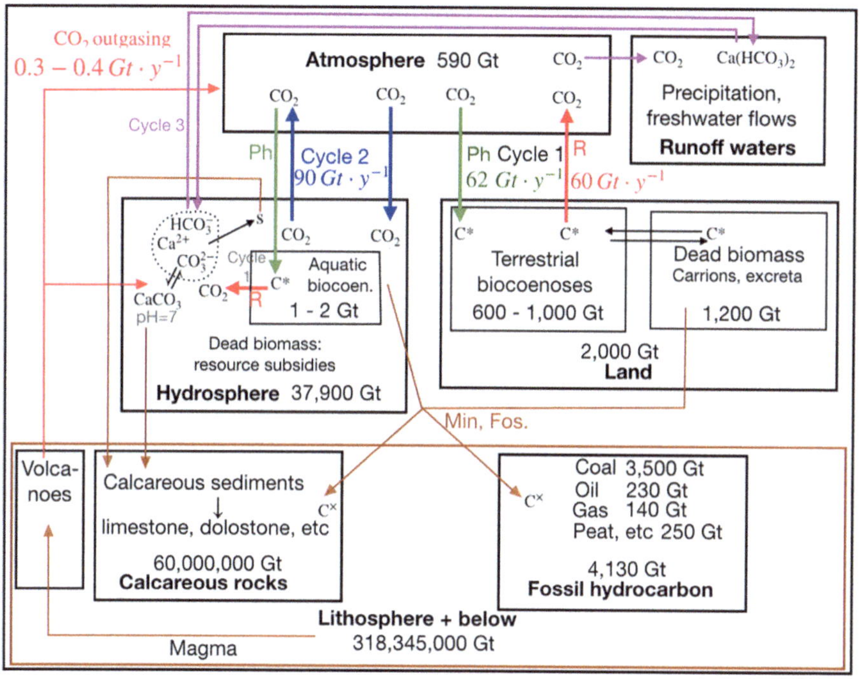

Fig. 10.5 Schema illustrating the complete carbon supercycle [12, 43]. Cycle 1, *red and green arrow* pairs, is in two pieces, one between the atmosphere and land compartments and one within the hydrosphere compartment. Cycle 2, the *blue arrow* pairs, regulates the constancy of the CO_2 flow; the latter nurtures locally cycle 1. Cycle 2 fosters cycle 3. Cycle 3 regulates calcium constancy, which impacts the calcium metabolisms of aquatic animals and has important side consequences on the sedimentation of calcareous formations. It has no direct implication on cycle 1. C* organic carbon, $C^{\times}$ inorganic carbon. The quantities in Gt represent carbon masses, *s* shells, tests and bones

nitrogen fluctuate side by side with one atom of another element (e.g. oxygen, hydrogen, carbon, argon and traces of other elements). The atoms of the mentioned elements are generally embedded in polyatomic molecules (e.g. N_2, O_2, H_2O, CO_2, CO); these structures and several rare noble gases that are monoatomic [Argon (Ar), Neon (Ne), Krypton (Kr)], overwhelmingly constitute the particles, subject to the Brownian motion, that together form the atmosphere. Moreover, dinitrogen molecules move with higher velocity than any other fluctuating particles constitutive of air, which explains why nitrogen atoms occupy almost 4/5 of the total volume of the atmosphere [11, p. 258-260].

The paradox lies in that nitrogen, in its gaseous form within the atmosphere, is chemically inert and, as such, cannot be used by most organisms. Moreover, very few living beings are able to return nitrogen into the atmosphere in the form of N_2. Nevertheless, nitrogen is absolutely essential for life as a vital element of crucial organic molecules, e.g. nucleic acids, proteins, ATP and NADP. The nitrogen atoms inserted in these molecules are called *organic nitrogen*.

Then, what mechanisms enable organisms to extract this so valuable nitrogen, albeit there is no other distal source of nitrogen than the atmosphere? We owe the Dutch microbiologist Martinus Beijerinck (1851–1931) the answer to this question [6]. The nitrogen atoms can only be used by organisms, thus biocoenoses, and returned to the AFS[21] as gaseous dinitrogen equimolecularly if they are inserted into primary producers within molecules like ammonia (NH_3) or ions like nitrites (NO_2^-) and nitrates (NO_3^-). Nitrogen atoms enter complex cycles where a small[22] proportion of inorganic nitrogen in the form of N_2 is sequentially transformed into ammonia (where the constitutive nitrogen is called *ammoniacal nitrogen*), nitrites and nitrates (where the constitutive nitrogen is called *nitric nitrogen*); a large portion of ammoniacal nitrogen is transformed into organic nitrogen thanks to an enzyme, the nitrogenase. Nitric nitrogen is retransformed via complex chemical mechanisms involving specialised bacteria, among the decomposers/mineralisers, into gaseous dinitrogen molecules, in such a way that dynamically, the quantity of nitrogen in the atmosphere remains globally preserved. Detailed and more advanced accounts of the process are given in relatively recent publications [5, 32, 48]; they are summarised in the next subsections.

10.6.2 Presentation of Some Nitrogen-fixers

The transformation of gaseous dinitrogen into ammonium is called *nitrogen fixation* and is caused mostly by nitrogen-fixing prokaryotes called *diazotrophs*[23] (58 bacterial and two archaeal genera; among the bacterial genera, 20 belong to phylum Cyanobacteria [50]). We will sketch a panorama on several examples taken from the literature.

Remark 10.1 Diazotrophs fall into different pairs of binary opposing categories that overlap.

1. Free-living vs symbiotic;
2. Aerobic vs anaerobic;
3. Phototrophic vs (chemoorganotrophic or chemolithotrophic);
4. Terrestrial vs aquatic;
5. Bacterial vs archaeal.

---

[21] AFS=Abiotic factor system.

[22] This small proportion is still enormous since it consists of the nitrogen atoms that are present in all living organisms and dead organic masses (fossil energy sources, calcareous sediments)!

[23] A small proportion of naturally occurring nitrogen fixation can be caused by abiotic phenomena like lightning.

I will draw hereunder a schema that exemplifies several types of diazotrophs (taxonomic examples from publications [32] and [5]).

From Fig. 10.6, we see that the diazotrophs are taxonomically very diverse. Indeed:

- Five phyla represent the domain Bacteria:
  - Actinomycetota with only one class Actinomycetia: one diazotrophic genus of the class is *Frankia*.
  - Bacillota with the sole class Clostridia: one diazotrophic genus of the class is *Clostridium*.
  - Cyanobacteria: only one class Cyanophyceae contains diazotrophs, e.g. the three genera – *Anabaena*, *Nostoc* and *Trichodesmium*.
  - Chlorobiota and its only class Chlorobia comprises a diazotrophic genus, *Chlorobium*.
  - Pseudomonadota (a recent term for Proteobacteria), which includes three classes: Alphaproteobacteria, Betaproteobacteria, and Gammaproteobacteria:

    Alphaproteobacteria contains two diazotrophic genera: *Rhodospirillum* and *Rhizobium*. .
    Betaproteobacteria comprises two diazotrophic genera, i.e. *Alcaligenes* and *Thiobacillus*.
    Gammaproteobacteria includes four diazotrophic genera: *Pseudomonas*, *Azotobacter*, *Methylomonas*, and *Chromatium*.

  - Thermosulfobacteriota with one class Desulfovibriona: this class contains the diazotrophic genus *Desulfovibrio*.
- The domain Archaea is represented by only one phylum, Euryarchaeota, with two classes Methanomicrobia and Methanococci: Methanomicrobia and Methanococci are respectively represented each by a single diazotrophic genus, *Methanosarcina* and *Methanococcus*.

10.6.3 Nitrogen Fixation

To this taxonomic diversity, we can add a gigantic diversity of ways of life within as well as between these groups (as explained by the categorisation sketched in Remark 10.1).

1. Some diazotrophs are free-living and practise their nitrogen fixation via soil bacteriophages and plasmids [48]—here, all the genera of Fig. 10.6 except *Rhizobium* and *Frankia*. The two last genera are symbiotic, carrying out nitrogen fixation within legumes via root nodules they generate [30] in a trade-off with the host for vital nutrients (see detail below). From this small sample, we can observe that the class Alphaproteobacteria comprises these two types of diazotrophs.

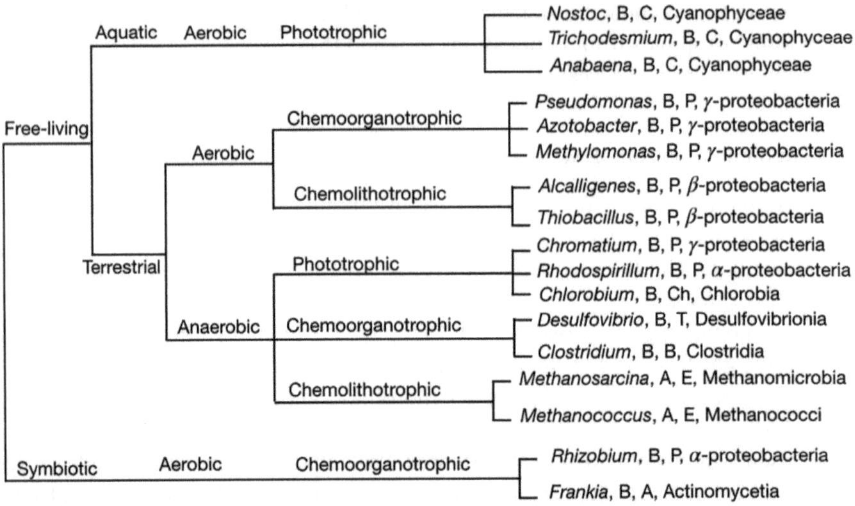

Fig. 10.6 Representation of several microorganism genera involved in nitrogen fixation. Each genus name is associated with the abbreviated names of its taxonomic domain (*B* Bacteria, *A* Archaea) and phylum (*A* Actinomycetota, *B* Bacillota, *C* Cyanobacteria, *Ch* Chlorobiota, *E* Euryarchaeota, *P* Pseudomonadota or Proteobacteria, *T* Thermosulfobacteriota) and the full name of its class (Cyanophyceae, ..., Actinomycetia)

2. The division between aerobic and anaerobic diazotrophs is not taxonomically clear-cut. Two classes of the Pseudomonata phylum comprise both types, e.g.:
   - Alphaproteobacteria with one aerobic () and one anaerobic (*Rhodospirillum*[24]) representative.
   - Gammaproteobacteria represented by a group of three aerobes (*Pseudomonas*, *Azotobacter*, and *Methylomonas*) and one anaerobe (*Chromatium*),

   On the other hand, all the genera of the Betaproteobacteria are aerobes (*Alcaligenes*, *Thiobacillus*).
3. Cyanobacteria is a phototrophic and oxic phylum.
4. Proteobacteria are divided into phototrophs (*Chromatium*, *Rhodospirillum*) and chemotrophs (the other Proteobacteria of the sample), shared between chemolithotrophs (*Alcaligenes*, *Thiobacillus*) and chemoorganotrophs (*Pseudomonas*, *Azotobacter*, *Methylomonas*). The methanogenic Archaea represented here are chemolithotrophs.
5. Mainly aquatic or humidity-related genera are represented in the phylum Cyanobacteria. The other taxa represented in the sample are mainly terrestrial and related to soil. But the two categories are not clear-cut.

[24] In certain condition, several species can switch to oxic respiration [38], e.g. *Rhodospirilllum rubrum*.

6. The sample comprises two representatives of the Archaea (*Methanosarcina, Methanococci*). The other genera are all bacteria. However, all these microorganisms are efficient nitrogen-fixers.

The chemical reaction of nitrogen fixation carried out by diazotrophs that need the contribution of at least 16 ATP molecules is

$$N_2 + 8\,H^+ + 8\,e^- + 16ATP \longrightarrow 2\,NH_2 + H_2 + O_2 + 16ADP + 16P \qquad (10.6)$$

It has been shown that the symbiotic genus *Rhizobium* comprises the most efficient nitrogen fixers among almost all the known diazotrophs. Since they only deal with members of the Fabaceae family, commonly known under the vernacular name *legumes*, the mechanism of nitrogen insertion in the other plants must also be evoked.

- In many cases, when these plants are around legumes, the exceeding quantity of organic nitrogen and nitric nitrogen is absorbed via the roots, a mechanism that has been long observed and analysed [32],
- In more recent research, an active and more direct integration of atmospheric nitrogen has been observed and analysed: the atoms penetrate the plant from the environment via soil bacteriophages, the plant's membrane passes and its plasmids [48] [see point (10.6.3) at the beginning of this section].

We will consider the case of the symbiotic diazotroph *Rhizobium* to exemplify a case of nitrogen fixation. The bacteria originally live freely in the soil, but are attracted by legumes via flavonoids[25] they release when the plants are in need for nitrogen. The microbes settle in the roots of the legumes they infest, inducing nodules on hairlike filaments barbing the rootlets of the host plant [30]. In these structures, they adsorb molecules of N_2; from then on, the transformation of atmospheric to ammoniacal nitrogen begins in the frame of a symbiosis between the host plant and the microbe, which consists of a trade-off between the *Rhizobium* and the legume. The former provides the latter nitrogen in exchange for carbohydrates. Ammonium compounds (NH_4OH) are elaborated in the nodules and conveyed into the leaves, where this molecule is transformed into ammonia (NH_3) thanks to an enzyme, nitrogenase.

- A proportion of the ammonia molecules enters the Calvin cycle (see the dark phase of the photosynthesis, cf. Sect. 9.4.2), which initiates the complex chemical reactions which eventually integrate the ammoniacal nitrogen within amino acids. At this stage, nitrogen becomes organic. In further steps of the metabolism, nitrogen becomes integrated into other vital organic molecules (e.g. nucleobases, ATPs, NADPs).
- The other proportion entering a process called nitrification is the subject of following subsection.

---

[25] Flavonoids are organic molecules involved in the metabolism of plants.

Notably, only a portion of the extracted inorganic nitrogen of the atmosphere is inserted in the vital molecules of all the biocoenoses of the biosphere.

10.6.4 Nitrification

The part of ammonia that is not invested in the elaboration of organic nitrogen is subject to a complex process, nitrification. It is the most common ammonia oxidation occurring in the presence of gaseous dioxygen. The latter consists of a series of chemical reactions performed by a number of specialised prokaryotes that change ammonia (NH_3) into nitrite (NO_2^-) and then nitrite into nitrate (NO_3^-). This stage ends with denitrification completing the cycle—the topic of Sect. 10.6.7. Nitrification proceeds in two phases:

Phase 1: The microorganisms carrying out the conversion of ammonia to nitrite are called *ammonia-oxidisers*, comprising species specialised in the oxidation of ammonia, e.g. (1) the bacterial genera *Nitrosomonas*, *Nitrospira* and *Nitrosococcus* [5] and (2) the archaeal genera *Nitrosopumilus*, *Nitrososphaera* and *Nitrosotales* [21]. Two following chemical reactions describe the process:

$$\begin{cases} NH_3 + O_2 + 2e^- \longrightarrow NH_2OH + H_2O \\ NH_2OH + H_2O \longrightarrow NO_2^- + 5H^+ + 4e^- \end{cases} \tag{10.7}$$

Phase 2: Nitrification proceeds with nitrite's oxidation, leading to nitrate's obtention. This phase takes place within prokaryotes called the *nitrite oxidisers*, which recruit other species of some of the same genera, e.g. *Nitrospira* and *Nitrosococcus*.

The process is described by the following equation:

$$NO_2^- + \frac{1}{2}O_2 \longrightarrow NO_3^- \tag{10.8}$$

The proportion of ammonia not transformed into organic nitrogen is efficiently processed by the decomposers/mineralisers in sustainable ecosystems,[26] where several of specialised prokaryotes proceed to denitrification.

[26] When ammonia is in excess in the environment, due to a disruption of the cycle, it can be toxic for the primary producers of the biocoenoses.

10.6.5 Anammox

Nitrification, as seen above, develops in oxic environments, but another type of ammonia oxidation, which is far from uncommon, takes place in anoxic situations; it has been called *anammox*, an acronym for "anaerobic ammonia oxidation" [34, 42]. Anammox is carried out by different bacterial genera belonging to the phylum Planctomycetes, e.g. *Candidatus* Jettenia, *Candidatus* Anammoximicrobium, Candidatus Anammoxoglobus, *Candidatus* Scalindua, *Candidatus* Kuenenia and *Candidatus* Brocadia [34]. Here, the oxidiser (electron acceptor) of ammonia is no longer oxygen but nitrate in a reaction that decomposes into two steps: (1) ammonia is modified in ammonium ion; (2) it is at the latter state that the oxider reacts. This reaction is described by the equation system:

$$\begin{cases} NH_3 + H_2O \leftrightharpoons NH_4^+ + OH^- \\ NH_4^+ + NO_2^- \longrightarrow N_2 + 2H_2O \end{cases} \tag{10.9}$$

In addition, we can see from this equation that anammox is simultaneously a denitrification process (see next subparagraph).

Anammox generally takes place in environments devoid or poor of oxygen, like in great depths of water bodies (marine and freshwater habitats), sediments and mangroves.

10.6.6 Ammonification

Carrions and excreta, during decomposition, release organic nitrogen, which is transformed into inorganic ammonia, by the same decomposers as the first cohort of diazotrophs mentioned in Sect. 10.6.3. To these ammonia molecules released from dead organic matter, we can add the ammonia molecule in excess left in the environment during nitrogen fixation. This process is called *ammonification*. These ammonia molecules are at the disposal of plants that can absorb this matter for their nitrogen fixation [5, 41].

10.6.7 Denitrification

Denitrification is a process which converts nitrate to dinitrogen gas (N_2), the inorganic form of nitrogen that is eventually released in the atmosphere. This phase completes the nitrogen cycle. This stage involves at each step (see enumeration below) different categories of denitrifying organisms (or denitrifiers) that are all chemoorganotoph,[27] e.g. bacteria (members of the genera *Bacillus*, *Paracoccus* and *Pseudomonas*) and benthic foraminifers (*Globobulimina*), *Nonionella* and *Stainforthia*) [37].

In sum, the reactions describing denitrification are as follows [5]:

1. A first category of denitrifiers reduces nitrates to nitrites, resp. $NO_3^- \rightharpoonup NO_2^-$.
2. A second category of denitrifiers starts from nitrites and, via several reactions, obtains a mix of nitric oxide and nitrous oxide resp. $NO + N_2O$ [17].
3. A third category of denitrifiers proceeds with succeeding reactions, eventually releasing inorganic N_2 in the atmosphere.

The complete redox reaction is:

$$2NO_3^- + 10e^- + 12H^+ \longrightarrow N_2^{\uparrow} + 6H_2O \tag{10.10}$$

10.7 The Cycle of Oxygen

10.7.1 Specificity of Circulating Oxygen

Atmospheric and hydrospheric circulating oxygen, mainly taking the form of gaseous dioxygen molecules at the disposal of the aerobic organisms for respiration, as well as oxygen contained in surface rocks (e.g. calcareous sediments and iron deposits), are of distal biogenic origin [18, 32]. Indeed, oxygen originally was the waste product of the first photosynthetic microorganisms that emerged on Earth ca 2.4 Ga ago, a phenomenon termed the Great Oxidation Event (GOE) [16]. The organisms responsible for that phenomenon were very probably ancestors of the present cyanobacteria, very probably already primitive cyanobacteria[28] [7]. All the living beings of that remote epoch [i.e. the beginning of the Proterozoic aeon (−2.5 to −0.542 Ga)], including the photosynthetic microorganisms in question, were necessarily prokaryotic anaerobic unicells,[29] obtaining the energy for their metabolism from glycolysis without the need for the presence of gaseous oxygen in their environment that was lacking.

---

[27] As a reminder, these organisms are supplied by organic carbon.

[28] Generally, based on the principle of parsimony, these ancestors are now considered cyanobacterial since cyanobacteria are the only organisms that can presently carry out oxygenic photosynthesis. Indeed, plants and algae perform this via their chloroplats, which are considered engulfed cyanobacteria [19].

[29] During the Archaean and the first part of the Proterozoic aeon, the primaeval atmosphere and hydrosphere of the Earth were devoid of free dioxygen; the latter was thus a lethal poison for the living beings of that time. Aerobic respiration was thus an adaptation of new organisms that appeared once dioxygen was sufficiently diffused due to the GOE.

Nowadays, all eukaryotes and most prokaryotes (bacteria and archaea) whose supramolecular organisations of respiratory systems have been identified practise aerobic respiration [23]. It is thus of the utmost vital importance that the cycle of oxygen takes place for these organisms.

It has been mentioned that the oxygen contained in surface rocks was biogenic. Indeed, most of the oxygen generated by the first cyanobacteria, because they were anaerobic, could not stay in the atmosphere but dropped on Earth's surface and was fixed in the rocks in the form of carbonates, copper oxides, iron oxides, sulphates or other oxides. All the organisms living in these early periods were adapted, for respiration, to different electron acceptors than oxygen. They lived in an anoxic "atmosphere rich in sulphur gases, with CO_2 partial pressure exceeding 2 bar that could have provided some warmth and protection against UV radiation" [28]. They were very probably chemolithotrophs: they depended on redox chemistry for their energy, and their electron donors were inorganic. The electron donors and acceptors were probably sulphates (ibid). For example, hydrogen sulfide (H_2S) could have been an electron donor and iron sulphide (FeS) an iron acceptor [40]. The living beings of that time thus rejected the gaseous dioxygen molecules as highly toxic wastes and precipitated them as iron oxides. This scenario would explain the vast extension of Archaean iron sesquioxide formation [32] and constitutes another indirect evidence of the reality of GOE.

From the GOE onwards, the evolution from primaeval almost anoxic atmosphere (with a negligible content of abiotic oxygen estimated at less than 0.2%) to the beginning of a genuinely oxic atmosphere that enabled aerobic organisms to emerge sustainably, 0.3 billion years were estimated necessary [10]. It was the elapsed time required for the gaseous dioxygen molecules to accumulate and sufficiently concentrate to elaborate an ozone layer thick enough to protect the emerging oxygenic organisms from the space's UV radiations. It was achieved when the concentration of O_2 in the air was 1% of the present, allowing the first aerobic phototrophs (probably cyanobacteria or related) to develop and evolve in the upper layers of water bodies [32, p. 52]. During the geological periods succeeding this primaeval period, the atmosphere and hydrosphere became increasingly oxic but the evolution was very slow; the relative concentration of oxygen in the air (ca 20% of its content), which is the present state, has been attained at the early Carboniferous, around 350 million years ago;[30] since that period, this figure has fluctuate, mainly according climatic conditions.[31]

[30] This datum originates from the Smithsonian Environmental Research Center, cf. website forces.si.edu/atmosphere/02_02_06.html.

[31] In the late Carboniferous and Upper Cretaceous, the oxygen rate reached 30% [4].

10.7.2 *The Oxygen Biogeochemical Cycle of Oxygen*

The two following properties characterise the biogeochemical cycle of oxygen:

1. The cycle brings the dioxygen of the air or water into the oxic part of a given biocoenosis.
2. The same cycle returns this dioxygen equimolecularly into the environment from which it was drawn so that the atmosphere or the hydrosphere remains oxygen constant.

The biogeochemical cycle of oxygen starts locally at the level of any given ecosystem and it completes globally. It starts with the well-known chemical equation that describes the quasi-equilibrium between oxygenic photosynthesis leading to glucose, the start of metabolism, and aerobic cellular respiration from glucose:

$$CO_2 + 6\,H_2O \rightleftarrows C_6H_{12}O_6 + 6\,O_2^{\uparrow} \tag{10.11}$$

- The biogeochemical cycle of oxygen starts with the reaction from left to right, describing oxygenic photosynthesis: each primary producer (plant, alga or cyanobacterium) takes in one molecule of carbon dioxide and six molecules of water from the surrounding fluid $\mathcal{F}$ (atmosphere or hydrosphere) and launches a chemical reaction that produces six molecules of gaseous oxygen, released into $\mathcal{F}$ and one glucose molecule. The six oxygen molecules liberated into $\mathcal{F}$ constitute the overwhelming source of the oxygen available to the aerobic organisms for their respiration, thus the start of the cycle.
- When the six molecules of a dioxygen molecule are released by one phototroph, the same amount of dioxygen molecules is taken out of $\mathcal{F}$ by one eukaryote or aerobic prokaryote by the converse phenomenon, aerobic respiration. The latter is described by the equation read from right to left. Indeed, the quantity of oxygen absorbed by the incriminated taxon consumes one molecule of glucose, which triggers the metabolic pathways. This process deprives $\mathcal{F}$ of six molecules of a dioxygen molecule.

At the level of one individual or a pair of individuals, the equilibrium described by Eq. (10.11) looks perfect. A problem arises when we extrapolate this balance to the entire oxic part of the biocoenosis. Indeed, the aerobic organisms comprise the aerobic primary producers and the aerobic consumers depending on them, whereas only the aerobic primary producers emit dioxygen. Since, theoretically, no added dixogen arises from other sources, there should be a depletion of dioxygen. However, the concentration of O_2 remains constant in the atmosphere and hydrosphere so long that humans do not interfere. It has been estimated that it takes ca 100 years for a dioxygen molecule used by respiration, thus disassembled, to be replaced by a newly generated dioxygen molecule [10, 19]. Since this process is continuous, the constancy of gaseous oxygen in $\mathcal{F}$ is conserved. Thus, though this situation appeared paradoxical, the apparent paradox has been explained.

1. It has been shown that about 0.1–0.2% of the organic carbon is not oxidised and is buried in sediments and that the burial of one mole of organic carbon liberates one mole of dioxygen [10]. This organic carbon remainder stems from buried organic matter. A reported case of release of O_2 in seawater is the result of the reaction between buried pyrite (FeS_2), sulphate ions (SO_4^-), ferric oxide and organic matter [19].
2. Atmospheric water vapour and nitrous oxide molecules can be broken down in a process called photolysis. High-energy UV radiation disassembles these molecules, and we have, as reactions
$$\begin{cases} 2\,H_2O + h\nu \longrightarrow 4\,H + O_2 \\ \quad \text{Photolysis of water} \\ 2\,N_2O + h\nu \longrightarrow 4\,N + O_2 \\ \quad \text{Photolysis of nitrous oxide} \end{cases}$$
The photolysis of water can occur in the thylakoids of cyanobacteria and the chloroplasts of green algae and plants [1, 9].

In sum, the quasi-equilibrium between the oxidative photosynthesis, added to the mechanisms described in points 1 and 2 explains the quasi-constancy of oxygen across the geological periods since the Carboniferous.

10.8 The Cycle of Phosphorus

The phosphorus cycle starts within the lithosphere amidst igneous rocks like apatites, which are groups of phosphate minerals. Mineral phosphorus is a limiting driver of biocoenoses since it accounts for less than 1% of simple bodies; thus, it significantly impacts productivity. Phosphorus is a vital element, crucial to nucleic acids and ATPs.

In the case of terrestrial ecosystems: Mineral phosphorus is washed away by lakes, rivers, or surface run-offs and introduced into terrestrial ecosystems, where the primary producers absorb it. Then, it circulates along the food webs across all the consumers of the biocoenosis. After the organisms die, the organic phosphorus is decayed by decomposers/mineralisers. Microorganisms retransform the element into mineral orthophosphates available to plants, and the cycle is complete.

In the case of aquatic ecosystems: Phosphorus penetrates aquatic ecosystems via run-off waters, notably rivers. The latter ones, rich in phosphorus, continuously feed the oceans with this element, which allows the development of phytoplankton. These species form the first level of a trophic web that encompasses all the successive aquatic trophic levels. When the organisms die, the small ones turn into marine snows and the larger into falls. Specialised prokaryotes transform the phosphorus of this organic matter from its organic into an inorganic state.

10.9 The Cycle of Sulphur

Sulphate (SO_4^{2-}) constitutes the main distal source of the sulphur cycle, which is crucial since its main element, sulphur (S), is a vital constituent of two amino acids, cysteine and methionine, which are extremely important for proteins. Indeed, these amino acids play a crucial role in forming the tertiary structures of the proteins, structured by disulphide bonds. It is found in all kinds of natural waters (freshwater, rainwater, marine water), thus at the disposal of plants, algae and cyanobacteria. It is incorporated, via trophic network, into all the species of any biocoenosis and released as abiotic factors by decayed corpses and excreta via the chains of decomposers/mineralisers. The last decomposers and mineralisers of the chain that are concerned with sulphur are heterotrophic bacteria liberate hydrogen sulphide (H_2S) from the proteins. Hydrogen sulphide are eventually transformed into sulphate by chemosynthetic bacteria, e.g. the genus *Thiobacillus*. The cycle is then complete [32, p. 59].

References

1. Begon M et al (2006) Ecology: from individuals to ecosystems. 4th edn, Blackwell Publishing.
2. Bergene T (1996) The efficiency and physical principles of photolysis of water by microalgae. Int J Hydrogen Energy 21:189-194, https://doi.org/10.1016/0360-3199(95)00071-2
3. Berner RA (1990) Atmospheric carbon dioxide levels over phanerozoic time. Science 249:1382–1386.
4. Berner RA, Landis GP (1988) Gas bubbles in fossil amber as possible indicators of the major gas composition of ancient air. Science 239:1406-1409, https://doi.org/10.1126/science.239.4846.1406
5. Bernhard A (2010) The nitrogen cycle: processes, players, and human impact. Nat Educ Knowl 3(10):25
6. Beijerinck MW (1901) Über oligonitrophile Mikroben Zentralbl Bakterol Parasitenkd Infektionskr Hyg Abt. II 7:561-582.
7. Blankenship RE, Hartman (1998) The origin and evolution of oxygenic photosynthesis. Trends Biochem Sci 23:94-97.
8. Brown GC, Mussett AR (1993) The Inaccessible Earth: an integrated view of its structure and composition, 2nd edn, Springer.
9. Bulak M et al (2022) Quantification of O_2 formation during UV photolysis of water ice: H_2O and $H_2O : CO_2$ ices. Astr Astrophys 657:A120, https://doi.org/10.1051/0004-6361/202141875
10. Catling DC, Claire MW (2005) How Earth's atmosphere evolved to an oxic state: A status report. Earth Planet Sci Lett 237:1-20, https://doi.org/10.1016/j.epsl.2005.06.013
11. Cox AN (2000) Allen's astrophysical quantities. 4th edn, Springer Science and Media Business, https://doi.org/10.1007/978-1-4612-1186-0
12. Falkowski PG et al (2000) The global cycle; a test of our knowledge of Earth as a system. Science 290:291-296, https://doi.org/10.1126/science.290.5490.291
13. Falkowski PG, Godfrey LV (2008) Electrons, life and the evolution of earth's oxygen cycle. Phil Trans R Soc B 363:2705–2716, https://doi.org/10.1098/rstb.2008.0054
14. Gaines RV et al (1997) Dana's new mineralogy: the system of mineralogy of James Dwight Dana and Edward Salisbury Dana. 8th edn, Wiley.

15. Garrone E et al (2017) How many chemical elements are there in the universe? a (not so) *Bohring* question. World J Chem Educ 5:20-22.
16. Gumsley AP et al (2017) Timing and tempo of the Great Oxidation Event. Proc Natl Acad Sci USA 114:1811-1816, https://doi.org/10.1073/pnas.1608824114
17. Hakata M et al (2003) Conversion of the nitrate nitrogen and nitrogen dioxide to nitrous oxides in plants. Acta biotechnol 23(2-3):249-257, https://doi.org/10.1002/abio.200390032
18. Ho RYN et al (1995) Biological reactions of dioxygen: an introduction. In: Valentine JS et al (eds) Active oxygen in biochemistry, p 1-36, Springer, Dordrecht.
19. Hocke K (2023) Oxygen in the Earth system. Oxygen 3:287-299, https://doi.org/10.3390/oxygen3030019
20. Hutchinson GE (1957) A treatise on limnology. John Wiley & sons
21. Jung M-Y at al (2023) Ammonia-oxidizing archaea possess a wide range of cellular ammonia affinities. ISME J 16:272-283, https://doi.org/10.1038/s41396-021-01064-z
22. Kittel C et al (1965) Mechanics. Berkeley physics course. Macgraw-hill, New York
23. Magalon A et al (2012) Supramolecular organization in prokaryotic respiratory systems. In: Poole RK (ed) Advances in bacterial respiratory physiology, Advances in microbial physiology, vol 61, pp 217-266, https://doi.org/10.1016/B978-0-12-394423-8.00006-8, Elsevier - ScienceDirect.
24. Mahan BH (1970) University Chemistry. Addison-Wesley, Reading (Mass)
25. Meinesz FV (2012) The Earth's crust and mantle, vol 1. Elsevier
26. Merrill GP (1897) A treatise on rocks, rock weathering and soils. MacMillan Co, New York and London
27. Mills SJ et al (2009) The standardisation of mineral group hierarchies: application to recent nomenclature proposals. Eur J Mineral 21:1073–1080, https://doi.org/10.1127/0935-1221/2009/0021-1994
28. Nisbet EG, Sleep NH (2001) The habitat and nature of early life. Nature 409:1083-1091, https://doi.org/10.1038/35059210
29. Peacock MA, Fuller RE (1928) Chlorophaeite, sideromelane and palagonite from the Columbia river plateau. Am Min 13:360–383
30. Price NPJ (1999) Carbohydrate determinants of Rhizobium-legume symbioses. Carbohydr Res 317:1-9.
31. Ralph J (2017) The most common minerals on the earth's crust - url:http://www.mindat.org/a/common_minerals
32. Ramade F (1982) Elément d'écologie, écologie appliquée: action de l'homme sur la biosphère (In French). 5th edn, Mc Graw-Hill, Paris.
33. Ramade F (2009) Eléments d'écologie: Ecologie fondamentale (In French) 4th edn, Dunod, Paris.
34. Ren Z-Q et al (2022) A review of anammox-based nitrogen removal technology: From microbial diversity to engineering applications. Bioresour Technol 363:127896, https://doi.org/10.1016/j.biortech.2022.127896
35. Revelle R, Suess HE (1957) Carbon dioxide exchange between atmosphere and ocean and the question of an increase of atmospheric CO_2 during the past decades. Tellus 9:18–27
36. Ricklcfs RE (2008) The economy of nature. 6edn, Freeman and co, New York.
37. Risgaard-Petersen N et al (2006) Evidence for complete denitrification in a benthic foraminifer. Nature 443:93-96, https://doi.org/10.1038/nature05070
38. Schultz JE, Weaver P (1982) Fermentation and anaerobic respiration by *Rhodospirillum rubrum* and *Rhodopseudomonas capsulata*. J Bacteriol 149:181-190.
39. Slocum G (1955) Has the amount of carbon dioxide in the atmosphere changed significantly since the beginning of the twentieth century. Mon Weather Rev 83:225–231, https://doi.org/10.1175/1520-0493(1955)083<0225:HTAOCD>2.0.CO;2
40. Southam G et al (2015) Geology, life and habitability. In: Schubert G (ed) Treatise on Geophysics, 2 edn, p. 473-486, Elsevier, Amsterdam (NL).
41. Strock JS (2008) Ammonification. In: Jørgensen SE, Fath BD (eds) Encyclopedia of Ecology, p. 162-165, Elsevier, Amsterdam (NL).

42. Strous M et al (1999) Missing lithotroph identified as new planctomycete. Nature 400: 446-449.
43. Terry Collins Assoc (2019) Scientists quantify global volcanic CO_2 venting; estimate total carbon on Earth. Peer-Reviewed Publication i AAAS EurekAlert, website: www.eurekalert.org/news-releases/736161
44. Udden JA (1914) Mechanical composition of clastic sediments. Bull Geol Soc Am 25:655–744
45. Valero A et al (2011) The crepuscular planet. A model for the exhausted atmosphere and hydrosphere. Energy 36:3745-3753, https://doi.org/10.1016/j.energy.2010.07.017
46. Von Cammerer S, Baker N (2007) The biology of transpiration. From guard cells to globe. Plant Physiol 143:3, https://doi.org/10.1104/pp.104.900213
47. Wentworth CK (1922) A scale of grade and class terms for clastic sediments. J Geol 30:377–392
48. Willison JC, Magnin J-P (2013) Role and evolution of endogenous plasmids in photosynthetic bacteria (chapter 8). Adv Bot Res 66:227-265, https://doi.org/10.1016/B978-0-12-397923-0.00008-4
49. Woodwell GM et al (1983) Global deforestation: contribution to atmospheric carbon dioxide. Science 222:1081–1086.
50. Zahran HH (1999) *Rhizobium*-legume symbiosis and nitrogen fixation under severe conditions in an arid climate. Microbiol Mol Biol Rev 63: 968-989.

Chapter 11
The Material Approach-II: The Physical Parameters

11.1 The Physical Parameters of the Environment

Chapters 9 and 10, respectively, treated the energetical and chemical approaches to ecosystems. The present chapter will discuss the abiotic factors of the environments we have yet to treat, i.e. physical parameters, which crucially determine the existence of the living beings composing the biocoenoses. We will now study some of the physical parameters, namely (1) temperature, (2) humidity, (3) light and (4) pH.

Like the chemical factors treated in Chap. 10, these physical parameters are directly exploitable by the populations of the biocoenoses and, as such, constitute first-order abiotic factors (see Sect. 2.3). Each abovementioned physical parameter takes on values within an interval between a minimal and a maximal value. A set of these intervals corresponds to the life conditions of the organisms forming a given biocoenosis [1].

Biocoenoses vary geographically, strictly depending on climatic parameters. Thus, climate determines the faunal, floral, fungal, and microbial composition of a biocoenose on a given surface of the Globe. Let us now proceed with the first physical factor mentioned: temperature.

11.2 Temperature

11.2.1 Different Adaptive Mechanisms to Temperature

The vital mechanisms of a living being are only possible when its internal water is liquid. This situation occurs when, at standard atmospheric pressure (1 atm =

Supplementary Information The online version contains supplementary material available at https://doi.org/10.1007/978-3-031-90656-5_11.

M. Salomon, *Integrative Ecology*, https://doi.org/10.1007/978-3-031-90656-5_11

760 mm Hg = 101.3 kPa), the temperature lies between 0° C and 100° C. Most organisms are mesophylic, i.e. they live in climatic conditions around 25° C (between 0° C and 45° C) at standard atmospheric pressure. The eukaryotes that can live at temperatures above 45° C are infrequent and inconspicuous and were unknown until ca two decades ago. Recent research has unveiled the existence of such eukaryotes [52]. Most organisms that live at temperatures higher than 45° C, often next to hot springs, are prokaryotes; these living beings are termed thermophiles.[1] Some species, called hyperthermophiles, were found next to hot springs or hydrothermal vents that reach temperatures far above 75° C [29, 45].

- Thus, at standard atmospheric pressure, in terrestrial habitats, the highest temperature where living beings are found exceeds 100° C. This record was recorded in solfataric fields, hot springs, and subterranean oil.
- At high pressure in deep marine habitats, next to submarine hydrothermal vents, extremophiles (see Footnote 1) can be found at very high temperatures, e.g. the methanogenic archaeon, *Methanopyrus kandleri* [46] observed at temperatures above 122° C.

To adapt to such high temperatures, the proteins of thermophilic prokaryotes show special adaptations that make them significantly different from the proteins of mesophiles. Indeed, mesophilic proteins denature at ca 45° C–50° C, whereas thermophilic proteins function at far higher temperatures.

On Earth's surface, the temperatures infrequently rise above 50° C, except in the geothermic springs and hottest deserts. On the other hand, cold temperatures below 0° C, thus below water's freezing point, are widespread and found in extensive stretches of the planet, covering 75% of the Earth's surface (mountainous and polar regions and deep sea) [21]. Many ectothermic species[2] are adapted to these very low temperatures. they are called *psychrophiles*. This situation looks paradoxical since the cells of any organism, including the psychrophilic ones, die when their inner water freezes. The reason is that ice crystals, which form during water freezing, have cutting edges that irremediably deteriorate the cell structures, interrupting the intracellular biochemical processes. This denaturation unavoidably leads the organism to its death.

[1] The organisms that can live at very high or very low temperatures, high pressure, or highly acidic or alkaline environments are called extremophiles. Thousands of these species have been recently discovered and analysed; the most extremophilic species are simultaneously thermophilic (adapted to very high temperatures) or psychrophilic (adapted to very low temperatures), barophilic (adapted to very high pressures), and acidophilic (adapted to very low pH) or alkaliphilic (adapted to very high pH) [47].

[2] Endothermic animals (mammals and birds) have sophisticated internal physiological mechanisms that render them far more autonomous to ambient temperature than the ectothermic species (all the organisms except the endotherms). Their body temperature is constant and far less tributary to the ambient temperature than the other species, the ectotherms. When they hibernate, their body temperature remains well above 5° C [19].

Psychrophilic species had to find adaptive strategies to respond to extreme cold's very harsh life conditions. Different mechanisms underpin these strategies:

1. Many species, mainly insects, prevent ice formation in their cells by generating a metabolism that maintains their body temperature above the freezing point [4]. This has been shown experimentally with the large-bodied Mediterranean dung beetle *Chelotrupes momus*, a species endemic to southern Spain. A sample of 15 individuals was placed in laboratory conditions, with air temperatures experimentally decreased from 15° C (the natural ambient winter temperature of the species) to −14° C. At the coldest temperatures, the insects maintained their body approximately at 6° C without warming up their bodies above this temperature, thanks to the role of their exoskeleton that equilibrates the absorption of infrared radiation and heat dissipation [25].
2. Other organisms have another strategy to survive the extreme cold. They let their body temperature be brought to the ambient one but simultaneously develop intracellular mechanisms within intracellular water that render them resilient against the lethal effect of the cold.

    - Some species prevent their inner cell water from freezing at 0° C by extruding the ice crystals out of it, e.g. the blossom weevil (*Anthonomus pomorum*) [43].
    - Other taxa, such as the Antarctican fishes or green algae like *Chlamydomonas*, develop tolerance to sea salt, which enhances the efficiency of osmoprotectants,[3] like glycerol, as an antifreeze agent [14, 36].

3. Supercooling consists of mechanisms that lower the freezing point of the inner cell waters below the standard water freezing temperature (0° C at 1 atm pressure) and has been observed mainly among insects of the temperate zone in winter, particularly in North America, where temperatures can be low, and in circumpolar aquatic ectotherms. Four mechanisms have been mainly reported in the literature:

    a. the dissolution of antifreeze proteins (e.g. the THPs that depress the hemolymph freezing point) [49],
    b. the elimination of the normal endogenous ice nucleators triggered at 0° C and their replacement by ice nucleators that initiate at temperatures below 0° C down to 25° C [24],
    c. the use of endogenous chemicals that launch supercooling, e.g. glycoproteins, [24],
    d. the toleration of partial freezing of the body so long that the inner cell water does not freeze (observed in eggs of a marine nematode) [54].

Generally, insects are better candidates for supercooling than vertebrates since this phenomenon seems more efficient with small-sized animals. It has been observed that the smaller the terrestrial arthropods tend to be, the more likely these organisms will be prone to supercooling [24]. However, some ectothermic

[3] This phrase means "a substance that lowers osmotic pressure".

vertebrates of cold temperate and arctic areas overwinter without damage thanks to supercooling, for example, the wood frog (*Rana sylvatica*) and the painted turtle (*Chrysemis picta*) that both live in Northern America beyond the Arctic Circle [10].

Two points are important to remember:

- In the psychrophiles studied, all the cold-proof mechanisms mentioned above are followed by an activity slowdown and even, in the most extreme cases, by a temporary vegetative state, awaiting the active phase accompanying the return of warmer conditions.
- Psychrophily consists of cold-proof mechanisms that fit ectotherms. Endotherms avoid death by freezing via other mechanisms, i.e. (1) migration, (2) hyperactivity related to body conformation, and (3) hibernation.

11.2.2 Temperature Optimisation: Constraints on Metabolism, e.g. on Proteins and, in Particular, on Enzymes

Temperature has divergent effects on vital processes.

1. Heat increases the kinetic energy of the molecules within the bodies and catalyses the metabolism's chemical reactions. When the temperature rises 10° C, the velocity of these chemical reactions doubles to quadruples, according to the species.
2. At high temperatures, proteins, notably enzymes, lose their stability. Beyond a given threshold, they become dysfunctional and eventually denature.

The optimal temperatures of an organism fall into an interval of equilibrium states corresponding to situations where proteins, particularly enzymes, sway to and fro between (1) the conservation of their conformation, on the one hand, and (2) their activity, on the other hand:

- Too low a temperature slows down the activity of the enzymes and impairs the vital chemical reactions of metabolism.
- Too high a temperature brings instability and returns us to point (11.2) just above.

11.3 Humidity: The Presence of Water in the Environment

Water in its liquid form is absolutely vital to all organisms. Its physical properties explain why this molecule is intricately related to the existence of living beings: water's thermic inertia, the link between water and soil, and air's water content.

11.3.1 Thermic Inertia of Water

Water buffers temperature variations and phase transitions (around the liquid phase), which ensures favourable life conditions (temperatures between 0 and 100° C, large proportion of water in its liquid phase). This property is largely due to H_2O's relatively high specific heat capacity.

- Water is temperature inert. To modify the temperature of 1 kg of this liquid by 1° C, much heat (4184 J) must be added to or removed from water.
- Water is resistant to phase transition, i.e., it withstands to (1) freezing, the passage from the liquid phase to the solid phase (ice), (2) condensation, the passage from the gaseous phase (water vapour) to the liquid phase or (3) evaporation, the passage from the liquid phase to the gaseous phase.
  - The energy of fusion, i.e. the energy needed for 1 kg of water to freeze at standard atmospheric pressure (corresponding to the temperature of 0° C), is 335 J (or 80 cal).
  - The energy of vaporisation, i.e. the energy needed for 1 kg of water to evaporate at standard atmospheric pressure (corresponding to the temperature of 100° C), is 2243 J (536 cal).
- Above 4° C, water's density is 1, but below that temperature, water's density decreases with decreasing temperature, which explains ice's buoyancy.
- Last but not least, water has a strong capacity (1) to dissolve inorganic compounds (mineral salt), chemical bodies fit to be absorbed by plants as nutrients and (2) to enable these chemicals, fundamental for the biocoenoses, to enter them. Indeed, water is a potent solvent that spontaneously dissolves mineral salts in the soil interstitial waters.

11.3.2 Link Between Soil and Water

Terrestrial plants draw the water they need from the soil. As mentioned above, this water is loaded with chemical elements essential for plants (and other primary producers) and the organisms depending on them (the consumers). Soil is a complex system composed of different interacting components, i.e. (1) a layer of decaying organic matter, (2) a community of biotic agents, (3) a network of interstitial water and (4) soil grains.

We will sketch a brief overview of these components.

Decaying organic matter:

Organic matter in the process of decomposition/mineralisation comprises humus (dead leaves and roots of plants), carrion, and excreta.

Biotic agents:

These organisms form a community of animals, fungi, plants, and microorganisms that fulfil the roles of burrowers, decomposers, and mineralisers. They carry out the continuous degradation, decomposition, and mineralisation of decaying organic material and are, as such, an essential part of the soil.

Interstitial water network:

This network concerns the circulating groundwater that feeds the primary producers, mostly terrestrial plants, with the dissolved mineral salts, which are vital for each terrestrial biocoenosis.

Soil grains

The water content of soil tightly depends on the sizes of soil grains. In increasing order of volume, the main soil grains are the following: (1) clay grains, (2) silt grains and (3) sand grains.

These soil grains originate from the bedrock and constitute a genuine soil skeleton with a solid capacity for retaining water molecules. Indeed, water has a strong potential for clinging to non-hydrophobic material like soil grains. The water loaded with vital chemical agents that penetrate the biocoenoses stems mainly from the water linked to the soil grains.

Soil granulometry, the study of soil grains, plays a crucial role in explaining the aspect of an ecosystem. A rule of geometry tells us that a sphere of surface S and volume V has a relative surface $\left(\frac{S}{V}\right)$ inversely proportional to its radius r. Indeed,

$$\frac{S}{V} = \frac{\cancel{4}\cancel{\pi}\cancel{r^2}}{\frac{\cancel{4}}{3}\cancel{\pi} r^{\cancel{3}}} = \frac{3}{r} \tag{11.1}$$

This result means that proportionally, the greater the radius is, the smaller the number of molecules that can stick to its unit surface will be. For this reason, if we consider equal clayey, silty or sandy surfaces, we can expect that, on average, a clayey surface will have a greater water content than a silty surface, and a silty surface will have a greater content than a sandy surface.

Consequently, the larger the soil grains are, the less dense the water molecules will be and the more arid the soil will be. Hence, sandy soils are less fertile than silty soils, which are less fertile than clayey soils.

11.3.3 Humidity

The content of water vapour in the air defines humidity. The more the atmosphere is humid, the more it can condense water in the form of rain. Humidity is a parameter essential to ecosystems and biome differentiations, as is the link of water with soil. It depends on multiple types of initial evaporations, such as (1) evaporation and evapotranspiration from plants, (2) evaporation from animals, (3) evaporation from rocks and (4) evaporation from vast water body stretches. The first three types of evaporation accompany heat radiation, and the last one is heat convection.

Evaporation and evapotranspiration from plants: Plants, particularly tall trees (and this salient in the rainforests), overwhelmingly contribute to significant ambient humidity through evapotranspiration.

Evaporation from animals:
Abounding animal biomass releases a non-negligible quantity of water through different physiological mechanisms (e.g. sweating and micturition). This water evaporates and can constitute a significant contribution to ambient humidity.

Evaporation from rocks:
Water trapped in soils (see Sect. 11.3.2 in the entry "soil grains") directly evaporates when the soil is naked. The more arid the soil is, the larger the soil grains will be, and the less water will evaporate. Rocks evaporate more water as their *hydrophilicity*, i.e. their water absorption capacity increases.

The three types of evaporation and evapotranspiration mentioned above accompany a short water cycle contrary to the type of evaporation that we will deal with, the evaporation from vast water body stretches (oceans and seas, lakes and rivers).

Evaporation from vast water body stretches:
This source of evaporation is overwhelmingly the largest contributor to the humidity of most ecosystems, with perhaps the notable exception of the tropical rainforests where the evapotranspiration of the high trees plays a considerable role in the ambient humidity next to the strong influence of monsoons. The type of hygrometry we are dealing with presently consists of the water transport under the influence of heat convection tightly depending on temperature, wind regime and *orographic* conditions[4] [7, 41].
The convection transport of water starts with air masses loading water vapour and thereafter rising above the water body stretches (the initial stage). The air masses are then driven by winds with different regimes according to their geographic locations (e.g. anticyclones, depressions, trading winds, monsoons). This initial convection transport of water is associated with a potential energy load (see the water cycle, cf. Sect. 10.4), which explains a potential capacity to launch precipitations. This potential energy naturally diminishes while the air mass movements advance due to gradual condensations, either as precipitations of weak intensity due to a natural loss of kinetic energy (according to the second law of thermodynamics) or in the form of heavy rains in contact with obstacles (e.g. mountains, violent updrafts owed to temperature contrasts or thunderstorms).
Various factors greatly influence the dynamics of the air masses loaded with humidity, temperature and orography:

[4] This phrase refers to influence of the presence of mountains and their impact linked to their altitudes and extensions.

Temperature:
The efficiency of the displacement of the water-loaded air masses, which I define as their sustainability and capacity to cover long distances, depends on the potential of the initial stage, i.e., the quantity of water that the air masses can load as water vapour then, which is called the *initial hygrometry*. The initial hygrometry is strongest at a mean temperature of 24° C–25° C, with a narrow minimum maximum deviation. The efficiency of the displacement of the water-loaded air masses diminishes rapidly. It becomes far weaker when the temperature is low (below 10° C on average, like in polar areas) or high (above 28° C on average).

Orography:
The volumes of the air masses driven by the prevailing winds are determined by their water vapour load. This feature of the air masses and the initial stages of their displacements set their altitudes. When they collide with mountain ranges, the humidity-loaded air masses driven by the prevailing winds frequently condense or deposit their water vapour, i.e. respectively precipitate their water in the form of (1) rain or (2) snow or hail. This phenomenon removes humidity from the displacing air masses that become drier, thus with less potential energy when they move beyond the mountain ranges, which explains the relative dryness of continental climates compared with the maritime ones. This factor is more pronounced as the mountainous range is wider.

11.4 Relationship Between Temperature and Rainfall Height

The annual precipitation height P in mm is negatively correlated to the mean annual temperature T in degrees Celsius when the temperatures and humidity degrees are not extreme (polar, desert or equatorial).

This trend is illustrated by an analysis I performed on data published by a team of Chinese researchers [50, Table 1] (cf. Table 11.1). From these data, I carried out data analyses (regression linear analyses, cf. Fig. 11.1), corroborating the following interpretation.

Four groups can be interpreted, but we must be cautious due to the small samples inferred. Two groups show a negative correlation between T and rainfall P:

1. The group comprising the European localities of temperate and subarctic latitudes shows an east-west cline with decreasing T and increasing temperature P. This would correspond to a semi-continental cold temperate climate. The regression fits the points, and the correlation test is significant.
2. The group comprising the central Siberian localities also shows a clinal gradation showing that when T decreases, P increases, but the cline is not clearly

Table 11.1 Table displaying the coordinates of 29 localities across continental Eurasia, their geographical coordinates and their temperature and rainfall height. *Lat* latitude, *Lon* longitude, *Alt* altitude (m), *T* mean annual temperatures (° C), *P* annual precipitation heights (mm), *loc* localities. *Be* Berlin, *Gr* Greifswald, *Ar* Arkona (Germany), *Kr* Krakow (Poland), *Ri* Riga (Latvia), *Es* Espoo, *Ro* Rovaniemi, *Ku* Kuopio (Finland), *Sp* Saint-Petersburg, *Mu* Murmansk, *Ka* Kalinin, *Mo* Moscow, *Ak* Arkhangelsk, *Ki* Kirov, *Pe* Perm, *Pc* Pechora, *An* Anderma, *Sa* Salekhard, *Kh* Khanty-Mansiysk, *Om* Omsk, Ba Barabinsk, *En* Enisejsk, Ir Irkutsk, *Ol* Olenek, *Bg* Bagdarin (Russia), *Wu* Wulumqu, *Qi* Qiqihar (China), *Ul* Ulaanbaataar (Mongolia)

Lat	Lon	Zone	Alt	T	P	loc
52°28'N	13°24'E	1	48	10	574	Be
54°06'N	13°25'E	1	2	9.2	635	Gr
54°41'N	13°26'E	1	42	9	564	Ar
50°04'N	19°51'E	1	205	9.4	647	Kr
56°58'N	24°04'E	1	3	5.6	661	Ri
56°11'N	24°50'E	1	30	6.6	706	Es
66°30'N	25°46'E	2	107	2.3	517	Ro
62°53'N	27°38'E	1	116	7.8	670	Ku
59°58'N	30°18'E	1	4	5.6	664	Sp
68°58'N	33°03'E	2	46	−0.2	489	Mu
56°54'N	35°54'E	1	31	8	649	Ka
55°45'N	37°34'E	1	157	5.4	695	Mo
64°35'N	40°30'E	2	13	1	548	Ak
58°39'N	49°37'E	1	164	7.1	653	Ki
58°01'N	56°11'E	3	161	2.1	616	Pe
65°07'N	57°06'E	2	56	−1.8	578	Pc
69°46'N	61°41'E	4	53	−7	443	An
66°32'N	66°40'E	4	16	−6.2	446	Sa
60°58'N	69°04'E	2	40	−1.3	563	Kh
55°01'N	73°23'E	5	94	1.6	401	Om
55°20'N	78°22'E	5	120	1.1	371	Ba
55°02'N	82°54'E	5	162	0.9	422	No
43°47'N	87°37'E	6	918	7.5	303	Wu
58°27'N	92°09'E	5	78	−1.5	491	En
52°16'N	10°21'E	5	485	0	445	Ir
47°56'N	10°59'E	7	1338	−0.3	249	Ul
68°30'N	11°26'E	8	220	−11.8	300	Ol
54°28'N	11°35'E	4	903	−5.3	438	Bg
47°23'N	12°55'E	9	147	4.3	581	Qi

interpretable as an ecogeographical gradation. The regression fits the points, and the correlation test is significant.

3. The groups, respectively, comprising the western and the Siberian Arctic localities, do not show a significant correlation between T and P. One could interpret this as a characteristic of polar regions, but again, one should be cautious because the samples are small.

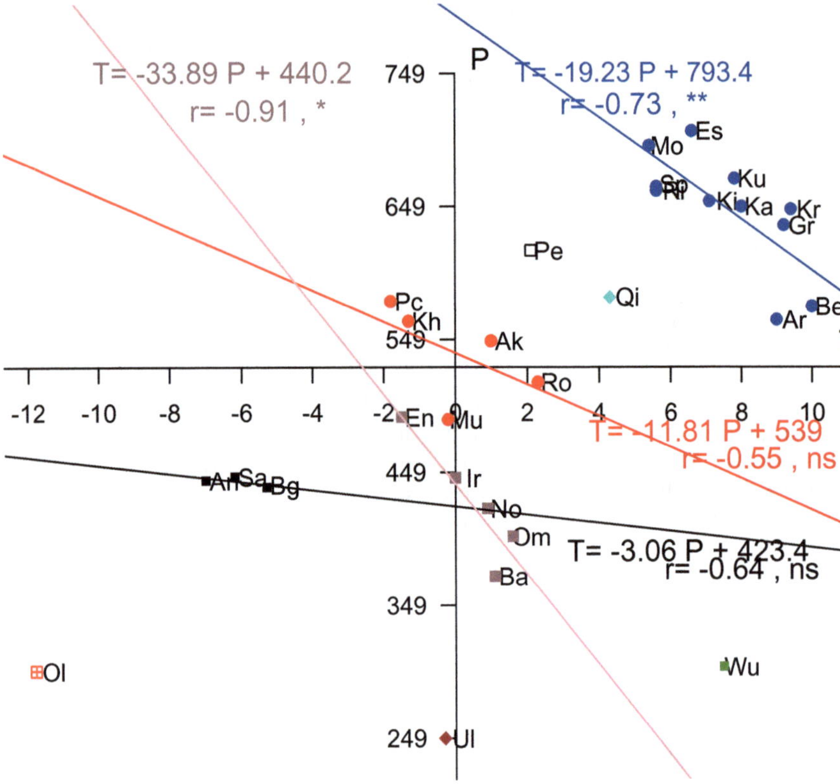

Fig. 11.1 Graphic showing the relationship between temperature T and rainfall height P across continental Eurasia. Each symbol represents a locality. The analyses (see above) highlighted nine clear-cut groups of localities: (1) the European localities (*blue bullets*), (2) the western Arctic localities (*red bullets*), (3) an isolated locality (Pe) (*white square*), (4) the Siberian Arctic localities (*black squares*), (5) the central Siberian localities (*brown squares*), (6) an isolated locality (Wu) (*light brown squares*), (7) an isolated locality (Ul) (*brown rhombus*), (8) an isolated locality (Ol) (*brown rhombus*), (9) an isolated locality (Qi) (*green rhombus*). The full forms of the locality names of the figure are in the legend of Table 11.1

11.5 Light

11.5.1 The Action of Light on Plants and Beyond

Light, a primordial energy source for most known ecosystems, is recognised as a double physical phenomenon. It simultaneously constitutes a train of electromagnetic waves and a flow of photons, i.e. energetic particles with a quasi-null mass [12, 16]. Light is associated with two important parameters: its energy and wavelength. Light energy is a decreasing function of wavelength, i.e., shorter wavelengths correspond to higher energies (Fig. 11.2).

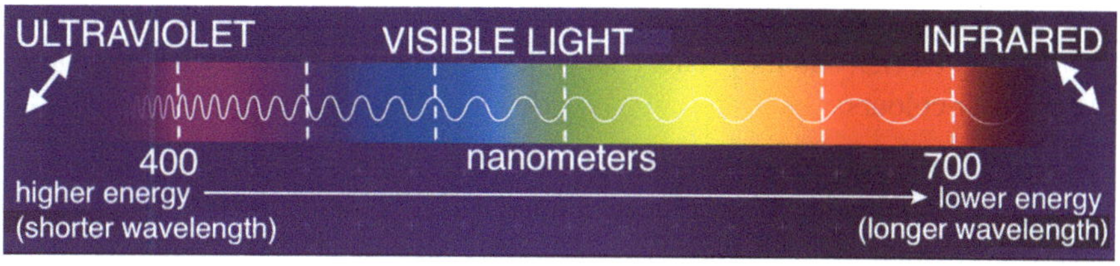

Fig. 11.2 Light spectrum with correspondence between colour, light energy and wavelengths. Image taken from the website of the Webb Space Telescope platform (https://webbtelescope.org/coopyright, image belonging to the Public Domain, no claim to copyright.)

For example, blue, having a relatively narrow visible wavelength, will be endowed with higher energy than red, which has a broader wavelength. Rays corresponding to wavelengths below 380 nm are called *ultraviolet* and are the most energetic light waves. X-rays belong to this category of waves; they are loaded with very high energy and can, above a given concentration, be very harmful to the biocoenoses. The stratosphere's ozone, a layer of the atmosphere located at altitudes between 10 and 50 km above ground level, absorbs most of this harmful UV light. The stratosphere is thus a screen protecting the lower strata of the atmosphere, and consequently Earth's surface, against much of the UV light. The portion of light reaching Earth's surface contains just a relatively reduced quantity of UV light, in addition to visible and infrared lights.

This portion of light concentrates all the energy taken in by most known biocoenoses via photosynthesis. The quantity of the intake varies between the different habitats of the planet:

- Water absorbs little of the visible light rays, which explains the relative colourlessness of this liquid; on the other hand, it absorbs all the infrared rays the first few metres below the surface. This phenomenon vastly explains that though photosynthesis is far more important in aquatic habitats than in terrestrial habitats,[5] water shows more diffused green patches in aquatic habitats, than in terrestrial habitats where the green colour is far more present.
- The pigments found in the leaves and stems of terrestrial plants are the main visible light absorbers and reflectors. (1) These chemicals take in the useful part of light energy, i.e., the energy diffused in the plants and potentially to the rest of the biocoenosis. (2) They reflect the light carrying the energy that is not absorbed. Among the main pigments, we can count chlorophyll and carotene.
  - Chlorophyll absorbs the red, a weak energy carrier, and the violet, on the contrary, a very potent energy carrier. It reflects the green, a weak energy carrier.

[5] Photosynthesis in aquatic habitats account for at least half of the total photosynthesis taking place on Earth, due to aquatic plants, algae, cyanobacteria and phytoplankton [17].

- Carotenes take in the blue, a potent energy carrier, and the green, a weak energy carrier, and they reflect the yellow, orange and red—three weak energy carrier colours.

These two last points explain the colours of the leaves we observe in the trees and shrubs of deciduous temperate forests and woodlands: green dominates in spring and summer, and yellow, orange, and red in autumn. These pigments are also found within the phototrophs of the aquatic environments.

Most importantly, pigments are the hubs where photons release the electrons that will launch the entire process of biosynthesis and metabolism within the photosynthesisers, with the emergence of the increase of free energy without violating the second law of thermodynamics, which will be transferred to the biotic agents of the biocoenosis through the trophic web. The leaves are thus the laboratories that generate the energy circulating throughout the entire biocoenosis.

11.5.2 Distribution of Light Across Time and Space: Seasonal and Latitudinal Effect

Two main factors explain the cause of the differences in light intensity observed in the various Earth's regions according to geographical coordinates and seasons [1]: the double rotation of our Planet and the tilt of the Earth's axis.

11.5.2.1 The Earth's Double Rotation

- The first rotation is the daily spinning of the Earth around its axis, which explains the diel cycle (periodical sequence of days and nights). By whirling around itself, the Earth displays a portion to the Sun during daytime; meanwhile, at the opposite of the Globe, which is not facing the Sun, it is nighttime.
- The second rotation is the elliptical revolution of the Earth around the Sun at constant velocity. It takes one year for the Earth to return to the same coordinates in the reference frame, the origin of which is in the Sun's centre. This period corresponds to approximately 365 daily spinnings.

11.5.2.2 The Tilt of the Earth's Axis

The tilt is relative to the incident rays of the Sun. The Earth describes an ellipse on a plane around the Sun, which is oriented so that the following phenomena are observed.

During the northern summer solstice: [6]

The Earth's axis is oriented so that the regions located north of the Arctic polar circle receive solar rays tilted in a manner that when the Earth rotates around its axis, this geographical zone is permanently hit by sunlight. The sun can be seen 24 hours a day. The light cone covers all the faces of the lit sphere portion. In the south of the Arctic polar circle, the cone lightens an increasingly smaller portion of the Earth when one proceeds southwards, resulting very quickly in the appearance of a day-night cycle, with a decreasing diurnal part: 16 hours per 24 hours at 45° N latitude, 12 hours per 24 hours at the equator, 0 hour south of the Antarctic polar circle.

During the two equinoxes:

The inclination of the Earth's axis is such that the incident sun rays of the light cone lighten half of the globe at each latitude. Here, the diel cycle divided into 12 hours daytime and 12 hours nighttime is observed everywhere on Earth.

During the southern summer solstice: (see Footnote 6)

Here, we have the converse of the northern summer solstice: 24 hours of daylight at the south of the Antarctic polar circle, 16 hours per 24 hours at 45° S latitude, 12 hours per 24 hours at the equator, and 0 hours (i.e. permanent night) north of the Arctic polar circle.

11.5.2.3 Seasonality

The Earth continues its periodical course from the northern summer solstice through the northern autumn equinox (or southern spring equinox) to the southern summer solstice and back through the northern spring equinox (or southern autumn equinox) to the northern summer solstice. This perpetual to-and-fro movement entails the seasonality phenomenon in distinct climate zones.

Cold climate zone around the poles

Seasonality is clear-cut in the high latitudes (above ca 65°), with a periodic succession of two distinct seasons—a very brief cool summer followed by a very long, icy winter. Hardly any intermediate seasons are observed between these two climate stages.

Temperate climate zone

In the middle latitudes (between 35° and 65°), seasons are clearly defined and last about three months (1/4 of the year's duration), with (1) relatively warm summers, (2) springs and autumns that are lukewarm, moist and often rainy and (3) cool to cold winters. Continentality emphasises these characteristics, whereas marine influences buffer the temperature range and spread the precipitations relatively uniformly and unpredictably throughout the year. The areas closest to the tropics are called *Mediterranean zones* and are characterised by frankly

[6] The northern (southern) summer solstice corresponds to the southern (northern) winter solstice.

hot and dry summers and winters that tend to be mild and rainy. However, occurrences of cold spells are not precluded.

Intertropical climate zone

Around the Tropics of Cancer and Capricorn, two situations arise between these two parallels.

- The initial hygrometry is high, i.e. above a given threshold. Then, the main determinants of the climate are the wind regimes and continentality. This situation generally occurs well inside the intertropical band. Here, we have two to four well-marked seasons, one or two dry seasons, and one or two wet seasons. Another essential factor around the equator line is the evapotranspiration of high trees. Within the tropical rainforest zones, a relatively narrow and disconnected belt around the equator, seasonality is hardly marked with permanent humid air maintained by evapotranspiration and heavy rains during evening hours at dusk.
- The initial hygrometry is low, i.e. below a given threshold. Then, the desert regime is the rule with extreme heat and quasi-permanent drought, albeit punctually but rarely, watered by diluvian rainfalls. This situation occurs within the continental masses around the Tropics for a reason we will expose below.

The altitudinal gradients

The seasonality type at low altitudes is similar to that of the surrounding plains, which defines the latitude's seasonality. At all latitudes, an observer who follows the altitudinal climate cline from the plain to the summit encounters a gradient of seasonality identical to the one he would see along a latitudinal gradient. The highest altitudes are subject to a seasonality type reminiscent of the cold climate zone.

11.5.3 How Light and Humidity Determine Climate Diversity

Definition 11.1 Luminous flux is the number of incident rays per surface unit.

Incident sunlight during its passage diffuses heat. The higher the luminous flux across the air, the higher the temperature. We will now describe the impact of this parameter on the different latitudinal zones.

In the cold climate zone around the poles:

The sun's rays hit the ground slantwise, thus with a weaker incidence than all the sun's rays that reach Earth's surface in other climate zones. The luminous flux is minimal; as a result, around the impact, the air loses its heat quickly. Hence, the explanation of why the cold stays at low altitudes, i.e., lower than in a hotter environment, is that cold air is heavier than warm air. Due to a physical phenomenon called Coriolis force, frequently occurring in the neighbourhood of the poles, many winds arise in the Arctic zone (Antarctic zone) and move, pushing cold air masses southwards (northwards).

In the temperate climate zone:

The closer a given area is to the equator, the less acute the angle between the sun's rays and the surface will be; consequently, the higher the luminous flux will tend to be, generating a greater quantity of warm air around the impacts of the rays. This warm air is lighter than the cold air originating from the air current descending from the closest pole; therefore, the warm air in question will tend to push the resident cold air upward. However, the cold air current continues to stream from the next polar area. A permanent conflict between warm and cold air ensues, which explains the presence of anticyclones and depressions. Still, the trend of air mass displacements remains the transfer of cold air towards the equator, which gets increasingly hotter as long as the air masses get closer to the tropics.

In the intertropical band:

The sun's rays are almost perpendicular to Earth's surface. The luminous flux is maximal, which entails the highest air warming around the impact. In the intertropical zone, two types of environments are seen in the intertropical band:

- the driest, hottest and most barren ecosystems on Earth, i.e. the tropical deserts.
- the most humid and fertile ecosystems with temperatures that, though high, are characterised by a heat buffered by moisture, with a decreasing gradient of humidity and increasing order of average annual temperature from the rainforests across deciduous forests and woodlands to savannahs and grasslands.

In the tropical deserts:

They are found across the continental masses, around the Tropics, an area that has been called the *Intertropical Convergence Zone*, abbreviated ICZ. The ICZ is where the trade winds and the Hardley cells converge [2, 38].

- The trade winds are the set of winds descending from one of the Poles summarily described in the last entry.
- The Hardley Cells or Hardley Circulation is a movement of air masses arising from the initial hygrometry around the equatorial circle (between 3°N and 3°S), which moves towards the Tropics.

Both wind systems unload their water content and meet in the ICZ, where their precipitation potential energy is almost cancelled, and the force of their movement considerably diminished; hence, the dryness of the air in this zone and the extreme heat that cannot be temperated by moisture—the temperatures are characterised by high monthly means and extreme daily drop between day and night (compared with all the other biomes): the monthly mean of nocturnal and diurnal temperatures are respectively 21° C and 32° C. Diurnal temperature often rise well above 35° C and can reach extremes close to 60° C [6].

In the other tropical formations:

The air, loaded with humidity originating from the Hardley Cells and, in some cases, from evapotranspiration, forms updraught areas: the air, when it rises, cools and reaches its dew point. The water vapour contained in the air condenses all the more quickly as the initial hygrometry is important. All this process explains the cloudy and rainy aspect of many intertropical zones. In these areas, humidity is constant; the air is concentrated with a high percentage of moisture, frequently saturated at 100% humidity close to the equator, resulting in a temperature buffered at 25° C–28° C on average.

11.6 Physical Parameters of Aqueous Ecosystem

We already discussed three physical parameters of Abiotic Factor Systems (AFSs), mainly in terrestrial habitats. The abiotic factors (AFs) studied were temperature, humidity, and light. Temperature and light are also major AFs in aqueous environments, to which we can add pH.

11.6.1 Temperature

The freshwater bodies:

An interesting presentation is given by an excellent review article [13] that I will summarise briefly:

The case of shallow waterbodies:

It has been established that the water temperature of streams, rivers and wetlands is highly correlated to the temperature of the air in contact with the surface of these waterbodies; this result has been shown with linear regression analyses [31] namely in New Zealand [22], England [11, 51], the United States [35, 44] and across 13 countries [32], i.e. United States, Europe (Belgium, Croatia, Czechia, Estonia, Finland, Germany, Macedonia, Norway, Poland), Asia (China, Japan), and Australia.

The case of deeper bodies (Figs. 11.3 and 11.4):

Within deeper and broader freshwater bodies (e.g. lakes), water temperature T is differentiated and is a function of depth $d \in \mathbb{R}^-$. Let us call $T(d)$ the temperature of a point in the lake at depth d. It has been observed that, generally,

- $T(d)$ is constant in the superficial layer called the surface layer or *epilimnion*, above a given depth threshold τ_e. $T(d) = \max(T) = T_{\max}$.
- T is constant in the deepest layer called the *hypolimnion*, below a given depth threshold τ_h . $T(d) = \min(T) = T_{\min}$.

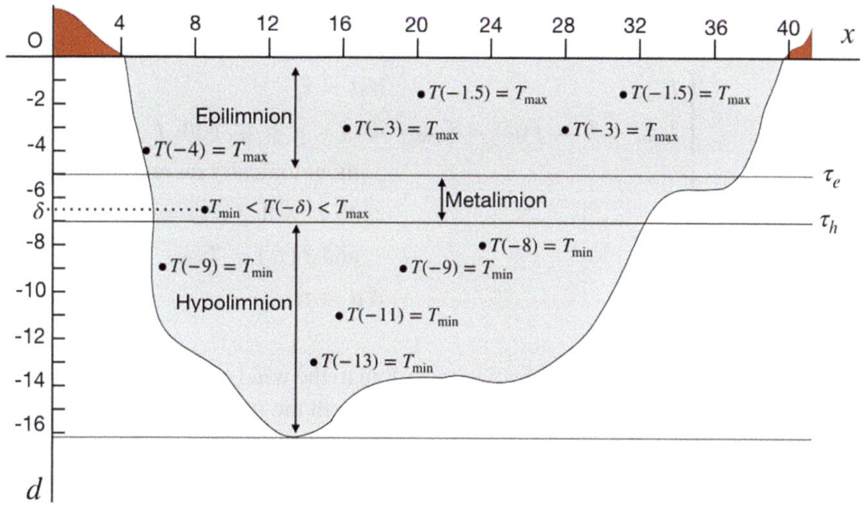

Fig. 11.3 Section of a lake with its three thermal layers, epilimnion, metalimnion and hypolimnion. The *grey surface* represents the section of the lake within a Cartesian diagram where the x-axis represents the distance along the longest geodesic of the lake between an arbitrary origin O and a point on the geodesic, and the y-axis the depth d (both in metres). The *black bullets* represent points associated with their temperature (in degrees Celsius) at a given depth. The brown surfaces represent the beginning of the banks at the opposite of the longest section

Fig. 11.4 Curve of temperature as a function of depth in a lakeWe took the example of Fig. 11.3, T temperature, d depth

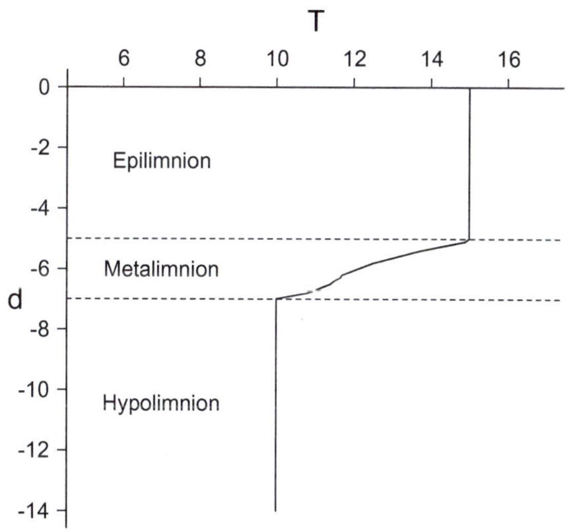

- Between these two layers, in a zone called *metalimnion*, $T(d)$ is a monotonic continuous increasing function between $T_{\min}$ and $T_{\max}$ ($T(d)$ decreases as d decreases, from $T_{\max}$ to $T_{\min}$).

Let us divide the lake into the three layers as in Fig. 11.3. In this example, $\tau_e = -5$ m and $\tau_h = -7$ m.

The variable T will be expressed as follows:

$$T = \begin{cases} T_{\max} & \text{if } d > \tau_e \\ T_{\min} < T = f(d) < T_{\max} & \text{if } \tau_h < d < \tau_e, \text{ with } f \\ & \quad \text{monotonically increasing from} \\ & \tau_h \text{ to } \tau_e \text{ with } f(\tau_h) = T_{\max} \\ & \quad \text{and } f(\tau_e) = T_{\min} \\ T_{\min} & \text{if } d < \tau_h \end{cases} \tag{11.2}$$

The width of the three layers varies according to the wind energy at the surface in spring, which induces the mixing of the heat from the sun and the proper heat of the water in the epilimnion. During late spring and summer, there establishes an equilibrium between these two types of heat, and this balance results in temperature $T_{\max}$. This mixing can sometimes extend beyond the photic zone, i.e., the zone that receives sunlight. At the level of depth threshold τ_e, in the metalimnion, there generally occurs a physical resistance to mixing, beyond which the heat accompanying the electromagnetic waves from the sun starts to have a decreasing effect on the water, which explains a decrease of T. Below the level of depth threshold τ_h, in the hypolimnion, the temperature of the water usually does not receive any influence of the sun, and no temperature mixing occurs, hence the constant temperature $T_{\min}$.

Hereunder, I present a graph largely inspired by [13, Fig. 2] (Fig. 11.4).

The marine waterbodies:

The marine waterbodies constitute the largest solar energy collector on Earth [42]. Only recently did scientists discover that (1) most of the solar system's energy reaching the Earth's surface was absorbed by the oceans and seas and (2) the way heat circulates in the marine environment.

About point (2):

The oceans are crisscrossed by vortical flows of different sizes, which are the main global and local heat conveyors across the oceans. These flows are phenomena that are reminiscent of and parallel with the atmospheric flows. The oceanic flows result from solar heat, the rotation of the Earth and the atmospheric flows [9]. Four categories are now recognised by the oceanographers:[7]

The major gyres:

These flows are the most extensive; they travel several thousand km. Five major gyres are now acknowledged: (1) the North Atlantic, (2) the South Atlantic, (3) the North Pacific, (4) the South Pacific and (5) the Indian Ocean ones.

[7] www2.physics.ox.ac.uk/collaborations/osmosis.

Approximately 40% of the heat transport is carried out by the major gyres, mainly in tropical and subtropical areas. It is largely tributary to wind-driven gyres. Abyssal mixing is negligible below 2000 m depth [18].

The mesoscale eddies:

These flows have lengths ranging between 10 and 300 km. They form lines of equal pressure, balancing between a move to low sea level and a move to high sea level (they are said to be geostrophic). Mesoscale flows are found in every part of the ocean, particularly in the middle and high latitudes. Notably, the mesoscale eddies generate smaller flows essential in transmitting heat, the submesoscale flows.

The submesoscale flows:

It concerns flows whose lengths range between 1 and 10 km. Though more modest in size, their role is important since they are responsible for vertical fluxes of heat that can reach deep sea bottoms and drive strong upward heat transfer from deep ocean levels to the surface. Vertical heat transport in the oceans caused by submesoscale flows appears to be a crucial regulatory mechanism of Earth's climate.[8]

The finescale flows

These very localised currents are associated with surface waves. An article has shown that in a specific case (the Yellow Sea in China), the effect on heat transfer is meagre and essentially consists of fluctuations that enhance the upward and downward mixing around the thermocline [33].

11.6.2 Light

11.6.2.1 Reminder on Some Physical Basics About the Propagation of Light in Water

Three main interactions exist between water and light: (1) reflection, (2) refraction and (3) absorption. All of these three phenomena co-occur. One part of a light beam hitting the surface of a water body is reflected. The rest of it penetrates the water body, first subject to being refracted and then to being absorbed.

Reflection:

The reflection of a light ray on a water body is a phenomenon where the moving photons of an incident ray emitted by the sun follow a sequence of events. (1) The ray travels through space and the atmosphere in a given trajectory. (2) This

[8] Beyond solving the thermal regulation problem, submesoscale eddies could relate two layers of many parts of the ocean: the superficial layer is characterised by an abundance of light and scarcity of nutrients. On the contrary, the deeper layers are dark but rich in nutrients. Such a situation is meanly found in the temperate and cold areas of the ocean, more than the intertropical areas [see Footnote 7].

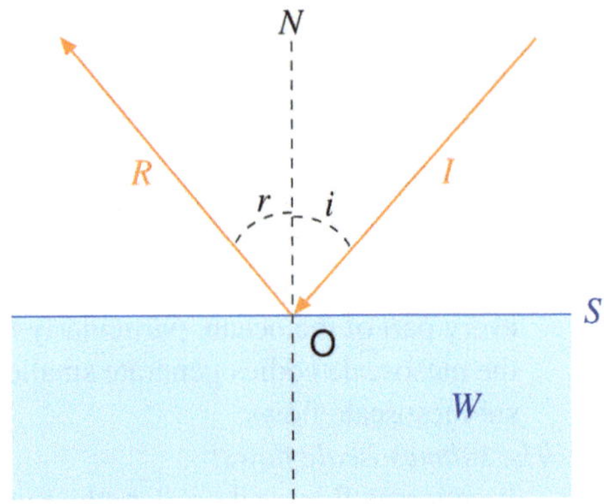

Fig. 11.5 Schema on the reflection of incident light ray *I* on water body W. *I* incident ray, *R* reflected ray (*in orange*), *N* normal line, *i* incident angle, *r* reflection angle, *W* water body, *S* water body surface

trajectory becomes linear at the approach of the water body's surface. (3) The ray collides with the surface of a water body. (4) The ray bounces back and remains linear. (5) The reflected line is symmetric to the incident line regarding the normal line *N* passing through O, the impact point of the incident ray (law of reflection). The incident angle *i* is the angle between the incident line *I* and the normal line *N*. The law of reflection stipulates that when the water body is still and homogeneous, without containing any particles, then $i = r$ (cf. Fig. 11.5).
Quantum physics explains that when excited by incoming light—an influx of photons—, any group of atoms departs from their ground state, where the electrons are associated with orbitals of their lowest energy orbitals at a given statistical distance from the atom nuclei. The electrons move then to higher energy orbitals, which are statistically at a longer distance from the nuclei than when they are in the ground state. During this phenomenon, the reflected light is emitted—the outcoming flux of photons sensed by the observer.
When all the sun's rays hitting an object are reflected, the "colour" of the reflected light is a mixture of all the wavelengths, thus white (in the air when no object absorbs it). White light covers much of the spectrum from infrared to ultraviolet. This passage is equivalent to the phrase "no light is absorbed". Air hardly absorbs light, but we will see that water absorbs much light.

Refraction:

The refraction of a light ray on a water body is a phenomenon where the moving photons of an incident ray emitted by the sun follow the following sequence of events (see Fig. 11.6). (1) The ray travels through space and the atmosphere in a given trajectory. (2) This trajectory becomes linear at the approach of the water body's surface. (3) The ray hits the surface and penetrates the water. (4) Then, the ray modifies its direction and velocity (see Fig. 11.6), the two phenomena being related for the following reasons:

- The light beam bends towards the normal line *N*. Indeed, both are transparent mediums; because water is denser than air, the light will cross water with a velocity reduced relative to its velocity through air, which explains the

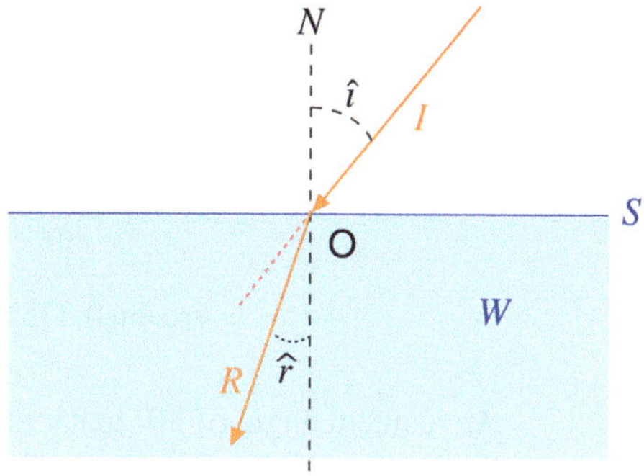

Fig. 11.6 Schema on the refraction of incident light ray *I* on water body W. *I* incident ray, *R* refracted ray (*in orange*), *N* normal line, $\hat{\imath}$ incident angle, $\hat{r}$ refraction angle, *W* water body, *S* water body surface

direction change of the ray closer to N. The refraction angle $\hat{r}$ is the angle between the refracted ray R and the normal line N.

- Hereunder, we will list the speed of light through different transparent mediums:

  Through vacuum:

  $c = 299,792,458\,\text{m/s}$.

  Through air:

  $v_1 = 299,702,457$ m/s.

  Comparing v_1 and c, we have $\frac{v_1}{c} = \frac{299,702,457}{299,792,458} \simeq 0.9997 \Rightarrow v_1 = 0.9997\,c \Rightarrow v_1 = \frac{c}{1.0003}$.

  This result means light travels in the air with a velocity that is 1.0003 times slower than in the vacuum. The value $n_1 = 1.0003$ is called the refraction index of light in the air.

  Through water:

  $v_2 = 225,000,000$ m/s.

  Comparing v_2 and c, $\frac{v_2}{c} = \frac{225,000,000}{299,792,458} \simeq 0.7505 \Rightarrow v_2 = 0.7505c \Rightarrow v_2 = \frac{c}{1.332}$.

  This result means light travels in water with a velocity 1.332 times slower than in the vacuum. The value $n_2 = 1.332$ is called the refraction index of light in water.

  Snell's law gives the refraction angle $\hat{r}$:

$$\frac{\sin\hat{\imath}}{\sin\hat{r}} = \frac{n_2}{n_1} \tag{11.3}$$

  This formula is valid for a transparent water body with minor surface roughness (flat lake or sea surface, almost no waves). If, for instance, the incident ray forms an angle of 60° with the surface $S, \widehat{i} = 30° = \frac{\pi}{6}$ rad. Hence,

$$\sin\widehat{r} = \sin\widehat{i} \times \frac{n_1}{n_2} = \frac{1}{2} \times \frac{1.0003}{1.332} = 0.375 \Rightarrow \widehat{r}$$
$$= \arcsin(0.375)\frac{180}{\pi} \simeq 22°$$

An incident angle of 30° generates a refraction angle of 22°.
Part of the refracted light is reflected with a "colour" which is not white, depending on the depth of the point where the reflection occurs.

Absorption:

The absorbed light within a water body is the part of the refracted light whose energy is changed into the potential energy of biotic and abiotic factors within the aqueous environments. The consequence of this phenomenon is the reduction of the colour spectrum. As a result, visibility is maximal just below the surface. It decreases down to total darkness below a given depth (a phenomenon called the *extinction of light* in a waterbody). On average, the water body absorbs 50% of the visible light reaching the surface within the first 10 m. Only 1% of the visible sunlight hitting the surface reaches the depth of 100 m.
The perception of visible sunlight at a given depth (in absolute value) z is measured with a variable called light intensity $I(z)$[9] measured with a light quantum sensor and K_d a constant [see formula (11.4)] assessed with a device called a Secchi disk, an old technique dating back from the nineteenth century[10] and still in use, clearly explained in English in a very clear exposition of the method [53]. Let I_0 be the light intensity at the water body's surface. According to the BBL[11] law describing the evolution of light intensity I as a function of depth z,

$$I(z) = I_0\, e^{-K_d \cdot z} \tag{11.4}$$

where K_d is a constant depending on the intrinsic property of the waterbody [like e.g. clarity, transparency level or homogeneity, type (for example, ocean, lake, river), latitude, altitude, salinity or pH]. This paramater is commonly used in phytoplankton studies [15].

[9] ...theoretically defined as the number of μmol photons $\cdot$ $m^{-2}\,s^{-2}$.

[10] Method [40] owed to the Italian Catholic priest and astronomer Pietro Angelo Secchi (1818–1878).

[11] BBL = Beer-Bouger-Lambert. The BBL law is an old formula, dating back from early in the Modern Age [3, 5, 23].

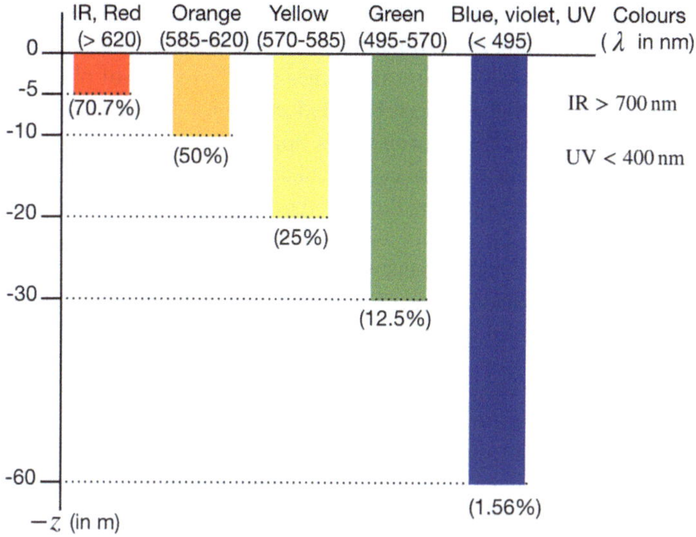

Fig. 11.7 Barplot showing the pattern of colour absorption at different depths in seawater. The x-line shows the wavelength of absorbed light, and the *y*-axis presents the depth of the water body; the abscissae λ corresponds to the wavelengths, and the ordinates $-z$ to the depths. The percentages in brackets are the visibility indices, i.e. the percentage of light intensity at the different depths relative to light intensity at the surface. Data collected in South China Sea, $K_D = 0.0693$

A relatively novel method called CLAHE (Contrast Limited Histogram Equalization[12]) has been applied to determine the depth range of the different colour absorption accurately [20, 26] (see Fig. 11.7).
Importantly, by the Planck-Einstein relation, the energy E carried by a photon is inversely proportional to its wavelength λ (cf. relation (11.5)).

$$E = \frac{hc}{\lambda} \text{ where } h \text{ is the Planck constant and } \lambda \text{ the wave length} \tag{11.5}$$

E is measured in J and λ in nm, $c = 2.998 \times 10^8$ in m.s^{-1} and $h = 6.626 \times 10^{-34}$ J · s.
The energy E_m of one mole of photons is

$$E_m = \frac{1.196 \times 10^8 \text{ J} \cdot \text{nm} \cdot \text{mol}^{-1}}{\lambda} \tag{11.6}$$

One mole is a sensible macrophysical magnitude to compare the energies of the different types of light rays according to their colour. Following the x-axis of Fig. 11.7, we can see a wavelength decrease, thus an energy increase, in the

[12] For a technical description of the method, see [55].

colour gradient between IR and IV through red, orange, yellow, green, blue, violet and UV, according to the relation (11.6). What interests us is the fraction of the solar energy received by the waterbodies at the disposal of photoautotrophic organisms; this part of the energy is called *photosynthetically active radiation*, abbreviated PAR. The PAR corresponds approximately to radiation from the visible light spectrum, and it has been calculated that, on average, it constitutes ca 45% of the total radiation hitting the waterbody's surface [15]. The area reached by the PAR is the photic zone.[13] As pointed out at the beginning of this entry devoted to light absorption, the energy associated with light rays constitutes the potential energy of physicochemical reactions related to photosynthesis and other critical chemical and biological processes in aqueous environments.

Light intensity and depth:
Light intensity decreases with decreasing depth (increasing in absolute value). The more energetic solar rays are, the higher their kinetic energy and the deeper the penetration will tend to be; simultaneously, they will tend to be scarcer. These two points are highlighted in Fig. 11.7

IR and Red:
The infrared waves[14] and the light beams corresponding to the red colour are absorbed **in the first five metres of the water column**. Compared to the rays of the other colours, the red light rays correspond to the longest wavelength and thus carry the weakest potential energy [as explained by formula (11.6)]. This potential energy is transformed into kinetic energy, the lowest among the solar rays penetrating into the waterbody, which explains the shallowest depth of red light penetration. They also constitute approximately 30% of the rays absorbed at the surface.[15]

Orange:
This colour extends from the surface **to 10 m depth**. The orange rays penetrate more deeply in the waterbody since, relative to the red rays, the wavelengths corresponding to this colour are shorter, and thus the kinetic energy of these rays stronger. Among the remaining colours between − 5 and − 10 m, the light beams corresponding to orange progressively disappear from the light spectrum. At 10 m depth, the light intensity is 50% of its initial value at the start of its penetration, meaning that half of the solar rays have been extinguished at the bottom of this colour column.

Yellow:
The solar rays, whose wavelengths correspond to yellow, penetrate **through the first 20 m** of the waterbody. The concentration of the yellow solar rays,

[13] According to the latitude, the photic zone can extend from 80 m (in the clear tropical oceans) to less than 10 m (near the poles).

[14] Infrared is absorbed within the upper 10 cm of the upper photic zone.

[15] At five metres deep and below, most red organisms appear blackish, but rays of shorter wavelength continue to be emitted, i.e. orange, yellow, green, blue and violet, which means that fishes or other organisms can still be observed with these colours.

on average, gradually decreases from maximal at the surface to null at 20 m depth. The wavelengths of the yellow light beams are shorter than those of the red and orange ones; relatively to the latter, the yellow light rays are endowed with a higher kinetic energy, hence their far lower penetration. At – 20 m, only 1/4 of the visible light that initially penetrated the surface is still present, entailing a visibility reduction and energy availability for the phototrophs of 3/4. Hence, light is very dim below this depth; the only colours perceived are green, blue and violet.

Green:

Following the decreasing wavelength gradient, the green solar rays come next, with thus enhanced kinetic energy, which pushes the presence of green further downwards (**to 30 m depth**). Only 12.5% of the sunlight penetrating the surface reaches the bottom of the column where the absorption of the green rays occurs. Here too, these rays gradually decrease from the surface to $z = -30$m, down to extinction. At this depth, solar energy has lost 87.5% of its initial value (at the surface). The aqueous environment is a bluish and purplish-blue extremely dim-lit universe.

Blue/violet:

Last but not least, the most profound water column absorbing visible light **reaches a depth of 60 m**. These sunlight rays have the highest kinetic energy among all the solar beams within the visible light spectrum, which explains why they gradually decrease their intensity from the surface down to −60 m, the extinction level. The environment is almost dark since only 1.56% of the sunlight on the surface reaches this depth. Still, surprisingly, we shall see that this extremely minimal quantity of solar energy is sufficient for photosynthesis to take place. In many oceans, phytoplanktonic organisms can be found in such depths and even lower.

Photosynthesis takes place in the ocean in the entire photic zone that extends between the surface and $z = -200$ m, that is, down to regions where solar light intensity is microscopic (9.56×10^{-7} J.mol^{-1}) but compensated by the immense energy these very few violet rays close to the UVs carry [28, 39] and the abundance of nutrients. In waterbodies like lakes and oceans, phytoplankton, constituting the main agents of photosynthesis, are distributed according to each species' needs. They are distributed into different layers of the photic zone: the shallower one, which receives much sunlight and relatively few nutrients, and the deeper one, which is hardly illuminated but nutrient-rich. Several of these species can migrate from one layer to the other. These layers' width and relative depths vary with different factors, e.g., latitude, season, turbidity and other ecological parameters. A thorough and detailed study on the subject in the Black Sea has been recently carried out by a Russian team [30], where they showed that season and turbidity, two related factors, played a significant role in the diverse migrations of the phytoplankton. Another article, performed on data collected in the Mediterranean Sea, pointed out two zones, quite deep and not well illuminated, that concentrate a maximum of phytoplankton biomass, thus a maximum of photosynthetic activities: (1) a layer between $z = -40$ m and $z = -50$ m , and (2) a layer between $z = -80$ m and $z = -100$ m [27].

11.6.3 pH

This physical abiotic factor is mathematically defined as

$$pH = -\log{[H_3O^+]} \tag{11.7}$$

where $[H_3O^+]$ is the concentration of hydronium ions, expressed in $mol \cdot L^{-1}$. This parameter is a good estimate of the impact of the concentration of hydrogen ions (H^+) on the quality of the soils in terrestrial environments[16] and of water in aqueous habitat. This concentration has been accepted by the scientific community since its invention by the Danish chemist Søren Peter Lauritz Sørensen (1868–1939), as reflecting the degree of acidity or alkalinity, in a gradient that runs from extremely acidic to extremely alkaline. I illustrate this gradient with Fig. 11.8.

- pH=7, the solution is said to be neutral.
- $pH < 7$, the solution is said to be acidic.
- $pH > 7$, the solution is said to be alkaline or basic.

The pH variable significantly determines the ecosystem type of a given Earth's portion.

11.6.3.1 pH in Terrestrial Habitats

An interesting article on the effect of pH on plant species richness throughout Brazil shows that pH determines the subbiome types [8]. This research, based on large samples (33,166 different plant species), showed the following results (data extracted from Fig. 5 of the article):

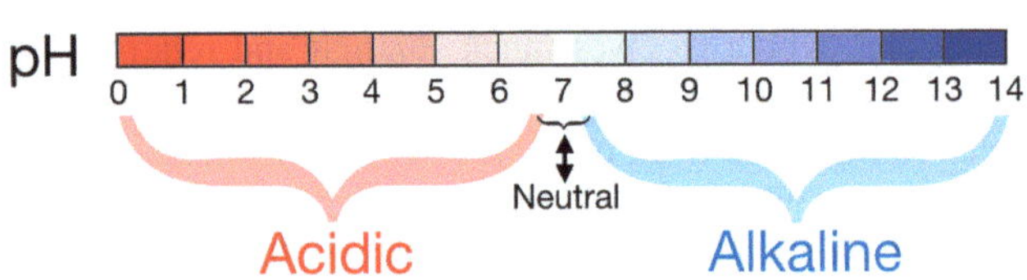

Fig. 11.8 A pH scale. The intensities of red and blue, respectively, depict the strengths of acidity and alkalinity

[16] As a reminder, soil pH means the pH of the interstitial soil water.

- Tropical and subtropical broadleaf rainforests are associated with a mean pH of ca 5.1;
- Tropical and subtropical broadleaf dry forests are associated with a mean pH of ca 5.9;
- Tropical and subtropical grasslands, savannah and shrublands are associated with a mean pH of ca 5.5;
- Flooded grassland sand savannahs are associated with a mean pH of ca 6.1;
- Deserts and xeric shrublands are associated with a mean pH of 6;
- Mangroves are associated with a mean pH of ca 5.2.

The distribution curves of the species richness as a function of pH were normal with comparable standard deviations. In all the cases mentioned above, the soil is slightly acidic. The mean pHs are different, with the dryer subbiomes being higher and thus less acidic than those of the moister subbiomes.

11.6.3.2 pH in Aqueous Habitats

The case of freshwater biomes:

The pH of the vast majority of freshwater bodies is, on average, equal to 7; in natural conditions, the pH of most rivers and lakes can fluctuate between 6 and 8, thus swaying their waters between slightly acid and slightly basic. In profound lakes, the deep waters are generally slightly alkaline ($7.5 < pH < 8.5$), whereas the superficial ones are slightly acid or neutral ($6.5 < pH < 7.5$). Such waterbodies shelter mesophylic organisms. However, a minority of lakes are highly alkaline (having pHs between 9 and 12) or highly acidic (with $pH < 5.0$). On some highly alkaline lakes, e.g. lake Natron in Tanzania, one can find very rich ecosystems comprising flamingoes, tilapia and specific algae that are resilient to alkalinity. Both highly alkaline and highly acid lakes are teaming with extremophiles, generally prokaryotes and, in particular, archaea.

An interesting report prepared by a collective specialised in water resources [37] lists thc vital pH intervals of different species, a situation that generates a Geasonian type of community (cf. Sect. 13.2.5)

The case of marinc biomcs:

Compared with freshwater bodies, the marine ones are significantly more alkaline, their pH being on average around 8.2, ranging between 7.5 and 8.5. Relative to freshwater, the high concentration of carbonate (see next entry) and the salinity explain this property of the oceans and seas.

On average, there is a pH difference between the different parts of the Global Ocean because pH depends on the dissolved organic carbon (DIC) concentration and alkalinity (Alk). At constant Alk, increasing DIC is accompanied by decreasing pH; and at constant DIC, when Alk increases, pH increases [48], thus: (1) The deep Pacific Ocean has a higher DIC, thus a lower pH ($7.9 < pH < 7.5$ [34]), than the deep North Atlantic. (2) The North Pacific has a higher Alk, thus a higher pH (close to 8.2–8.3 at the top 100 m [34]), than the Atlantic Ocean.

The deep Pacific Ocean is more acid globally than the deep Atlantic Ocean. This is due to the "pattern of the deep ocean circulation" [48]: the deep water of the Global Ocean is formed in the North Atlantic and around Antarctica and gradually spreads across the different oceans. Since the Pacific has existed for a longer period than the North Atlantic, the deep water with its organisms has taken a longer duration to fill the former than the latter; mechanically, via respiration and decomposition mechanisms, the deep Pacific has become more acid than the deep Atlantic (for reasons explained in the next entry).

Factors impacting the water's pH:

- An increase in the concentration of carbon dioxide induces an increase in acidity. CO_2 is released in water by respiration and decomposition of corpses and excreta and reduced by the photosynthesis of the primary producers (aquatic plants, algae, phytoplankton). The explanation is simple. The carbon cycle responsible for the first steps of the sedimentation of carbonate rocks occurs in the aqueous environment; one of these steps is the formation of carbonic acid (H_2CO_3), which loses its hydrogen ions (see Eq. (10.3), Sect. 10.5.4).
- Rain and snow are, on average, slightly acidic, with a pH around 5.6. It is an element that tends to acidify water bodies that are naturally neutral or slightly basic.
- Decomposing organisms, mainly decaying plants, are also an acidifying factor.

Quizzes and Problems

11.1 In the South China Sea ashore to Malaysia, 50% of the initial light intensity disappears at a depth of 10 m.

1. If I_0 is the light intensity when the light beam penetrates into the water surface and $I(z)$ the light intensity at depth $-z$, give the formula of the BBL law.
2. Calculate the constant K_d of the South China Sea.
3. Calculate the visibility index at the bottom of the water columns where the red, yellow green and blue are absorbed.

11.2 Planck-Einstein relation states that the energy of a photon is $E = h\frac{c}{\lambda}$ where E, h, c and λ are respectively expressed in J, $J \cdot s$, $m \cdot s^{-1}$ and nm. As a reminder, in these units, $c = 2.998 \times 10^8$ and $h = 6.626 \times 10^{-34}$. Moreover, in the SI, $J = kg \cdot m^2 \cdot s^{-2}$. Verify the relation (11.6).

11.3 This problem refers to Fig. 11.7.

1. Establish the gradient of energy intervals of the different visible colours per mole, from IV to UV.
2. What do you infer from this gradient?

References

1. Barbault R (2003) Ecologie générale. Structure et fonctionnement de la biosphère (in French). 5th edn, Dunod, Paris.
2. Barry RG, Chorley RJ (2009) Atmosphere, weather and climate. 8th edn, Routledge, London & New York.
3. Beer A (1852) Bestimmung der Absorption des rothen Lichts in farbigen Füssigkeiten. Ann Phys 162:78-88.
4. Bentz BJ, Mullins DE (1999) Ecology of Mountain Pine beatle (Coleoptera: Scolytidae) cold hardening in the Intermountain West. Environ Entomol 28:577-587, https://doi.org/10.1093/ee/28.4.577
5. Bouguer P (1729). Essai d'optique sur la gradation de la lumière [Essay on optics, concerning the attenuation of light] (in French), Claude Jombert, Paris.vspace1.5 mm
6. Britannica, The Editors of Encyclopaedia (2016) Tropical and subtropical desert climate. Encyclopedia Britannica, 14 March 2016, www.britannica.com/science/tropical-and-subtropical-desert-climate.
7. Çengel YA, Ghajar AJ (2015) Heat and mass transfer. Fundamentals and applications. 5th edn, McGraw Hill Education.
8. Crespo-Mendes N et al (2019) Relationships between plant species richness and soil pH at the level of biome and ecoregion in Brazil. Ecol Indic 98:266-275, https://doi.org/10.1016/j.ecolind.2018.11.004
9. Colling A (2004) Ocean circulation (vol. 3). Heinemann and the Open University, Milton Keynes, UK
10. Costanzo JP, Lee RE Jr (2013) Avoidance and tolerance of freezing in ectothermic vertebrates. J Exp Biol 216:1961-1967, https://doi.org/10.1242/jeb.070268
11. Crisp DT, Howson G (1982) Effect of air temperature upon mean water temperature in streams in the north Pennines and English Lake District. Freshwater Biol 12:359-367, https://doi.org/10.1111/j.1365-2427.1982.tb00629.x
12. De Broglie L (1924) Recherche sur la théorie des quantas, Doctorate dissertation (in French). Annales de Physique (extrait). Masson, Paris.
13. De Stasio BT et al (2022) Temperature as a driving force in aquatic ecosystems. In: Mehner T, Tockner K (ed) Encyclopedia of inland waters, p. 257-269, https://doi.org/10.1016/B978-0-12-819166-8.00062-1
14. Devries A (1982) Biological antifreeze agents in coldwater fishes. Comp Biochem Physiol A:Physiol 73:627-640.
15. Domingues RB, Barbosa AB (2023) Evaluating underwater light availability for phytoplankton: mean light intensity in the mixed layer versus attenuation coefficient. Water 15:2966, https://doi.org/10.3390/w15162966
16. Einstein A (1905) Über einen die Erzeugung und Verwandlung des Lichtes betreffenden heuristischen Gesichtspunkt. Ann Phys 322:132-148
17. Falkowski PG, Raven JA (2013) Aquatic photosynthesis. Princeton UP
18. Ferrari R, Ferreira D (2011) What processes drive the ocean heat transport. Ocean Model 38:171-186, https://doi.org/10.1016/j.ocemod.2011.02.013

19. Geiser F (2001) Hibernation: endotherms. In: Encyclopedia of Life Science, p. 1 - 8, Macmillan Publishers.
20. Hitam MS at al (2013) Mixture Contrast Limited Adaptive Histogram Equalization color models for underwater image enhancement. Conference paper, https://doi.org/10.1109/ICCAT.2013.6522017
21. Irwin JA (2020) Overview of extremophiles and their food and medical applications. In: Salwan R, Sharma V (eds) Physiological and biotechnological aspects of extremophiles, p. 65-87, https://doi.org/10.1016/B978-0-12-818322-9.00006-X, Academic Press. Get rights and content
22. Johnson FA (1971) Stream temperatures in an alpine area. J Hydrol 14:322-336.
23. Lambert JH (1760) Photometria sive de mensura et gradibus luminis, colorum et umbrae [Photometry, or, On the measure and gradations of light intensity, colours, and shade] (in Latin), Eberhardt Klett, Augsburg (DE).
24. Lee JR Jr et al (1996) Regulation of supercooling and ice nucleation in insects. Eur J Entomol 93:405-418.
25. Lobo JM (2024) Hotter-is-not-better: A study on the thermal response of a winter active and nocturnal beetle. J Insect Physiol 153:104602, https://doi.org/10.1016/j.jinsphys.2023.104602
26. Luo M et al (2019) An effective underwater image enhancement method based on CLAHE-HF. J Phys: Conf Ser 1237:032009, https://doi.org/10.1088/1742-6596/1237/3/032009
27. Marañón E et al (2021) Deep maxima of phytoplankton biomass, primary production and bacterial production in the Mediterranean Sea. Biogeosciences 18:1749-1767, https://doi.org/10.5194/bg-18-1749-2021
28. Marra JF et al (2023) Parameters for the depth of the ocean's productive layer. Front Mar Sci 10:1052307, https://doi.org/10.3389/fmars.2023.1052307
29. Merino N et al (2019) Living at the extremes: extremophiles and the limits of life in a planetary context. Front Microbiol 10:780, https://doi.org/10.3389/fmicb.2019.00780
30. Mikaelyan AS et al (2021) Seasonal evolution of deep phytoplankton assemblages in the Black Sea. J Sea Res 178:102125, https://doi.org/10.1016/j.seares.2021.102125
31. Mohseni O, Stefan HG (1999) Stream temperature/air temperature relationship: a physical interpretation. J Hydrol 218:128-141, https://doi.org/10.1016/S0022-1694(99)00034-7
32. Morrill JC et al (2005) Estimating stream temperature from air temperature: implications for future water quality. J Environ Eng 131:139-146, https://doi.org/10.1061/(ASCE)0733-9372(2005)131:1(139)
33. Nie L et al (2023) The influence of ocean processes on fine-scale changes in the Yellow Sea cold water mass boundary area structure based on acoustic observations. Remote Sens 15, 4272, https://doi.org/10.3390/rs15174272
34. Park K (1966) Deep-Sea pH. Science 154(3756):1540-1542, https://doi.org/10.1126/science.154.3756.1540
35. Pilgrim JM et al (1998) Stream temperature correlations with air temperatures in Minnesota: implications for climate warming. J Amer Water Res Assoc 34:1109-1121 https://doi.org/10.1111/j.1752-1688.1998.tb04158.x
36. Raymond JA et al (2020) Glycerol is an osmoprotectant in two Antarctic *Chlamydomonas* species from an ice-covered saline lake and is synthesized by an unusual bidomain enzyme. Front Plant Sci 11:1259, https://doi.org/10.3389/fpls.2020.01259
37. Robertson-Bryan Inc (2004) Technical memorandum: pH requirements of freshwater aquatic life - Website: www.waterboards.ca.gov/waterrights/water_issues/programs/bay_delta/deltaflow/docs/exhibits/bigbreak/dscbb_exh5.pdf.
38. Rodrigues SC, Rodrigues GSDSC (2009) Geomorphology and biogeography of tropical deserts. Tropical Biology and Conservation Management-Volume IX: Desert Ecosystems, 12.
39. Ryther JH (1956) Photosynthesis in the ocean as a function of light intensity. Limnol Oceanogr 1:61-70, https://doi.org/10.4319/lo.1956.1.1.0061
40. Secchi PA (1864). Relazione delle esperienze fatte a bordo della pontificia pirocorvetta *L'Imacolata Concezione* per determinare la trasparenza del mare. Il Nuovo Cimento, 20(1):205-238, https://doi.org/10.1007/BF02726911

41. Sibarani DA et al (2021) Experimental study of seawater evaporation using natural convection principles. J Phys: Conf Ser 1898:012029, https://doi.org/10.1088/1742-6596/1898/1/012029
42. Siegelman L et al (2020) Enhanced upward heat transport at deep submesoscale ocean fronts. Nat Geosci 13:50-55, https://doi.org/10.1038/s41561-019-0489-1
43. Sømme L (1999) The physiology of cold hardiness in terrestrial arthropods. Eur J Entomol 96:1-10.
44. Stefan HG, Prud'homme EB (1993) Stream temperature estimation from air temperature. J Amer Water Res Assoc 29:27-45, https://doi.org/10.1111/j.1752-1688.1993.tb01502.x
45. Stetter KO (2002) Hypothermmophilic microrganisms (ch. 11). In: Horneck G, Baumstark-Khan C (eds) Astrobiology. The quest of the conditions of life, pp. 169-184, Springer.
46. Takai K et al (2008) Cell proliferation at 122° C and isotopically heavy CH_4 production by a hyperthermophilic methanogen under high-pressure cultivation. Proc Natl Acad Sci USA 105:10949-10954, https://doi.org/10.1073pnas.0712334105
47. Verma D et al (2024) Developments in extremophilic bacterial genomics: A post next generation sequencing era. Ecol Genet Genom 32:100255, https://doi.org/10.1016/j.egg.2024.100255
48. Voelker C (2018). Re: Does the pH of the Atlantic, Pacific and Indian oceans differ? Retrieved from: https://www.researchgate.net/post/Does-the-pH-of-the-Atlantic-Pacific-and-Indian-oceans-differ/5b0adad210569f02850f202f/citation/download.
49. Walker VK et al (2001) Surviving winter with antifreeze proteins: Studies on budworms and beetles. In: Denlinger et al (eds) Insect timing: circadian rhithmicity to seasonality, p. 199-211, Saunders, https://doi.org/10.1016/B978-044450608-5/50048-9
50. Wang T et al (2020) A climatological interpretation of precipitation $\delta^{18}O$ across SIberia and Central Asia. Water 12:2132, https://doi.org/10.3390/w12082132
51. Webb BW et al (2003) Water–air temperature relationships in a Devon river system and the role of flow. Hydrol Proc 17:3069-3084, https://doi.org/10.1002/hyp.1280
52. Weber AP et al (2007) Metabolism and metabolomics of eukaryotes living under extreme conditions. Int Rev Cyt 256:1-34, https://doi.org/10.1016/S0074-7696(07)56001-8.
53. Wernand MR (2010) On the history of the Secchi disc. J Eur Opt Soc 5:10013s, https://doi.org/10.2971/jeos.2010.10013s
54. Wharton DA (1994) Freezing avoidance in the eggs of the antarctic nematode *Panagrolaimus davidi*. Fundam Appl Nematol 17:239-243.
55. Zuiderveld K (1994) Contrast limited adaptive histogram equalization. Graphic Gems IV:474-485.

Chapter 12
The Biosociological Approach-I: Species, The Relays of Matter and Energy

12.1 Organisms: The Relays of the Energy and Matter Dynamics Through the Ecosystems

The agents of energetic flow (see Chap. 9, p. 187) and material cycling (see Chap. 10, p. 221) within ecosystems are the organisms building up the living world (or the system of biotic factors) [140].

- On the one hand, all living beings—if we only consider the cellular organisms—share many similar structures and functions.
  - They are made up of one or more cells—complex open systems containing a cytoplasm, i.e. a colloidal substance containing 80% water and a great number of particles called ribosomes, and one or several right-twisted helicoidal macromolecules—a nucleic acid enclosing their hereditary material, the DNA, structurally very similar from one organism to another. We will not enter into the details of the molecular biology of transcription and translation. However, let us remind roughly that in each organism, a DNA molecule contains functional portions called genes that induce proteins within the ribosomes mentioned above, distributed into two categories according to two different functions. The first type of proteins are chemical catalysers called enzymes whose function is to accelerate the metabolic cycles, constructing the chemical components necessary for life (e.g. carbohydrates, lipids, and other organic molecules); the second type of proteins build up the backbones of many structures internal to the cell(s).
  - Organisms are separated from the outer world (the universe less themselves) by a layer (a skin or a membrane). This fundamental trait defines the individuality of each organism. They are sensitive, adaptive, and, for some

---

Supplementary Information The online version contains supplementary material available at https://doi.org/10.1007/978-3-031-90656-5_12.

M. Salomon, *Integrative Ecology*, https://doi.org/10.1007/978-3-031-90656-5_12

of them, motile. They can all be generated from a previous generation, grow, reproduce, age and die.

- However, on the other hand, the living beings are immensely different. This differentiation has been taking place for the last ca 4 billion years. Indeed, the living world is an immensely diversified system. The organisms that are the elements composing it vary tremendously for their sizes (from 10^{-4} mm for the smallest bacteria in diameter to more than 8×10^4 mm in height for the largest trees), inner and physiological structures and complexities (from the "simplest" prokaryotes to the most differentiated animals or plants), as well as their trophic statuses and modes of reproduction.

A common feature of all organisms is that they are necessarily related to other living beings. At an individual level, this relation can be beneficial, neutral or harmful (impairing or lethal). Each organism is an individual living in a population, i.e. a group that can be defined as a trophic and reproductive unit at a given time t, generally belonging to the same species (a concept that will be studied later). Thus, two organisms in contact can belong to the same species (they will experience intraspecific interactions) or to different species (in which case, their interactions will be termed interspecific).

- Members of the same species share identical trophic needs and locate themselves in a given position in a trophic network whose function is to transfer energy and cycle matter through the ecosystem. These populations act thus as relays of this energy and material. Section 13.4 will study this topic.
- A population can also be seen as an evolutionary community where speciation can take place, fostering the evolution of species. Both interpretations of a population are interrelated. The different elements of any trophic network emerged via a speciation event at a given time t during evolution.
- Each population comprises individuals of the same species whose life histories (birth, growth, reproduction, ageing, and death) are studied through demography, population ecology, genetics, and mathematical and statistical tools.
- Populations form communities (biocoenoses), i.e. species related by all the interactions described in Chap. 4.

Now, we will treat a property of the biotic parts of the exosystem, their multiscality.

12.2 The Biosphere Is Multiscale

The biotic agents organise themselves temporally and spatially. We will first deal with the temporal aspect. Communities are components of ecosystems described by compartment models (cf. Sect. 6.2, p. 130) that evolve with time. Biological systems, and among them, ecological ones, are metasystems [136]—systems emerging from one of their component systems with new properties. Indeed, the historicity of ecosystems has two faces:

- their statics or proximal level that depicts their structure and functions directly observed during a short period δt;[1] and
- their dynamics or distal process that has led the ecosystems to existence, reconstructed from their remote past to present by means of investigations from clues, logical reflections and modelling. This dynamic falls into two categories: (1) the ecological successions that constitute relatively short-term dynamics that construct ecosystems from the simplest form to more elaborate and intricate ones corresponding to relative and provisory equilibriums; and (2) the long- to very long-term evolutionary dynamics explaining the establishment of the biotic and abiotic agents of the ecosystems (since their origin).

Like many biological systems, ecological systems share as a unique characteristic the two features mentioned above, the first one pointing out the necessary functioning of the biological machinery to allow the bricks of these systems, i.e. organisms, to live, reproduce and interact, and the second one highlighting the very long and continuous development necessary for ecological systems, a. o—ecosystems, to exist. There is a tight relationship between space and time. Indeed, short-term processes tend to occur in nearby territories, generally during a period compatible with the times of continuous observations. In contrast, long-term processes have stages in far wider range areas and have been established for far longer, generally studied via historical reconstruction.

12.3 The Spatiotemporal Scales of Communities (Biocoenoses)

The species or groups of species building up a community will be studied at different spatiotemporal scales. But one should beware of two acceptations of the term "scale" when applied to surfaces, the one of the biogeographers that we will use and term spatial scale, and the one of the cartographers or map scale.

A small spatial scale corresponds to a small portion of the earth, like a locality or a small natural area that a single field observer can thoroughly cover during one stay (e.g. a 5000 km^2 track of desert or forest, a lagoon). It corresponds to a large map scale, measured with fraction 1/n where 1 cm on the map represents n cm in the field. For example, in a 1/25,000 scaled map, 1 cm on the map corresponds to 25,000 cm in the field, that is 250 m. The closer the representation to the actual area is, the smaller n and the greater the fraction 1/n is. On the other hand, a large spatial scale corresponds to a vast portion of the earth—sometimes huge areas like an entire continent or a whole biome—and will correspond to a small map scale. If we apply the cartographers' fraction $1/n$ criterion, n will correspond to a great length, and

[1] ...a period much shorter than the lengthy dynamics that corresponds to the time needed to construct the ecosystem observed.

the fraction will be small. For example, 1 cm on the map representing 100 km, i.e. 10,000,000 cm, corresponds to a map scale of 1/10,000,000, a very small number indeed.

12.4 The Proximal (or Short-term) Processes

Proximal processes occur during a brief duration of time in comparison with distal processes. Short-term processes take place during about the life span of organisms, that is, between one hour and several hundreds of years depending on the species, whereas long-term processes can last between 10,000 and some tens of million years [42].[2]

Thus, proximal processes can be approximated to instantaneous events compared to the distal processes' span. Proximal processes ensure the maintenance of a biological/ecological system at a given epoch; existence conditions must be met during a period Δt compatible with the biological needs of the component organisms; if they fail, the biological/ecological system collapses and ceases to exist.

Proximal processes are nested interdependent compartments consisting of (1) mechanisms of a morphophysiological nature, and (2) adaptive strategies.[3] Both ensure the maintenance of the individual organisms—a requisite for the constitution of the member population of a community. The appropriate functioning of the interacting taxa in ecosystems—as a result of their fitness in their ecology, genetics and population dynamics—thus occurs at the individual or intraindividual level and ensures the continuity of the existence of the biocoenosis, which is challenged at every instant (during short periods); and this is why these traits and their dynamics form the proximal processes.

Living beings[4] are made up either of one cell (unicellular organisms) or of many cells[5] (multicellular organisms). Each cell contains smaller components (organelles, vesicles, nuclei, ribosomes, macromolecules, and small molecules, atoms, electrons, protons, etc.) at lower integration levels. Appropriate biochemical and biophysical conditions must be met for matter exchange and energy transfer within these subcompartments, which are also performed at every instant. In sum, we see that maintenance at the individual and population levels, both functionally related, belongs to proximate processes.

[2] Theoretically, much older relictual ecosystems are to be discovered in the bedrock sediments or hydrothermal vents in the form of biogenic signatures (e.g. Carbon 13/carbon12 ratio), perhaps dating back to the origin of life, 3.7 billion years ago [94].

[3] Adaptive strategy: this concept applies in a given species to behaviour patterns or ecophysiological features of individuals or groups that are adequate responses to environmental pressure to ensure their survival or that of their progeny.

[4] ...except for viruses that are biological agents, thus organisms, but not considered as true living beings by every biologist.

[5] Up to billions in complex animals.

The distribution of many plant species can be explained by very simple responses to a limited number of environmental factors, as exemplified by the common wild madder (*Ruba peregrina*) and the small-leaved lime (*Tilia cordat*) . One single parameter or the crossing of two parameters can explain these plant species distribution [3, p. 45].

- The common wild madder *Ruba peregrina*, a European species, is associated to 4.5° C January isotherm.[6] This species grows in areas where temperatures are equal to or above 4.5° C in January; this plant has not been observed outside this area.
- As far as the distribution of the small-leaved lime is concerned, two parameters were crossed, explaining the presence of this tree.

  – the minimal temperature observed of the coldest month, and
  – the maximal temperature of the hottest month.

The correlation between ecological parameters and spatial distribution in animals is not as straightforward as in plants. In the zoological kingdom, a whole complex of factors can explain the differences between the ecological requisites and the distribution areas, which can already diverge between related species groups. This appears obviously when these related taxa, with similar sizes and niches, are endowed with different ecological tolerance degrees. Cox and Moore [19, p.41] give the example of two species of insects of the order Odonata, the common blue damselfly *Enallagma cyathigerum* and a dragonfly, the ruddy sympetrum *Sympetrum sanguineum*).

- *Enallagma cyathigerum* abounds in diverse freshwater habitats, namely pools, ponds, lakes and rivers preferably when the current flow is not too strong [19]. The species favours a habitat composed of a large variety of floating plants (in terms of diversity and species richness). In addition, the pH and temperature ranges of its habitat are very large. This explains why the distribution area of *Enallagma cyathigerum* is very extensive throughout the temperate zone of the Ancient and New World. The limit of its distribution at the southern rim of the holarctic (area covering the temperate periarctic zone in the northern hemisphere) can be explained by the lack of adequate water surfaces there. The species have thus weak ecological requirements explaining its very wide and pervasive distribution; it is said to be *eurytopic*.
- The dragonfly *Sympetrum sanguineum* has a far more limited distribution, though the distribution area seems very large, extending throughout Western Europe. However, its habitat is restricted to water-filled ditches and stagnant-water ponds where several species of reedmace (e.g. *Typha lastifolia*) and horsetails (e.g. *Equisetum*) grow. The dragonfly *Sympetrum sanguineum* has thus strong

[6] Isotherms at a given month are, on certain maps, curves connecting localities where the same mean temperature has been recorded.

ecological requirements explaining its disjointed and piecemeal distribution. Such a species is termed *stenotopic*.

12.5 Ecological Successions: The Proximal Dynamics Process

Now, I will address the dynamics of communities and a fortiori biocoenosis, termed ecological successions.

Definition 12.1 In a given territory, an ecological succession is a spontaneous dynamic towards the complexification of a community of autotrophic species (and associated heterotrophic species depending on them) through time. This complexification eventually tends to a stable equilibrium called climax.

Here, the autotrophic species are the motor of the dynamic. By speaking of them, most authors refer to plants, in which case the heterospecific species, depending on the latter, are animals or fungi. However, we can address the topic in the prokaryotic world as well. Ecological successions are composed of successive stages that replace one another and are characterised by a specific biocoenosis. The successive biocoenoses of an ecological succession form a temporal sequence or sere [103].

Corollary to Definition 12.1, we can give a mathematical expression of ecological succession. Let n and N be two natural integers such that $1 \leq n \leq N$, and a temporal sequence of N successive biocoenoses where

- if $n < N$, the n-th biocoenosis is replaced by a $(n + 1)$-th one that is denser and more complex; and
- if $n = N$, the n-th biocoenosis is not replaced,

then the ecological succession becomes stable at stage N and reaches the climax or a *climactic equilibrium*.

The mode of action of such ecological successions is the elaboration of increasingly rich, complex and deep soils[7] along the successive stages. Soils thus evolve while accompanying the successive stages until the eventual climax is reached. In the most extreme cases among terrestrial habitats, an ecological succession starts from a ground devoid of any eukaryotic organisms—a situation that can occur in recently-emerged islands or any area that has become empty of any eukaryotic life following fire, volcanic eruption, flooding, glaciation, or any other life-destroying phenomena, either natural or caused by man.[8]

---

[7] As a reminder, soils are complex systems involving many factors, notably (1) a complex chain of species ensuring the process of mineralisation and numerous abiotic factors, (2) the mechanic actions of plants on the bedrock that inject abiotic elements among the abiotic factors, (3) different elements of live or more or less decaying dead organic matter, and (4) a skeletal frame composed of grains of various sizes (elemental parts of sand, silt and clay).

[8] The first modern ecologists interested in the dynamic of ecosystems put forward their theory of ecological succession from these extreme cases [14, 16, 18, 127]. In those epochs when

In addition, an ecological succession can unroll via intrinsic processes or under the influence of external factors. A staunch debate took place in the 1930s about what could be considered intrinsic processes or influence of external factors [96], but finally, the theory put forward by Tansley has been adopted, considering that the autotrophic part of the biocoenosis is the motor of the succession. Intrinsic processes have been termed autogenic succession and the influence of external factors, allogenic succession [126, 127] (cf. Definition 12.2):

Definition 12.2 (Autogenic and Allogenic Succession)

- An ecological succession is termed autogenic if the passage from one stage to the next is only due to the action of the autotrophic species (*e.g.* the plants) on the habitat.
- However, an ecological succession is said to be allogenic if the habitat alteration is due to other factors (abiotic factors like wind, temperature, humidity, etc, but also biotic agents like non-autotrophic microorganisms, fungi, animals or man).

Ecological succession is thus a mix of autogenic and allogenic processes. Now, we will proceed with another characteristic of ecological succession. The latter can develop either from a plantless habitat or from an environment partially degraded in a given stage comparatively to a previous stage—which stage is always degraded comparatively to the climactic equilibrium. In both cases, they will complexify [35, 146], eventually approaching the climax. Hence the following Definition 12.3:

Definition 12.3 (Primary and Secondary Successions [13, 120])

- A succession is said to be primary if its first stage is a habitat composed of bare rocks only sheltering a biocoenosis of prokaryotic microorganisms (bacteria or archaea).
- On the other hand, a succession is called secondary if it started with a degraded habitat.

Both succession types alter with time and tend to a climax. We will now exemplify the scenario of a simple autogenic and primary succession—a model that is quite old, having been proposed in the eighteenth century [21, 31, 146].

Stage 0: In the beginning, there is no soil but just bare rock sheltering only microorganisms (mostly prokaryotes but also some fungus species) [15, 29, 59, 137].[9]

Stage 1: The first macroorganisms that colonise these sites are very simple and primitive small-sized multicellular eukaryotes (mosses, lichens) and some

---

the destruction of natural habitat was far less frequent and massive than it is nowadays, these were virtually the only evolving ecosystems that one could witness relatively easily. In general, undisturbed habitats are climactic by nature. Under present anthropic pressure, there are fewer and fewer of them.

[9] Until recently, the authors publishing on ecological successions considered that stage 0 corresponded to lifeless bare rock [143].

minute fungus species, forming the primaeval biocoenosis whose photoautotrophic elements have no root system. A very elementary and shallow soil is forming, resulting from the interaction between the biomass of these colonising organisms and the microorganisms *in situ*.

Stage 2: In the course of generations, the soil deepens and becomes enriched, thus favouring the growth of annual herbaceous plants. This new community gradually replaces the former one mainly composed of mosses and lichens, and even frequently eliminates them. These populations of herbaceous plants exert a stronger mechanical action on their substrate, which in turn enriches the soil, entailing the increase of plant biodiversity.

Stage 3: The enrichment and soil deepening process proceeds, causing the succession continuation. Stage 2 plants are partially or totally replaced by a flora composed of perennial herbaceous plants with stronger and more complex roots and small-sized vineous plants.

Stage 4 … stage $N - 1$: The ligneous community develops with time, becoming increasingly dense, high and complex. The landscape evolves from a low meadowlike one-stratum type of vegetation to a multistrata system, with more and more levels and increasingly closed foliage cover. A model of this development [17], based on a specific typology of shrub formations (cf. Table 12.1), runs as in Fig. 12.1. The process continues until the climax, represented by stage N (see hereunder). Figure 12.1 illustrates the fact when the ecological parameters are not optimised—for example, when the pH of the soil, the hygrometry or the temperature of the ecosystem deviates from some optimal values, climax can be represented by a habitat that is less complex and less rich than a dense forest. From that pattern, a network of scenarios can be drawn (Fig. 12.1).

Stage N: Climax can be reached at whatever stage from stage 0 and onwards, which means that there exist habitats on the Earth where macroscopic life is absent and others, on the contrary, that are covered by primary dense forests teeming with macroscopic species (plants, fungi and animals). Any habitat has a climactic equilibrium between these two extremes, which corresponds to the maximal photoautotrophic potential of an ecosystem. In terrestrial ecosystems, the latter is equivalent to the biomass of the plants. Climax determines the recovery potential of an ecosystem in a primary succession, which is mainly tributary to the abiotic factors (AFs) in the environment of this ecosystem, in particular soil AFs.

Table 12.1 Typology of the shrub formations proposed by Costermans [17]

	Height	
Foliage cover	< 2 m	2–8 m
Very sparse (< 10%)	Low open shrub land (LOS)	Tall open shrubland (TOS)
Sparse (10–30%)	Low shrubland (LS)	Tall shrubland (TS)
Mid-dense (30–70%)	Open-heath (OH)	Open-scrub (OS)
Dense (70–100%)	Closed-heath (CH)	Closed-scrub (CS)

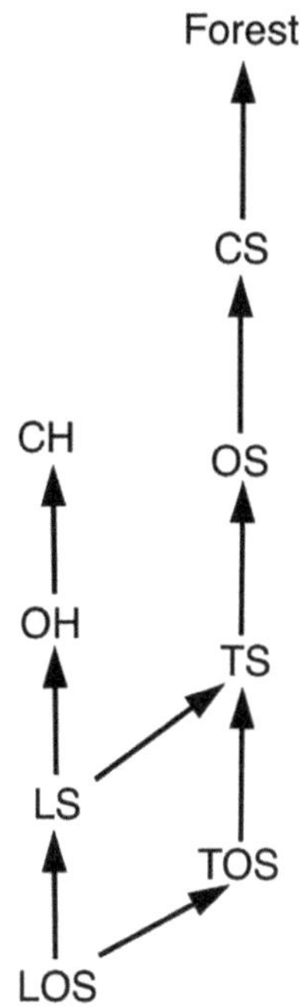

Fig. 12.1 Network of possible scenarios for a succession : a scenario is a sequence of adjacent arrows. When a shrublands turns into heathland, due to the acidity of the soil, their evolution into forest is hardly probable, hence the absence of arrows linking the left and right sequences of the figure from OH and CH

Each plant association in a given stage of ecological succession is related to a community of heterotrophic species (animals, fungi, phagocytic unicellular eukaryotes) and to regulatory microorganisms, building up together a biocoenosis (cf. Chap. 2). These ecosystem components are links in the trophic network. The biocoenoses thus fully reflect each stage of the ecological succession and necessarily evolve with the stages.

12.6 Elements of Evolutionary Ecology: The Distal Dynamics Process

Compared to proximal processes and ecological successions, distal processes unroll for a period that is 2–10 orders of magnitude longer. Long-term processes result in the relative constancy of the species constituting an ecosystem and the functional relationships between them. This relative constancy is perpetually set anew by reproduction and the replacement of the generations within the biocoenoses and by geophysical mechanisms such as subduction and volcanism, for instance, within the abiotic factor system. Long-term processes constitute a perpetuation of succeeding short-term processes. Seemingly constant, they are, in fact, sequences—mostly gradual, smooth and long, but sometimes dramatically discontinuous and fast-changing—of successive, often-contingent proximal processes. Distal processes are the objects of paleoecology, a science whose purpose is to reconstruct the ancient ecosystems from the origin of life through the succeeding geological eras until the present. The sequence of the reconstructed series of ancient ecosystems from the origin until the present is a huge topic beyond this present book, based on comparative genetics, palaeontology, molecular biology, and morphological biology, and can broadly be the subject of another book.

The emergence of the species, together with the mechanisms and processes of this phenomenon, constitute the subject of evolutionary ecology. Indeed, the motor of this process is an ecological factor, *natural selection* [24], i.e. the response of the organisms' vital and reproductive capacities to environmental hazards. This is why a school of thought appeared in the 1930s, the *synthetic theory of evolution* (also called the *modern synthesis*), which strived to find the ecological factors that could lead to an evolutionary process. The primordial role of the ecological parameters in evolution has not been focused on by most evolutionary biologists for the last forty years. This mainstream considers that classification based on cladograms should be the main topic of evolutionary studies. It bases its conception on the hypothesis that the nodes of the cladograms are formal but not actual ancestors, the latter and their geographic origin being unknowable. This argument is based on the fact that finding the missing links in the fossil record is so improbable that it is scientifically sound to reason in terms of cousinhood rather than direct ascendancy. Known taxa are always processed as "cousins", even if they are fossils from a remote past. Hence, they treat current evolutionary mechanisms as speciation scenarios. Though most of these scientists are not opposed to this type of research, they consider this topic minor.

Despite this strong trend, field studies on the mechanisms of speciation still belong to current research, with continuous discussions on (1) the necessity of a barrier for speciation to occur [77], (2) or whether evolution processes gradually [87], by jumps [129], or a mix of the two [54].[10] This research plays a significant role in understanding the genesis of biodiversity. Multidimensional analyses and artificial intelligence are now involved in speciation studies, carried out far more efficiently than in the pre-computer era when most of the tools describing speciation were forged. Species are structured in populations that (1) either do not show any significant variation throughout their entire distribution area (the interpopulation variability being inferior to the intrapopulation variability), (2) or present genetic variations without evolutionary effect, (3) or show patterns of current speciation, We will now overview some basics of evolutionary biology and biogeography to understand how life developed and formed biocoenosis. For the understanding of most readers, I will remind some basics of evolutionary biology and biogeography.

[10] An interesting and critical review summarises all these aspects [40].

12.7 The Distal Development of the Biosphere

Contrarily to the classical *scala naturae* [101, 103] depicting this history as a long process towards "visible" ecosystems, with biocoenoses formed of multicellular eukaryotic species constituting the biomes, I take a more general stand. The overwhelming part of the actual biodiversity comprises unicells, the least complex organisms, and nearly all invisible to the naked eye. Despite the apparent uniformity of their shapes and colours, they are biologically so diverse that they constitute the three primary divisions of life: the domains Bacteria, Archaea and Eukarya [149, 150]. Domains constitute the highest-ranked taxonomic category in modern classification, apparently counterintuitive: e.g., Human and Amoeba belong to the same domain, Eukarya, whereas three rodlike microbes like Trypanosoma, Halobacterium and Bacillus are respective members of the three domains, being respectively a eukaryote, an archaeon and a bacterium. Unicells can build up complex biocoenoses interacting between themselves and with abiotic factors they need, precisely like more conspicuous ecosystems. Due to their inconspicuousness, these ecosystems were too often overlooked. However, they constitute an essential part of the biosphere, mainly determining the ecological equilibrium of our planet.

12.7.1 Development of the Biosphere: An Evolving, Spontaneous and Stochastic Process in Permanent Instantaneous Thermodynamic Disequilibrium

The *biosphere* is the system comprising all the existing terrestrial and marine ecosystems. Since its origin, this supersystem has become increasingly complex with the increase in the number of living beings, *species richness* and *biological complexity*. Several evolutionary thresholds were of the utmost importance.

1. During the first half of the history of life (from −4 to 2.2 Gys[11]), the only forms of life must have been *anaerobic* prokaryotic microorganisms, i.e., unicellular organisms (1) without a nucleus and any distinguishable organelles, (2) with only one circular chromosome floating in their cytoplasm, and (3) breathing with other electron acceptors than gaseous dioxygen. These organisms would very probably remind us of our present anaerobic bacteria or archaea and were the only organisms that could be adapted to the environment of the primaeval biosphere, which was partially reconstructed from laboratory work and the analyses of the most ancient rocks [85, 89]. In that remote epoch, the oceans and atmosphere hardly contained the gaseous dioxygen, which is so prevailing yet and so vital for eukaryotic life and many prokaryotes nowadays. Life was probably constrained in aqueous environments, first in restricted pools, then

[11] 1 gigayear (Gy) = 1 billion years.

in oceans. At that time, these habitats would very likely contain a mix of methane (CH_4), water (H_2O), ammonia (NH_2) and molecular dinitrogen (N_2) [85]. The vast majority of scientists now hypothesise that life appeared in one geographic point, so to say, in one original life-inducing pool, then expanded into other spatially-related chemically similar pools, and eventually conquered the primaeval oceans that acquired progressively the chemical characteristics of the primary pools favourable to the emergence of life. A small number of authors, though, the proponents of the so-called Gaia theory, question this hypothesis and propose that Earth has been a planet uniformly favourable to life everywhere, in the remote past as well as at present [70].

2. A major evolutionary step occurred about 2.4–2.2 Gys ago with the appearance of the first prokaryotes carrying out a form of photosynthesis or chemosynthesis, releasing thus dioxygen molecules into their surroundings ([100])[12] in large amounts. This event has been termed the Great Oxidation Event and could have induced the extinction of many primaeval anaerobic organisms. Many present bacteria practise this type of chemical reaction, namely groups like cyanobacteria, and it is assumed that in the past, the prokaryotes that released this molecule were also cyanobacteria. This set of events first occurred in the hydrosphere where life was concentrated; gaseous dioxygen spread afterwards into the atmosphere. This set of events entailed the emergence of new forms of life increasingly adapted to this new type of environment—organisms able to use the O_2 molecules as electron acceptors, i.e. the aerobes or living being practising aerobic respiration, that is, breathing with oxygen.
3. The next step in the process of biosphere complexification was the emergence of the first eukaryotes— unicells containing a nucleus enclosing a fixed number of linear chromosomes and definite organelles. These organisms were called protozoans in former taxonomic classifications. All known eukaryotes are obligate aerobes, and it is still now widely acknowledged that these organisms practised aerobic breathing since their appearance;[13] which means that they would necessarily have emerged after step (2). The first eukaryotes were very probably endosymbiotic organisms, consisting of bacteria and archaea engulfed in a larger cell of unknown origin [22, 108]. They very probably invented sexual reproduction—a revolutionary event that considerably magnified the genetic potential and induced ecophenotypes of these organisms compared to those of the prokaryotes, with a formidable impact (1) on the sizes, weights, aspects and many other kinds of phenotypic variability, (2) on the ability to generate stable multicellular organisms, and (3) on the efficiency of resource exploitation.
4. The following evolutionary stage of the biosphere could have been the emergence of the different lineages of multicellular organisms among the eukaryotes,

[12] **Exercise**: were these organisms anaerobic or aerobic?

[13] Recent publications [7, 41] suggest that multicellular fungi already existed before the Great Oxidation Event, or date back at least to that event. This paper thus favours the pre-existence of the eucaryotes to the Great Oxidation Event.

i.e. *plants*, most of the *fungi* and *animals* (see Footnote 13). These are by far the most voluminous and complex living beings, constituting the only visible part of the living world to the naked eye. These organisms have undergone a spectacular evolution and development for the last 2 billion years [83]. The multicellular eukaryotes dramatically improved the efficiency in exploiting energy and material resources by adopting two modes of nutrition, i.e. (1) production where organisms practise biosynthesis, taking in all their nutrients in the form of inorganic molecules, and (2) consumption, where organisms carry out feeding, a process involving ingestion, digestion, assimilation and excretion from organic matter.

Among the multicellular eukaryotes, plants are the sole *producers*; almost all practice biosynthesis exclusively, and the other two groups are exclusively *consumers*.[14] A hyperspecialisation concerning nutrition and energy intake then took place between the different kingdoms of the multicellular eukaryotes, photolithoautotrophy in plants, and organoheterotrophy in fungi and animals. The trophic network emerged then and has developed to become the very intricate tree-like organisation it is now.

Before that period, the trophic relationships were much simpler, consisting either of direct feeding on organic or inorganic molecules found in the environment (as is generally the case in prokaryotes) or phagocytosis, a mode of nutrition where large unicells, generally eukaryotes, incorporate and digest smaller unicells, generally prokaryotes. These relationships did not form trophic networks. By contrast, we can realise the importance of this step.

5. Gradually, the multicellular eukaryotes and also the relationships between them increased in complexity. In animals, fungi and plants, it took different forms.

   - Within different lineages, animals were subjected to cephalisation, acquisition of bilateral symmetry, pair sensory organs, and the development of increasingly sophisticated and concentrated nervous systems. They formed societies that developed a wealth of feeding strategies and interrelationships unequalled in the rest of the Living World.
   - Fungi form the only group comprising unicellular and multicellular organisms, though most of them are multicellular. They have developed a diversity and complexity level in their reproductive process and life cycle that has been unsurpassed in other multicellular eukaryotes.
   - Plants have developed increasingly complex vascular and reproductive systems, more efficient shapes and organs for photosynthesis, and increasingly sophisticated life cycles.

[14] There are notable exceptions. Among the producers, we can mention the carnivorous plants that add some organic nutrients obtained from predating some small animals or eukaryotic unicells to the nutrients normally taken in by photosynthesis and also the case of sea slugs that engulf chlorophylls of the algae they feed on—the chlorophylls remaining intact and functional in the body of the animals, enabling them to photosynthesise their nutrients during starvation periods [98].

The increase in organic complexity went with a rise in the diversity of the habitats. In return, the latter could contain an increasing number of organisms and forms of life that were more and more diversified. The evolution of the biosphere, as superficially depicted by the five points just above, comes with an increase in the material and energetical balance, which is paradoxical thermodynamically. The biosphere is a materially and energetically open system in permanent thermodynamic disequilibrium at each time t, contrary to ordinary physical systems, despite its being also a physical system. In the long run, a biosphere and all its compartments, the ecosystems, are also subjected to the three laws of thermodynamics.

12.7.2 *Biogeographical Events as a Driver to the Distal Processes*

12.7.2.1 Let Us Set the Stage: Biogeography, as a Science, Is One Consequence of the Discovery of the Great Scientific Travellers

Biogeography is the science that deals with all the geographic aspects pertaining to the geographical distribution of the organisms on Earth. This discipline has existed since the eighteenth century and has evolved tremendously following the huge scientific advancements achieved from that time through the nineteenth until the present. It is about the distribution of the organisms on Earth, but the scope and concepts of this discipline have gained in accuracy and efficiency with time (see below). Biogeography is based on the *Linnean concept of species*. Before it, the concept of species was not very clear in the minds. The scenario of Genesis, where God created every type of animal and plant ca 6000 years ago and was not tied to any specific geographical area, was widely accepted, even in scientific circles. This philosophical attitude, completely obsolete in science nowadays, is called creationism. It was the common philosophical and scientific paradigm before the advent of the different theories of evolution (or evolutionism) during the nineteenth century (see Part II).

As a reminder (see Sect. 7.3), Linnaeus was the inventor of the Linnean species just mentioned—the oldest modern concept of species still used and that runs as follows: A species is defined by a single specimen deposited in a museum collection called the type of the species, identified by a binominal name, and described in a publication (article or book). Whether an individual, in the form of a museum specimen, pertained to a species or not was left to the appreciation of an expert appreciating the similarity between the individual and the type. The conclusions were then extrapolated to live organisms. Linnaeus was a devout Protestant and very prone to accept most of the Christian dogma; however, he was forced to admit from the evidence collected by the great scientific travellers that the constancy of the species between different parts of the globe was not observed. To make a compromise between his religious credo and the facts, he invoked Noah's ark to explain that when the waters receded, the animals adapted to different altitudes of the mountain—a kind of parable used by the author to evoke the diversity observed without involving the geography and even less the possibility that species were not fixed.

I will now summarise the successive concepts of biogeography; the former concepts were never totally replaced by the later ones but refined. All these concepts, even the oldest ones, can contribute to analysing ecological distal processes, which is our aim here.

12.7.2.2 Predarwinian Concept

The oldest known biogeographical essay is attributed to Georges Louis Leclerc de Buffon (1707–1788) [9][15][32–34], and was about

> the study of migration as the causal explanation of the distribution of faunas.[16]

In this book, Buffon established that animal species had a distinct geographical origin, their members having dispersed from these geographic centres to specific distribution areas. Buffon's explanation was compatible with his creationist belief. He was convinced that the ideal specimens of each animal species were found close to their "geographical centres". Through a process of migration and degeneration, similar races emerged, occupying different distribution areas. This principle was called *Buffon's law* by his contemporaries, a theory based on the firm belief of a universal degenerescence of the races coupled with dispersion among animals and man alike.[17] In addition, the increasing discovery of fossils, associated with his 'law', convinced Buffon that the Earth was much older than the 6000 years commonly believed by his contemporaries; he proposed for the planet, thus life, an age of several hundred thousand years.

With his theory of degeneration and migration, Buffon adapted creationism to his novel idea that a group of species dwell in specific distribution areas—he was thus the inventor of the concept of faunal distribution. In the eighteenth century and the beginning of the nineteenth century, the first biogeographic works after Buffon's, notably those of Augustin de Candolle and Alexander von Humboldt, could not have an evolutionary interpretation[18] and applied Buffon's law that was at that time the most convincing explanation of the distribution of living beings. Candolle and Humboldt were the first biogeographers who applied Buffon's vision of biogeography to plants and realised that the plant species, too, but also

---

[15] Buffon and subsequent authors who performed biogeographical work did not use the term 'biogeography' which was coined much later, by the end of the nineteenth century.

[16] Translated from French by the author.

[17] The terms 'species' and 'races' were indifferently used by the author in the sense of morphological types.

[18] This was the situation before the works of Lamarck [67] and Darwin [24] who introduced evolution as a central concept of biology.

entire communities of definite plant species, had specific distribution areas—they developed the concept of floral distribution.

12.7.2.3 First Biogeographical Theory After the Darwinian Revolution

The implication of the modern evolutionary approach to modelling distal ecological processes was immense. The idea that each species was born in a specific locality and that migration was the principal explanatory cause of its geographic distribution allowed later authors favourable to evolutionary thinking [from Lamarck (1744–1829) and Darwin onwards] to assign a geographical origin to the species at their nascent state. These biologists kept the idea of migration but replaced the causal principle of Buffon's law with evolutionary mechanisms instead of the old idea of degenerescence (see Sect. 12.7.2.2). This reasoning was impossible without the idea of all species deriving from ancestor species, thus of phylogeny. This term, coined by Haeckel (1834–1919) [50], designates a kind of genealogical tree[19] where an ancestor species is related to daughter species (nascent species derived from it), each becoming in turn ancestor species of another generation of species, and this recursively, from probably a primordial species dating back to the Archean (the oldest geological era that sustained life) that was the common ancestor to all present living beings, extremely probably a prokaryote. The motor of such an evolution is reproduction and by this mean, the transmission of characters from an organisms to its offspring, and this, resursively during many generations. Characters, which are relatively labile, can change with time, leading to new species.[20] This is the basis of evolutionary thinking, mainly explaining the history of organisms until now.

The first evolutionists (Darwin, Wallace, Haeckel) adopted Buffon's law (see Sect. 12.7.2.2) . They applied this hypothesis to set up the evolutionary theory in the following way. If A, a nascent species at time t, originates in a given locality ℓ, its descendants would travel around this point and constitute a distribution area. Any locality of this distribution area, in turn, could potentially constitute the geographical origin of a daughter species. The phenomenon would have been gradual and smooth. Figure 12.2 illustrates this vision of speciation.

[19] An idea already suggested by Darwin [24].

[20] Other modes of gene acquisition than vertical gene transmission from ancestor to descendant, the motor of phylogenesis (the biological process underlying phylogenies), seemed to have also played a great role, e.g. lateral gene transfer, endosymbiosis and chimerism. Eukaryotes are hypothesised to have been generated via endosymbiosis; they very probably originated during the Archaean in a world inhabited exclusively by prokaryotes, from the endosymbiosis between a cell of unknown origin and prokaryotes, engulfed but not phagocyted by that cell. This theory, yet widely accepted, posits that the cell machinery is of archaeal origin, and the cell membrane, the mitochondria and the chloroplasts are of bacterial origin. As a result, the explanation of evolution via phylogeny only, i.e. the construction of a hypothetic tree of life explaining all the biodiversity on earth, has been seriously challenged and debated [28, 66, 105, 108, 117].

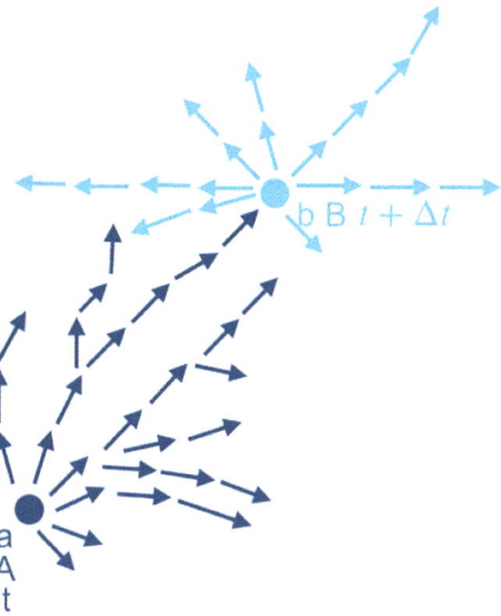

Fig. 12.2 Biogeographical model odel of the first evolutionists, fitting gradual speciation. Dispersal of species **a** (*in dark blue*): the ancestor species represented by a *dark blue disk* appears at time t in location A, and then migrates with time from that position. The migration is represented here by a kind of "vectorial field" whose elemental oriented segments are coloured in *dark blue*, with the sense depicting the march of time. Each arrow symbolises the discrete spread of a group of species **a** during a period of indefinite duration, generating other groups among which subgroups will spread likewise, symbolised by succeeding arrows. In the course of time, under the pressure of natural selection, the organisms gradually vary at different rates between the groups, several of them faster and statistically in one direction so as to depart from the morphophysiology of ancestral **a**. One of the groups located at time $t + \Delta t$ in position B has accumulated enough divergence to separate from ancestral **a** and become a new species, i.e. **b** (nascent **b** represented by a *light blue disk*). The same process unrolls with **b**; all the symbols are re-used for **b** with the same meaning but in *light blue*

The first evolutionary theories were gradualistic, explained by the following scenario. Species **a** spreads gradually from and around its origin centre A according to the scenario depicted in the legend of Fig. 12.2. At a certain point of the migration, after gradual transformations, **b** has accumulated so many differences that it has become a nascent species. A speciation event thus occurs in locality B that generates species **b**. In turn species **b** scatters from and around locality B, its origin centre, likewise. And this, recursively. This was the type of evolutionary biogeography and speciation reconstruction commonly accepted, a vision based on the theory of blending inheritance—the paradigm at that epoch, before the advent of genetics that totally disproved this theory. In the view of these authors, the "degenerescence of the race" as proposed by Buffon (see Sect. 12.7.2.2) within the species could explain why several individuals were unfit, thus counterselected by natural selection. The basis was good but had to be ameliorated, and that was the job of the historical biogeographers in the middle of the twentieth century.

In the frame of the old vision of the first evolutionary biogeographers, I can already propose an explanatory model for some of the features making the distal process of an ecosystem (see Fig. 12.2 and its legend). The species' distribution areas result from a kind of "vectorial fields", as shown in this figure, analogous but not identical to the accurate vectorial fields studied in physics. Contrary to these, the biogeographical "vectorial fields" cannot be mathematised deterministically since they are the consequence of very complex behaviour (or tropisms) linked to contingent events and not the consequence of material properties. The biogeographical "vectorial fields" are composed of "vectors" circumscribed by a front that is the observed limit of the distribution area, the "vectors" representing extant or extinct demes of the distribution areas. Each "vectorial field" of a species corresponds in this model to a specific colour. The overlap of vector fields of different colours models the distal process of a biocoenosis if we accept the partially true theory of the first evolutionists. This model will be refined.

12.7.2.4 Historical Biogeography: Its Implication to the Modelling of Distal Processes

Many evolutionary biologists of the twentieth century involved in biogeographic studies, namely the authors belonging to the school of thought named evolutionary systematics or synthetic theory of evolution, accepted the gradualism and hypothesis of geographic centres and migration proposed by the former evolutionary biogeographers. However, the advent of genetics at the beginning of the 1900s forced them to reject the theory of blending inheritance. Among the famous biologists leading that school of thought, we can cite a.o. Mayr [74], Simpson [122], Darlington [23], Dobzhansky [27] and Lewontin [69]. These authors found another explanation for the occurrence of speciation events. Species were gene pools normally remaining homogenised when nothing prevented the genes from travelling free across the species, i.e. the profile of the alleles would remain statistically constant throughout the range. For speciation, barriers were needed to cause a divergence between the profile of alleles of two demes, a requisite to splitting a species into two ones. In addition, these authors did not reject the interest of searching the missing link among fossils or even rarely among extant species[21] and of trying to reconstruct its hypothetical origin centre (see Sect. 12.7.2.3). The biogeography they performed is called historical biogeography [34]. Until the 1970s, these authors still admitted the old, now obsolete theory of continental fixity, positing that the present continents were always at the same place and constituted an immutable stage on which the species evolved.[22] Only one mode of species diffusion was overwhelmingly acknowledged at that time, active transportation, which exists in two forms: (1)

[21] The missing link being the common ancestor of related species whose origin centres are known.

[22] Curiously, the works of Suess (see Sect. 8.4) and of Wegener were still not recognised and accepted only fifty years ago.

locomotion (walking and running, flight, swimming, etc) in the case of motile animals or microorganisms, and (2) vectorial transport, i.e. a diffusion where the organisms are conveyed by motile animals, winds, streams, etc, in the case of plants, fungi, and sessile animals and microbes. The organisms tend, at each generation, to scatter in nearby territories, resulting in a continuous spatio-temporal displacement of their populations. In these cases, there are adaptive controls and feedback by organisms of their spread.

Contrary to active transportation as described above, passive transportation is defined as an abrupt and rapid displacement mode of the individuals relative to the distribution area of their ancestors, due to events they cannot control. It is not an adaptive process. Only two passive transportations are invoked in historical biogeography: (1) the transports causing the sets of "insular" speciations and (2) rafting theory explaining how two relatively related taxa can existing different continents very distant from one another, e.g. the monkeys of the Old World and those of the New World.

- The setup of "insular" speciations results from ecological successions that can take place, for example

  - when an oceanic island emerges,
  - after a catastrophic event like the mass extinctions of the end of the Permian or the end of the Cretaceous, or
  - when local extinctions, like those caused by volcanic eruptions, earthquakes, tsunamis, landslides, etc occurs.

  Here, passive transportation may involve conveyance by streams or currents, vagrancy of flying organisms due to storm and chance, or probably transports of small animals or other organisms on floating logs or masses of intertwined vegetation called rafts driven by oceanic currents. Indeed, floating vegetation rafts with small animals and fungi on them driven by currents are regularly observed nowadays, and this phenomenon may seriously be involved in the distribution of faunas. This hypothesis has been called the rafting theory. Organisms that succeeded in adapting and reproducing in their new insular habitats became geographically isolated and relatively rapidly differentiated and speciated, a phenomenon called *adaptive radiation* (cf. Sect. 12.7.6).
- Rafting theory also explained why similar taxa were found in different continents. It is this theory that convincingly explains the presence of monkeys in South America, as was mentioned above; their ancestors very probably were transported from Africa to the New World via rafts at the beginning of the Tertiary (and we will see below that the acceptance of a new paradigm in the late 1970s will add credits to the raft theory concerning this example).

12.7.2.5 Phylogeography and Continental Drift

This is the most recent explanation of the geographical distribution of species and ecosystems. The different distribution areas evolved parallelly to the species. Consequently, the different distribution areas are likewise processed with a tree representation. The phylogeographic tree is drawn in parallel to the phylogenetic one. In the frame of this practice, the continent drift, now accepted as factual, is involved and explains many ecological facts, namely, the existence of the "live fossils" in isolated ecosystems and how faunas and florae became specific of their different biogeographical regions [20, 33].

12.7.3 An Example: The Evolution of the Equids, Revealing the Evolution of the Paleoenvironments

Let us take the example of the natural history of the family Equidae, or horse family, since the Eocene [128]. The remarkable evolutionary traits of this group of mammals are (1) the reduction of the number of fingers from four to one, (2) the complexification of the teeth structure, and (3) the biogeographical evolution of the group. The evolution of this group of mammals tells a whole story about their successive paleoenvironments; in return, the modification of the morphology of the equids had an impact on the rest of the total biocoenosis in which these species lived.

- At the beginning of the *Eocene*, about 60 Ma ago, the oldest representative—*Hyracotherium*, also named *Eohippus*—lived in a supercontinent reuniting Europe and North America. Its remains were found on both continents. (1) This animal was very small compared to modern equids, with a length of 50–60 cm and a height of 20 cm at withers. It weighed 9 kg on average. (2) It carried *brachyodont*[23] teeth, i.e. low-crowned teeth, numbering 44 and hypothesised adapted to eat soft leaves, soft and hard fruits, and plant shoots [124]. (3) The forelimbs of the genus were four-toed, and the hindlimbs three-toed.
- During the Eocene, the Northern Atlantic Ocean started to form and progressively widened, isolating Northern America from Europe; the two continents continuously drifted apart during the Tertiary and reached their present positions in the late Pliocene (ca. −2.6 Ma) [134].

  Meanwhile, the horse family evolved strictly in North America with numerous genera of succeeding fossils. Figure 12.3 illustrates the evolution of the equids on a small sample of these genera.

[23] Brachyodont teeth, i.e. low-crowned teeth, have limited growth. Once they attain their maximal size, they slowly diminish due to continuous abrasion. This type of tooth is adapted to relatively soft food that contains little abrasive elements.

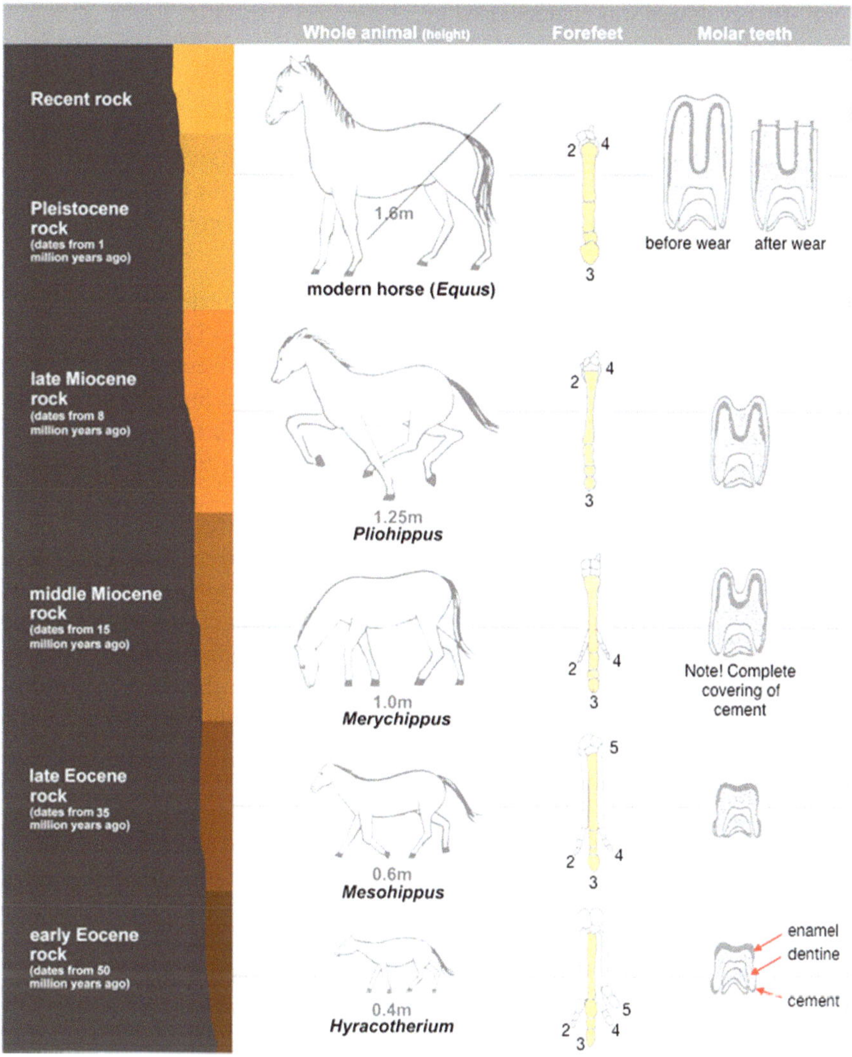

Fig. 12.3 Example of a distal process—a lineage in the horse family since the Eocene, highlighting the evolution from standard mammalian *pentadactyly* and brachiodont dentition towards *monodactyly* and hypsodont dentition specific to modern equids (courtesy of Wikimedia Commons)

This evolution was characterised by:

- a continuous size increase,
- a transformation of the shape of the teeth, from brachiodonty to hypsodonty,[24] with, in parallel, an increase in the number of the crests,
- a reduction of the four subequal toes in *Hyracotherium* (see above) to one very solid column-like toe terminated with a developed nail, the hoof, the other toes reduced to quasi-absence in modern equids. We show hereunder four steps of the evolution.

- *Mesohippus* was a genus of the lineage dating back to the late Eocene and early *Oligocene* (37–32 Ma ago). Compared to *Hyracotherium*, (1) its size was larger with a length of 90 cm and height of 60 cm at withers on average. The weight of a *Mesohippus* was 38 kg on average. (2) The teeth are still low-crowned, but the set of teeth contains a gap before the front teeth and six grinding teeth. Its diet was probably mainly composed of a mix of leaves and grass [82]. (3) The four limbs were terminated with three toes.
- *Miohippus* was a genus that existed during the *Oligocene* between 32 and 25 Ma before the present. (1) Its length varied between 90 and 120 cm, and its height was 60–90 cm at withers. Compared to *Hyracotherium* and *Mesohippus*, the body was significantly heavier (ca 40–55 kg). (Website: equipepper.com/2023/04/21/the-evolution-of-horses.) (2) No clear information is found in the literature about the height of the teeth crowns, but it has been noticed that the upper molars had one supplementary crest relatively to the counterparts of their ancestors. (3) The limbs were equipped with three still subequal digits.
- *Merychippus* lived during the *Miocene*, 16–5 Ma ago. (1) The animal has a length of 1.20 m and a height of 1 m at withers. Its weight lay between ca 70 and 100 kg. (2) The teeth are hypsodont, adapted to the grassland that appeared and became extensive during that geological period. The head started to look similar to that of modern equids. (3) The limbs were equipped with three digits, but only the middle one touched the ground, the others starting to reduce. Morphologically, this protohorse began to have the look and gait of a modern horse.
- *Pliohippus* was the first horse-like genus of the lineage and lived at the end of the Miocene[25] between 15 and 5 Ma ago. (1) On average, its size was 2 m in length and 1.20 m in height at withers. Its weight lay between ca 150 and 200 kg. (Cf. website uchytel.com/Pliohippus.) (2) The premolars and molars are hypsodont with a relatively complex tooth structure. However, these teeth are more curved than those of modern horses [8], which tends to prove that *Pliohippus* is not a direct ancestor of the modern equids. These teeth were adapted to grazing. (3) According to the species, the limbs are either one-toed like the modern horses, or

[24] In hypsodonty, the teeth are high-crowned and continuously growing—an adaptation to a mode of nutrition where the teeth are abraded by gritty food, e.g. hard plants like meadow grass, which contains a great amount of silica.

[25] ...and not during *Pliocene* as it was previously thought and suggested by the denomination.

three-toed with one column-like central toe and two lateral toes that are visible but vestigial and non-functional [147].

- Last but not least, the modern horse *Equus ferus* and all the other current equids, belonging to the same genus *Equus*, appear approximately 3 Ma years ago, thus at the Pliocene, a geological period extending from 5.33 to 2.58 Ma ago. If we take, as a representative of the group, the domestic horse *Equus ferus caballus*, (1) its length is 147–240 cm, and height 1.42–1.73 mm at withers. Its weight is around 500 kg on average. (2) The teeth are hypsodont with very complex structures, adapted to grazing. (3) The limbs terminate with one column-like toe terminated by a hoof.

A land bridge emerged linking America to Asia between the mid-Miocene and early Pleistocene, allowing a part of the Equidae to migrate from America to Asia. All the current species of the horse family, with no exception, stem from these migrants. The Equidae thrived in the Old World (Asia, Europe, Africa) while the family progressively declined until complete extinction in the New World (North America).

The ancient horses of the New World were extinct about 10,000 years ago, coinciding strangely with the arrival of humans in America. All the present equids originate from a handful of equids that migrated from America to Asia through the Bering Strait about 800,000 years ago [92, 141]. This example shows how, with time, considerable changes have occurred that can explain the biodiversity observed now.

We have treated above a case of macroevolutionary process; we will now make an overview of the elemental events that, in their succession, construct such macroevolutionary processes. These elemental events are called *microevolutinary mechanisms*, of which I will designate two types, polymorphism and (sub)speciation.

12.7.4 Polymorphism

An interesting case where two types of individuals are very distinguishable phenotypically and sometimes genetically is polymorphism. Sympatric morphs are groups of individuals that are phenotypically but not genetically different from the rest of the population. The existence of different morphs is called polymorphism, a phenomenon interpreted as a response to different local or seasonal environmental parameters. Distinct morphs appear fortuitously and are the dramatic expression of interacting alleles; they do not impair the interbreeding ability of these individuals with the rest of the species. Like all other population members, morphs can be considered as interindividual variates. They can build up separate groups as a consequence of segregation due to phenotypic preference, but by no means are they involved in a speciational process. They live locally in sympatry with the rest of the species. Well-known examples are the black panther within the leopard (*Panthera pardus*) or the bridled guillemot within the common guillemot (*Uria aalge*). Here, the interactions between the individuals or the social units are intraspecific and depend on the spatial organisation of the species (see Sect. 13.1.6) and on the sex of the individual.

In both situations, intraspecific competition for food resources, when the latter are not limited, is generally weak. The existence of different morphs of the common guillemot was shown as responses to different local or seasonal environmental parameters (e.g. temperature or other parameters of the sea surface); see an interesting review on the subject [65]. Some cooperative behaviour traits that maintain and strengthen the coherence of the colony are the rule in some species, e.g., the side-blotched lizard [123]. In several situations, intraspecific competition has been observed, notably the sparsely- and densely rakered whitefish *Coregonus lavaretus* in the lakes of northern Norway [1].

Now, we will treat the problem of the emergence of nascent subspecies and species, which is the beginning of the evolutionary process. This is the mechanism that generate the bionoenoses since the origin of life.

12.7.5 The Barriers

Speciation occurs when a novel norm appears in an isolated subpopulation of a population. Different types of isolation mechanisms (or barriers) exist between the group that undergoes the norm change and the rest of the population that conserves the original norm, i.e., the (1) behavioural, (2) geophysical (habitat or geographical), (3) temporal, (4) mechanical, and (5) genetical barriers [75]. The first four barrier types are involved in premating isolation mechanisms; the genetic barrier is the basis of postmating isolation mechanisms.

Behavioural barriers:

These isolation mechanisms consist of behaviour patterns specific to each group that render impossible communication that could to lead to mating between members of different groups. The beginning of the shift of behaviour patterns leading to behavioural barriers is sensorial preferences that can evolve into isolation mechanisms. Let us take as an example within passerine taxon, the species complex formed by the common and the Iberian chiffchaffs (resp. *Phylloscopus collybita* and *P. ibericus*) in southwestern Europe. The sexual interactions between the two taxa show a partial premating isolation caused by the following behavioural barrier: *P. collybita* ♀♀ are sensitive to, thus attracted by the song of *P. ibericus* ♂♂, whereas conversely *P. ibericus* ♀♀ do not respond to, and thus are not attracted by the song of *P. collybita* ♂♂. This behaviour pattern has considerable consequences for the pairing; it induces assortative mating [112].

Geophysical (habitat) barriers:

This phrase means a physical impossibility for the sexes to meet due to a physical barrier like, for example, (1) distance, (2) a stretch of land for an aquatic species, (3) a body of water for terrestrial organisms, (4) mountain ranges and valleys or (5) climatic impossibilities, e.g. glaciation or desert formation. Several examples of species, within which such a geophysical barrier isolates distinct populations, are displayed in the literature [75]. One of the habitat barriers is the oceanic barrier, from which Mayr draws examples in birds, like the following ones: Fifty-three passerine species comprise populations specific to different islands of the central Solomon archipelago, in particular the white-eye *Zosterops rendovae*; these populations are efficiently isolated by the sea arms between the islands. Very probably, the early ancestors of the Geospizines of the Galápagos Islands and of the Hawaiian honeycreepers would have undergone a similar subspeciational process after they colonised their respective archipelagos.

Temporal barriers:

Two populations can be isolated by a temporal barrier, either seasonal or circadian.

Seasonal barriers are common among plants and invertebrates, in particular insects and aquatic invertebrates. A detailed analysis of a seasonal barrier on an Indonesian ladybird beetle *Henosepilachna diekei* has been carried out by a Japanese research team. Two races of this species coexisted in the same territory, respectively associated to plants that do not grow in the same season [72].

Day-night barriers within species or species complexes have been documented, e.g., in the fisher (*Pekania pennanti*), a North American mustelid [64] or in two closely related antelopes, the Peters' Duiker (*Cephalophus callipygus*) and the Bay Duiker (*C. dorsalis*) [38].

Mechanical barriers:

Anatomical incompatibilities can appear within a population that renders copulation or the meeting of gametes impossible. This phenomenon is exemplified by the evolution of genitalia in insects [121], or flower structure in plants, particularly in angiosperms where pollinators play an important role in reproduction [47].

The barriers mentioned above are the main factors explaining speciations. Two great categories of speciations fundamentally differ in their ontological development: (1) *insular speciations* and (2) *continental speciations*. On the one hand, insular speciation is a process mainly initiated by geophysical barriers, either obstacles or unfavourable environments. On the other hand, continental speciation is more complex as far as its incipient development is concerned; it involves simultaneously manifold barrier types, among the ones depicted in Sect. 12.7.5.

12.7.6 Insular Speciations

Insular speciations are generally the consequence of adaptive radiation, which occurs when an "insular" environment happens to be quasi-empty[26] or highly fragile. This situation is observed, for example, when volcanic islands emerge from the oceanic bottom or when islands or ecological islands (like table mountains) undergo massive extinction events due to catastrophic phenomena (e.g. volcanic eruptions, hurricanes and typhoons, earthquakes, or extensive or violent anthropogenic disturbances). Insular speciations develop in three successive stages: invasion, colonisation, and genetic differentiation [115].

Invasion:
The invasion of habitat occurs when an ecological equivalent does not or weakly populates the latter or when the latter has a less offensive dynamic. Indeed, the dynamic of the invasive species is, as a general rule, more active, frequently resulting from a more severe natural selection than that of a native ecological equivalent in its insular environment, which is relatively more protected, with weaker competition and predation pressure.

Colonisation:
This stage corresponds to the expansion of the population across the entire available area, provided that the territories gained by the invading species are suitable for the latter.

Genetic differentiation:
It occurs when the invaded environment differs from the invading species' initial environment. This situation entails the alteration of their adaptive characters, which allows their carriers to survive, reproduce and transfer these favourable traits to their descendants.

When the islands are small, the weak number of the "colonists" that they can carry statistically leads to genetic drift, i.e. a rapid loss of alleles and enhanced mutations, with, as a result, a speciation rhythm that is more rapid than on large stretches of land (mainlands, continents, large islands). The larger an island, the larger the number of individuals it can sustain, and the more its environment will tend to resemble a continental one, and the slower the speciation rate.

In archipelagos, significant environmental differentiations can develop between their constitutive islands with time. This will entail accelerating phenotypic differentiations in the descendants of the invading species, hence a rapid speciation rate (above all when there is a *founder's effect*[27]). This has been broadly verified by a large body of literature, in particular on Darwin's finches (*Geospiza sp*) [24, 45, 95, 102, 138].

[26] I use the phrase "quasi-empty" in the sense that the environment is only inhabited by microorganisms.

[27] This is so, due to the weak initial number of a dynamic population.

The acquisition of adaptive traits, which accelerated speciation in island species, is manifold. We can see this phenomenon when comparing the morphologies of insular and continental forms of closely related taxa, which are often different, e.g. the systematic differences in the wings of related butterflies observed by Alfred R. Wallace (1823–1913) [144] (op. cit. in [44]). The size of insular forms can be smaller or greater than the size of their continental counterparts from which they originate; in birds, their bills, wings and/or tarsi can be longer or shorter [44]. Remarkably, in insular environments, the evolutionary tendency towards apterism or flightlessness is commonly observed in species descending from primitively volant taxa, namely in pterygote insects [142] and in birds [152]. Such an evolutionary trend is characteristic of the insular world as the physical constraints of islands impose it on these environments. Indeed, islands are often lands of small areas encompassed by a liquid surrounding of immense area, frequently swept by winds and storms. One can already detect this trend among efficient insular flyers. This is even truer with islands that are far from continents. Indeed, these islands are much smaller than those close to them; they are thus relatively isolated from resource subsidies and have reduced accessibility to invading species, two vital features for establishing insular ecosystems.

For these reasons, compared to more extended land stretches, the following properties have been statistically observed in insular ecosystems:

- The species of remote islands that succeeded in their settlement and colonisation have a smaller carrying capacity, favouring the K strategy (i.e. a reduced number of offspring per generation compensated by a longer life span, thus the possibility to repeat reproductions relatively manyfold).[28] Consequently, we can predict smaller offspring per female [71] or slower generation rates [39].
- Due to these demographic constraints, intra- and interspecific competition is less intense, which allows a greater plasticity in the phenotypical and behavioural responses to environmental pressures and vital needs.
- Sexual dimorphism tends to be less pronounced [106].
- Bird plumage tends to be drabber [44].
- The communication signals responsible for the social stability and reproduction, thus the sustainability of the species, have a wider spectrum.

  All these characteristics are due to broader reaction norms in insular species that lack the strong predation and competition pressures universally prevailing in continental biocoenoses, though other types of environmental factors, mainly abiotic, press on the insular populations (e.g. volcanism or cyclones). The last point explains the trends mentioned above in volant animals or a number of functional specific traits in plant species (limited dispersal abilities, larger seeds in smaller numbers, larger number of ecotypes) [91]. In sum, in the islands, above

[28] The alternative demographic strategy is the r-strategy consisting for a species to invest in the production of a large number of offspring more than a long life span. The r-strategy species are often short-lived.

all in the small ones, fewer animals are efficient flyers or runners, and fewer plants have developed offensive defence systems like venoms glands or thorns.

12.7.7 Continental Speciations

There are two broad categories of continental speciations:

Sympatric speciations:
This vast category comprises situations in which a nascent species emerges within the parental population without a geophysical barrier.

Allopatric speciations:
A geophysical barrier isolates a small population from a large one, allowing the former to diverge genetically from the latter. In this process, the glaciations of the Pleistocene are frequently invoked as the geophysical barrier.

We have to keep in mind that different schools of thought underlie these categories, which are, in fact, hypothetic and frequently debated. These hypotheses are not necessarily incompatible, and different mechanisms could have occurred during evolution.

12.7.8 Sympatric Speciations

Different types of sympatric speciations have been proposed in a rich body of literature, mentioned and cited in [115] for points (1)–(5): (1) parapatric speciation, (2) speciation by hybridogenesis, (3) speciation by polyploidy, (4) Robertsonian raciation, (5) speciation by heterochrony, and (6) speciation by genetic duplication.

12.7.8.1 Parapatric Speciation

Regular geographical variations of phenotypical or genotypical traits are observed within a species, or even a complex of closely related species. These variations can be gradual, brutal or piecemeal. Here, the concept of cline comes in, a notion we owe to the British biologist Julian S. Huxley (1887–1975) [55, 56]. Parapatric speciation is explained in terms of cline evolution.

Definition 12.4 Cline is a trend, globally increasing or decreasing, of phenotypical or genetic variables along a topographical line, a succession of localities (or transect).

Cline theory, according to Huxley and followers [10, 36, 37], says that a cline develops in three stages: (1) smooth cline, (2) stepped cline and (3) steep cline [115].

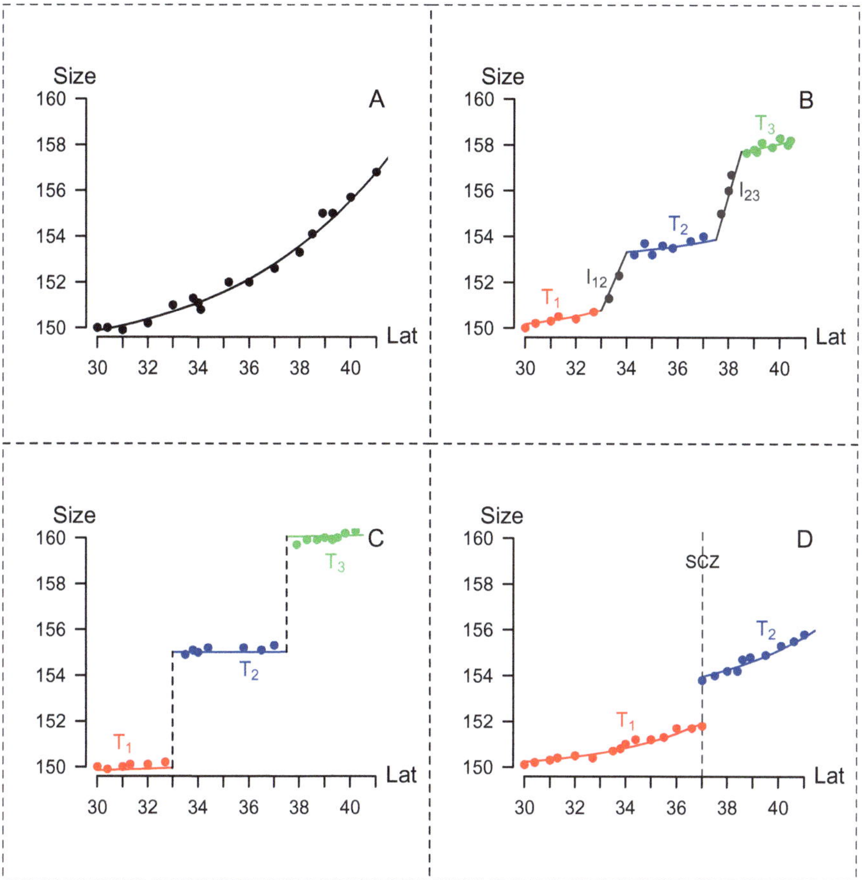

Fig. 12.4 Cline dynamics. The x-axis corresponds to latitudes (in °N), and the y-axis is the measured size of an organism (in mm). *Bullets* = individuals of the same taxon along a transect across its range. *Curves*=regression polynomial (RP) of the observed points across a taxon's range. (**a**) *Smooth cline*: (*black bullet*) = individuals of a monotypic species across its range. (**b**) *Stepped cline*: three subspecies: T_1 represented by (*red bullet*) and *red* RP, T_2 by (*blue bullet*) and *blue* RP, and T_3 by (*green bullet*) and *green* RP; I_{ij} hybrid zone between T_i and T_j represented by (*gray bullet*) and *grey* RP (the individuals can belong to the red, blue or red × blue taxon). (**c**) *Steep cline*: same legend as subfig. (**b**), with the disappearance of the hybrid zones I_{ij}. Parapatric speciation achieved. (**d**) *Broken cline*: same legend as subfig. (**c**), secondary contact zone after a geophysical isolation. *SCZ* secondary contact zone. Allopatric speciation achieved

Smooth cline:

This initial stage corresponds to the spatial variation of conspecifics (cf. Fig. 12.4a). It corresponds to a situation where a species is monotypic and settled in an ecological homogeneous area. When the latter is vast, no geophysical barriers hamper or block the individuals that, with time, tend to move and spread continuously. The ecological parameters vary gradually and continuously, entailing a correlative gradual and continuous phenotypical and genetic variation of the individuals. If we take chunks of equal size of a transect along the cline, the mean values of the processed variable change monotonously, but the variance remains statistically equal.

Stepped cline:

Here, a number of subspecies are increasingly individualised, with the statuses first of weak subspecies, then of strong subspecies, and eventually of semispecies (quasi or vice species) [5] (cf. Fig. 12.4b). These different types of subspecies or incipient species are separated in quasi-exclusive areas called allopatric areas by hybrid zones that evolve with time from intergradation zones to introgression zones.

Definition 12.5 (Intergradation Zone) An intergradation zone between two conspecific taxa, T_1 and T_2, is a sympatric zone where most individuals of the species are hybrids, and only a minority are pure breeds of T_1 and T_2. Such an integration zone separates the neighbouring allopatric zones of the taxa.

At the beginning of the process, the crossing between the individuals is theoretically random between the taxa, though individual preferences and cultural factors cannot be excluded. [This was amply shown in passerine birds like the yellow-rumped cacique in Panama (*Cacicus cela vitellinus*) [135] or the Eurasian chaffinch (*Fringilla coelebs*) [104] in western Europe]. We speak of *panmixy* or panmictic crossing. With time, assortative mating takes place when the allopatric ranges of the taxa become increasingly different ecologically. The individuals produced by mixed crossing are soon more ill-adapted than the pure breeds, in the allopatric areas where they appear accidentally as well as in the sympatry zones, where they are still more ill-adapted than in the allopatric ranges. Soon, the intergradation turns into an introgression zone.

Definition 12.6 (Introgression Zone) An introgression zone between taxa T_1 and T_2 is a sympatric zone where most individuals are pure breeds. Here, mating is assortative, leaving as products few hybrids. We witness here a reproductive isolation that is of two kinds. Premating isolation mechanism results from behaviour between the sexes, namely sexual preferences and choices. Postmating isolation mechanisms intervene after the pair formations, where two types of causes can be involved, either reduced fitness of the young hybrids or genetic incompatibilities.

Eventually, the young's lack of fitness can be due to the maladaptation of the hybrids to the existing conditions. At this point, taxonomists can decide that the two taxa are nascent species with an incipient niche differentiation, and we fall into the next stage, *steep cline*. In a stepped cline, there is an increase in the variance of the cline variable during the evolution of its physiognomy—between the incipient stage, close to a smooth cline, and the final one, close to a step cline.

Steep cline:

The allopatric zones are contiguous and ecologically distinct, sheltering distinct paraspecies T_1 and T_2 (cf. Fig. 12.4c). Parapatric speciation is achieved.

Broken cline:

This situation reveals an allopatric type of speciation [section (cf. Fig. 12.4d] (continued, cf. Sect. 12.7.9, p. 328).

12.7.8.2 Speciation by Hybridogenesis

Definition 12.7 (Hybridogenesis) Hybridogenesis is the phenomenon described as follows. A male m_A of taxon A crosses with a female f_B of taxon B; this crossing produces a female f'_B of taxon B, genetically identical to her mother (f'_B is thus a clone of f_B).

This happens during fecundation in the amphimixis stage when the chromosomes of the father (f_B) do not reach the equatorial plate and are thus eliminated. Consequently, the father does not have a genetic influence on the offspring; its intervention, though, is necessary since the reproductive process is launched by the mechanical contact and penetration of the spermatozoon into the ovum. Backcrosses are generally hampered or reduced.

This mechanism is a case of parthenogenesis induced by fusion. It launches speciation by hybridogenesis, which will be illustrated with the example of the edible frog [*Rana (esculenta)*, Amphibia, Anura]. This species complex comprises three taxa R (*ridibunda*), L (*lessonae*) and E (*esculenta*). It has been shown that *esculenta* is the hybrid between a female *ridibunda* and a male *lessonae* (i.e. E = ♀*R* × ♂*L*). These non-assortative pairs produce viable and fertile *esculenta* offspring of both sexes. However, a hybridogenesis phenomenon occurs when a male *ridibunda* crosses with a female *lessonae*; this crossing only produces female *lessonae* that are clones of their mothers (i.e. ♀L = ♂R × ♀L). It has been shown that when a population of edible frogs comprises, in sympatry, the taxa E, R and L, if there is no recruitment of L by migration, the *lessonae* taxon is counter-selected until it disappears. At the same time, the other taxa, E and R, are favoured and eventually fixed, forming a new equilibrium, but it has been shown that this equilibrium is unstable. Indeed, if E is not recruited by migration, the density of *esculenta* decreases until this taxon eventually disappears. Finally, only *ridibunda* remains. From this complex process, we can definitely conclude that E, L, and R are nascent which species, *Rana esculenta, R. lessonae and R. ridibunda*, which form a species complex [30].

12.7.8.3 Speciation by Polyploidy

In polyploidy, a reproducing individual A, whose chromosome number per somatic cell is $2N$, gives more than one gamete, for example two gametes, to its partner B, whose gamete contains N' chromosomes, to constitute the zygote during fecundation. If $N = N'$ as it occurs in the same species, the product is a zygote containing $3N$ chromosomes, thus a trisomic individual, characterised by impaired fitness and sterility.

Between two individuals of closely related species, if $N = N'$ and N' is odd, a genetic barrier prevents the taxa from producing offspring (the homologous chromosomes cannot match adequately), and we have a postmating isolation mechanism after only one generation. If N' is even, the hybrids and their descendants will not be able to reproduce indefinitely. Examples exist in molluscs and insects. A special case is a variant of polyploidy that occurs between two closely related interfertile species crossing normally: Let A and B be such species, respectively characterised by 2N and 2 N' chromosomes. During fecundation, in the fusion stage, $N + N'$ chromosomes combine and duplicate; the somatic cells of the hybrids thus contain 2 (N + N') chromosomes and will form a new species. This species, obtained by polyploidy, does not back-cross freely with either parental species. This type of speciation is observed in plants; it has been well-documented in studies on cultivated grasses (80% of these plant species have been obtained by polyploidy [12]).

A particular case of speciation by polyploidy is speciation by hybridisation, where two sympatric species A and B, interfertile in F1, cross to produce A × B, allopolyploid individuals[29] showing hybrid vigour. The hybrids do not back-cross with the parental forms, and having a stronger fitness than the latter, successfully compete with them, eliminate them progressively them, colonise their distribution area, and eventually replace them. A botanical example illustrates this type of speciation, i.e. the emergence and development in western France of *Spartina anglica* as cross-products of two ancestral species, a local one *S. maritima* [49] and an introduced one *S. alterniflora*. Speciation by hybridation was also observed among Darwin's finches, namely between *Geospiza fortis*, *G. scandens* and *G. fortis × scandens* on the small Galápagos island of Daphne Major [46].

12.7.8.4 Robertsonian Raciation

Within a given species, several individuals can undergo identical chromosome recombination, changing their chromosome topology and numbers similarly. If these organisms and their descendants are viable and indefinitely interfertile, they constitute a sustainable subpopulation called Robertsonian race, showing various degrees of isolation from the rest of the population. Such a situation has been reported with a population of domestic mice (*Mus musculus domesticus*) living in a valley of northern Italy (High Valtellina). The domestic mouse typically possesses a karyotype of $2\,N = 40$ chromosomes; a Robertsonian race called the Poschiavo race has a karyotype of $2\,N = 26$. In the contact zone, many mixed pairs (normal species × Poschiavo race) are not sterile despite the lack of topographical chromosome homology. The products are viable but show reduced fertility and, as a result, a

[29] Allopolyploid individuals are individuals that have several complete sets of chromosomes stemming from different species.

reduced gene flow between the domestic mouse and the Poschiavo race. Though genetically very close, their evolutionary destinies seem inevitably distinct, and we can consider this type of chromosomal raciation a form of speciation [97].

12.7.8.5 Speciation by Heterochrony

Heterochrony is a type of speciation that involves altering the maturation process during the ontological development of taxa within an evolutionary lineage.

- When a species reaches its sexual maturity at a stage that corresponds to the juvenile stage in its ancestors, we speak of *neoteny* or *paedomorphosis*.

  The best-known case of neoteny is that of the axolotl (*Ambystoma mexicanum*), a critically endangered amphibian living in two volcanic Mexican lakes, i.e. the lakes Xochimilko (in Mexico City) and Chalco. This species does not complete its metamorphosis and becomes sexually mature in the tadpole stage [62], with the physical anatomy of the larva of a salamander, notably external gills that force it to be strictly aquatic [107]. This article describes the physiological evolution that led the ancestral salamander to block its ontologic development at the juvenile stage at adulthood: it was, very probably, a complex process that involved the thyroid hormone.

  Another case of neoteny is illustrated by the modern human (*Homo sapiens*). Paedomorphic traits are obvious if we compare our species to non-human primates or the other extinct species of humans that lived in the Pleistocene. Neotenic characters are manifold,[30] among which" (1) the globular skull, (2) the thinness of skull bones, (3) the reduction of the brow ridge, (4) the large brain, (5) the flattened and broadened face, (6) the hairless face, (7) hair on the head, (8) larger eyes, (9) ear shape, (10) small nose, (11) small teeth, (12) small upper and lower jaws. Many of these traits would correspond to the features of a last-stage foetal chimpanzees." Plastic behaviour that lasts a lifetime is also a neotenic trait. In other primates, this characteristic is generally observed in the young. Hairless bodies and genitals, specific to humans (no penis bone, presence of hymen, forward-facing vagina) are also juvenile traits. [2, 43, 86]
- Papers on fossils have shown cases of *peramorphism* in a temporal series of Cambrian trilobites [57, 78] and dinosaurs [79]. Here, the descendant species reaches sexual maturity at a hypothetically more advanced ontological stage than its ancestors.

  In a recent paper [11], the skull shapes of neotropical leaf-nosed bats (Chiroptera, Phyllostomidae) were compared, which were shown to be correlated with the nutrition mode of the animals. The skulls of frugivores had structures and biometries related to a more advanced ontogenical development than the skulls of insectivores. The fossil records of bats worldwideare highly favourable

[30] I copied the text below and citations from the English wikipedia on neoteny.

to the hypothesis that insectivory is ancestral relatively to frugivory. This example and another one still on bats [90] illustrate cases of peramorphosis.

12.7.8.6 Speciation by Gene Duplication

Definition 12.8 (Gene Duplication) *Gene duplication* is the process of duplication a DNA portion corresponding to a gene within a chromosome. This occurrence results from two types of events: (1) several types of errors in DNA replications that are corrected by repair machinery or (2) fortuitous capture by selfish DNA portions.

Readers who wish to have more details on these genetic topics can refer to different publications [48, 61, 84, 153]. When gene duplication is massive and differential, it can have evolutionary consequences for the carrier, e.g. the emergence of new species within a population. Recent publications showed that this phenomenon played a crucial role in speciations within salmonids [25, 63, 93].

12.7.9 Allopatric Speciation

12.7.9.1 Development of Allopatric Speciation

Allopatric speciation[31] functions as follows. Geographic isolation may (1) either persists, (2) or be followed by a phase of *secondary contact*, i.e. contact of two forms issued from the same stock, then isolated and eventually returned in contact. Here, a population with a conservative norm has been replaced by a set of populations with distinct novel norms acquired previously in allopatry zones, and coming into contact in a suture zone, generally a narrow sympatry belt. This is a current case of speciation, a dynamic process that according to the observed biology of the sympatric populations, can be assigned diverse *taxonomic ranks* reflecting their differentiation:

1. the rank of conspecificity (when the populations are *morphs* or *weak subspecies* [26, 76]),
2. a rank intermediate between conspecificity and heterospecificity (when the populations are *strong subspecies* (or *megasubspecies*), *quasispecies*, *vicespecies* or *semispecies* [5, 26, 76]).
3. heterospecificity (when the populations are distinct species (*paraspecies* and eventually *synspecies*).

Figure 12.5 illustrates this process already described in one of my previous publications [116]. In this figure, I represent the analysed populations as a set of

[31] This type of speciation has also been called geographic speciation. Other modes of speciation exist and have been well documented in the literature [115].

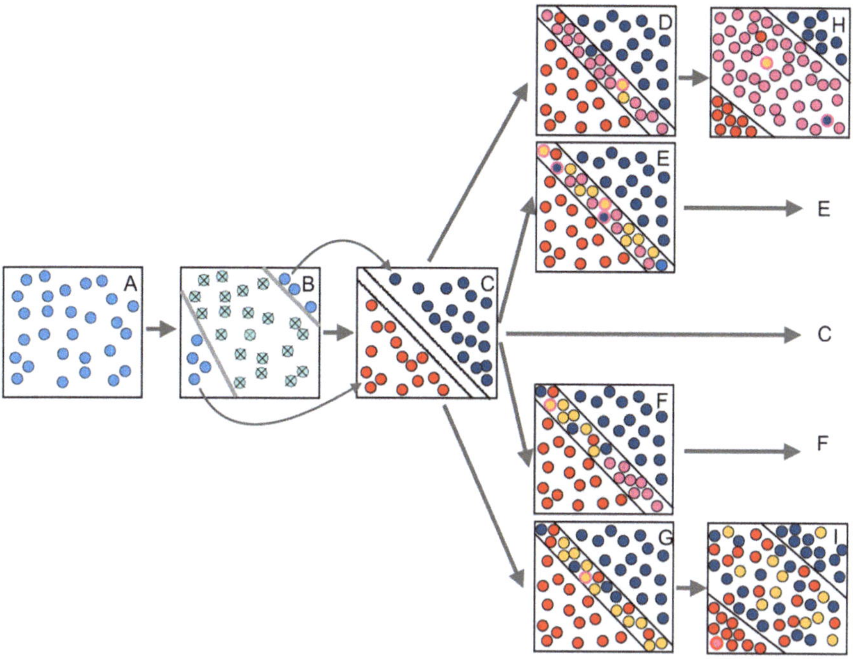

Fig. 12.5 Allopatric speciation. *Filled circles* of various colours = demes. (**a**) *Light blue* = demes in an ancestral species through its range. (**b**) After a catastrophic event, extinction of most ancestral demes (barred pale cyan). Surviving demes in peripheral areas = *light blue*. (**c**) Surviving demes of (**b**) evolve (curved arrows) into 2 new allopatric taxa respectively containing *dark blue or red* demes. From (**c**), 5 outcomes: (1) taxa remaining allopatric. (2) The other ones involve a narrow hybrid belt: (**d**) Megasubspecies. Unstable hybrid belt containing many demes with only hybrids, expanding to a vast intergradation zone (**h**). (**e**) Quasispecies. Stable hybrid belts. Limited and often unbalanced hybridisation between the taxa. (**f**) Vicespecies [5, 88]. Stable hybrid belt. A group of demes behaves like in (**e**) and another one, like in (**g**). (**g**) Paraspecies. Heterospecificity. Very little gene flow beween the taxa. Still territorially competitive. Eventually, belt turns into vast sympatric area with descendants becoming species coexisting territorially (synspecies) (**i**)

demes depicted by a filled circle whose taxonomic identity is assigned by a given colour, for example, 'red'. Then, I called such a deme 'red deme'.

I will now comment on the scenario of an allopatric described in Fig. 12.5.

An ancestral species **a** (the set of light blue demes) sharing a given breeding area (subfig. a) is widely destroyed, particularly in the centre of its distribution area, by a catastrophic event (the cycles of glaciation of the last ice ages are often invoked) (subfig. b); only a couple of small populations in peripheral zones manage to survive, i.e. $\mathbf{a_1}$ and $\mathbf{a_2}$), subject to genetic drift and allelic changes. According to the amount of genetic change, the descendent populations of $\mathbf{a_1}$ and $\mathbf{a_2}$ [respectively$\mathbf{b_1}$ (red bullets) and $\mathbf{b_2}$ (dark blue bullets)] may conserve the initial norm or see their norms diverge. Populations $\mathbf{b_1}$ and $\mathbf{b_2}$ increase but remain isolated, unable to colonise improper environments (e.g. places that are too cold or devoid of the necessary resources). Populations $\mathbf{b_1}$ and $\mathbf{b_2}$ have become dense allopatric taxa (subfig. c). Stage C can evolve into five biogeographical situations that can be observed at present:

1. Populations $\mathbf{b_1}$ and $\mathbf{b_2}$ remain allopatric. Every case of divergence between them can have been reached, but no direct evidence can be made from ecological or behavioural observations.[32]
2. Populations $\mathbf{b_1}$ and $\mathbf{b_2}$ become parapatric. They come into secondary contact and overlap in a narrow hybrid zone; according to their genetic, phenotypical and ethological level of divergence, they turn out to be megasubspecies, semispecies (either quasispecies or vicespecies) or paraspecies. We will now overview these cases.

    (a) Populations $\mathbf{b_1}$ and $\mathbf{b_2}$ are *megasubspecies* (or strong subspecies, a synonymous term) (subfig. d). They are not differentiated enough from one another to have erected a barrier preventing gene flow between them and remain thus conspecific and form a population. Strong subspecies are the taxa of the lowest level. They interbreed perfectly but show constancy in their difference and tend to live in complementary areas. This situation reveals a tendency towards competition for food resources, which explains the evolution towards allopatry.[33] This is explained by the beginning of a behavioural barrier sufficiently constant to maintain a global ecophenotypic distinctness.
    Strong subspecies are exemplified by many cases, e.g.in birds, (1) the carrion crow (*Corvus corone corone*) and the hooded crow (*C. c. cornix*) [99], (2) the western coal tit (*Periparus ater melanolophus*) and the eastern coal tit (*P. a. aemodius*) [151].

    After a limited number of generations, one expects to see the following.

    - An overwhelming majority of demes are composed only of hybrids $\mathbf{c} = \mathbf{b_1} \times \mathbf{b_2}$ (pink demes circled black).
    - Besides, a small minority of the demes are composed otherwise. They can be made up of: (1) one parental form only and no hybrids (red or dark blue demes circled black), (2) one parental form only and hybrids (red or dark blue demes circled pink), (3) both parental forms only and no hybrids (yellow demes circled black), and (4) both parental forms only and hybrids (yellow demes circled pink).

    Here, the population norm is conserved and very probably not significantly different from the ancestral population depicted by subfig. a. Very naturally,

[32] This biogeographic evolution, widely observed, can be predicted by a computerised Monte-Carlo simulation [68].

[33] Allopatry is the biogeographic situation of two taxa whose distribution (breeding) zones are geographically complementary and (1) exclusive except in the contact zone, when the word “allopatry” is taken *senso latu*; (2) and totally exclusive, thus spatially isolated, when it is taken *senso strictu*.

the initial intergrade narrow hybrid belt depicted in subfig. c continuously evolves into a broad intergrade zone populated by taxon **c** (pink demes circled black) that has mostly resulted from the fusion of taxa $\mathbf{b_1}$ and **b2** by interbreeding, a dynamic that also proceeds when the intergrade zone widens. At term, this situation unrolls at the expense of $\mathbf{b_1}$ and $\mathbf{b_2}$ whose allopatry zones decrease and dwindle at the extreme of the overall distribution area.

(b) Populations $\mathbf{b_1}$ and $\mathbf{b_2}$ are *quasispecies* (subfig. e). Here, gene flow is restricted: the taxa are partially isolated. Mating tends to be assortative. The isolation is often asymmetric, the heterogametic pair being generally counterselected in assortative matings (Haldane's rule).

- Demes containing only hybrids (pink demes circled black) are a minority scattered in the hybrid belt.
- Most demes consist of parental forms with or without hybrids (yellow demes circled black or pink) or demes with only one parental form with or without hybrids (dark blue or red demes circled black or pink).

Thus, the hybrid belt remains a narrow strip approximately constant, separating the vast and relatively stable allopatry zones of species $\mathbf{b_1}$ and $\mathbf{b_2}$. Genes can be exchanged between $\mathbf{b_1}$ and $\mathbf{b_2}$ in small amounts and even travel across the allopatry zones. This phenomenon is called gene introgression. Here, we witness a clear differentiation of the population norms, revealing incipient speciation. Many colleagues, mostly phylogeneticists, consider these taxa already species. The International Commission of Zoological Nomenclature (ICNZ) tends to countenance this point of view in their decisions nowadays. A well-studied case of such a situation is the current speciation within a bird taxon in southwestern Europe, belonging to the passerine genus *Phylloscopus*. This group of warblers, scattered in the Old World (from Central Africa through Eurasia to Eastern Asia), is represented by about 50 species very similar morphologically but highly distinct acoustically, their songs and calls being trustful signatures of their taxonomic identities [114]. The taxon in question is a complex of quasispecies, (1) the common chiffchaff and (2) the Iberian chiffchaff, diagnostically distinguished by their songs and calls [110–112], but so poorly distinguishable morphologically that they were long considered simple acoustic forms by most ornithologists, without any taxonomic implication. Several authors, though, considered them as distinct subspecies [6, 73, 133, 139] and named the Iberian chiffchaff *P. c. brehmii* after the description and denomination by the German ornithologist Eugen F. von Homeyer (1809–1889) in his book chapter on the birds of Portugal. This description was revealed later to be erroneous [118]. Consequently, the name of the Iberian chiffchaff was later changed to its current name, *Phylloscopus c. ibericus*, the name actually given to an Iberian chiffchaff defined by its songs [118, 125, 132, 148]. German authors [130, 131] already suspected the two acoustic forms were taxonomically more differentiated, basing their arguments on acoustic, behavioural and biogeographical considerations. From 1982 onwards, many new arguments convinced the scientific community to consider the taxa a distinct species, though I would call them nascent species or quasispecies:

- The Iberian chiffchaff emits calls that differ from the common chiffchaff calls [109–111].
- A thorough analysis of the songs of the Iberian chiffchaffs, the common chiffchaffs and the mixed-singers, taking into account different contexts, i.e. intraindividual variations related to the presence of a female or competing males or the effect of seasonal change, has been carried out with statistical multivariate analyses. This study showed that (1) the difference between the taxa is always present, whatever the context, the three categories of singers have specific and distinct calls (so that the males[34] and females can be distinguished), and that the mixed-singers, forming a minority of the singers, were overwhelmingly hybrids [4, 53].
- Subtle habitat preferences have been shown between the common chiffchaff and the Iberian chiffchaff in the sympatry zone. (1) The latter tends to have a narrower spectrum of altitude than the former. (2) Besides, in hilly and mountainous areas, *P. collybita* has no preference concerning the exposition of their habitat, whereas *P. ibericus* tends to breed on slopes exposed southwards, southwestwards and westwards.[111, 113]
- Mating is assortative in the sympatric zone. Most pairs are *collybita* × *collybita* or *ibericus* × *ibericus*, and the mixed pairs are quasi-exclusively ♂*collybita* × ♀*ibericus*. Thus, partial premating isolation is evidenced here. Moreover, hybrids and early backcrosses are less fertile than the "pure" taxa [4, 53], which adds postmating isolation. The gene coding for cytochrome b diverged between the common chiffchaff and the Iberian chiffchaff by 4.6%, which shows a very restricted gene flow, and already an indication that the two taxa have reached complete heterospecificity, a situation similar to the taxonomic status of modern human vs Neanderthals [53, 80].

  Furthermore, multivariate statistical analyses show systematic biometrical differences between the males of the two taxa. The Bergman rule is observed within each taxon but not between the southern populations of *P. collybita* and the northern ones of *P. ibericus* [118]. Their migratory behaviours are different [58].

(c) In several situations, populations $\mathbf{b_1}$ and $\mathbf{b_2}$ behave like *megasubspecies* [see case (a)] in a portion of the hybrid belt (of surface $\mathcal{A}_1$) and like *paraspecies* [see case (d)] in the rest of the hybrid belt (of surface $\mathcal{A}_2$). Above a given threshold θ, depending on the biology and dynamic of the organisms studied (thus case by case), we fall into two different situations. If $\frac{\mathcal{A}_1}{\mathcal{A}_2} \geq \theta$, we fall

[34] As a reminder, in most oscines, only males sing actively, but both sexes call.

into case (a); otherwise, the biogeographical and evolutionary consequences are similar to case (b),[35] in which case we say that populations $\mathbf{b_1}$ and $\mathbf{b_2}$ form two vice species.

(d) Populations $\mathbf{b_1}$ and $\mathbf{b_2}$ are *paraspecies* when they are reproductively isolated but still remain *parapatric*,[36] i.e. sympatric only within a narrow belt that separates their *allopatry zones* (subfig. g). They do not or rarely hybridise, and when this happens, the hybrids $\mathbf{c} = \mathbf{b_1} \times \mathbf{b_2}$ have a reduced fitness. However, at the social unit level, each taxon territorially excludes the other, as within a species, since (1) both have the same ecological requirements, competing for the same limited resources, and (2) the ethological barrier relative to premating behaviour may, in some cases, not be totally achieved, with rare products, enough to propagate genes of the other taxon to the dwelling one. This might explain the weak gene introgression sometimes observable in sympatry as well in allopatry.
In my opinion, this stage is the first to correspond to the differentiation of evolutionary destinies, which can evolve into the stage depicted in the subfig. i. In the latter, the species have different ecological and reproductive requirements, enabling them to live in syntopy, thus widely in sympatry, in a zone that will get increasingly broader, relegating the former allopatry zones in decreasing peripheral areas.

12.7.9.2 Barrier Invoked: The Glaciations of the Pleistocene

A succession of glacial events, glaciations separated by periods called interglacials (cf. Table 12.2), occurred in the Holarctic region during the Quaternary. These glacial events were periods during which ice caps covered vast areas; for example, in Europe, it could expand southwards as far as Northern Germany or the London basin, causing the climate to be extremely cold in most parts of the Holarctic. During *interglacials*, climate softened in these parts, becoming warmer and moister and sometimes reaching almost subtropical temperature peaks.

12.7.9.3 Impact of the Pleistocene Cycles on the Evolution of Tropical Biomes

A very probable and well-evidenced scenario explaining the evolution of the tropical rainforest biome during the Pleistocene has been proposed [52], which runs as

[35] For this reason, Jörgen Haffer lumped the categories "megasubspecies" and "vice-species" into the concept of semipecies [51].

[36] In the case of parapatry between two taxa, their respective distribution areas are complementary, with a border between them that can theoretically be a line or an overlapping belt (which is generally the case)—the last situation often being called stasipatry [60, 145].

Table 12.2 The last glacial events (in bold) and interglacials

Glacial events and interglacials	Time periods
Günz	600,000–540,000
Günz-Mindel	540,000–480,000
Mindel	480,000–430,000
Mindel-Riss	430,000–240,000
Riss	240,000–180,000
Riss-Würm	180,000–120,000
Würm	120,000–10,000
Present	10,000–

follows. The more intense the cold in the temperate region was (corresponding to the glacial events), the stronger the dryness of the *tropical biomes* at the same period tended to be. It was at that epoch that the tropical rainforest shrank to its greatest extent. Conversely, climatically softer periods, i.e. the interglacials, corresponded to relatively high temperatures in the temperate biomes, entailing a wetter climate in the tropical biomes favourable to the expansion of the rainforests. The latter had their maximal extent when humidity peaked in the tropical zone, and when the interglacial in the periarctic areas was hottest. An example of the phenomenon is the evolution of the Amazon forest studied by Haffer and Prance [52].

12.7.9.4 Impact of the Quaternary Cycles on the Suture Zones

The presence of suture zones, narrow belts of territories that separate or used to separate the distribution (or breeding) areas of closely-related species or speciating taxa, is largely explained by the theory of allopatric (or geographic) speciation [74]. The mechanisms of allopatric speciation entail the presence of suture zones revealing former refuge areas and the subsequent evolutionary dynamics of closely related taxa sharing common ancestry during the Pleistocene, as explained in Fig. 12.5. This phenomenon can be observed in many species of pluricellular eukaryotes (animals, plants or fungi). We can mention many taxa complexes in Europe like for example the common-Iberian chiffchaff complex (*Phylloscopus collybita-ibericus*), the carrion-hooded crown complex (*Corvus corone-cornix*), the house mouse complex (*Mus musculus-domesticus*) or the viviparous lizard complex (*Zootoca vivipara*) . The following scenario is classically invoked.

Initially, these taxa would have formed one single population that shrank, a number of survivors taking refuge in few small but favourable habitats during the glacial events at their harshest. These individuals were isolated for several generations, forming populations sufficiently small to be submitted to genetic drift.

In many cases, these populations came into contact again, sufficiently differentiated genetically but still close enough ecologically,

1. not to be able to hybridise freely
2. but at the same time to exclude each other geographically via species competition and diverse unknown demographic events.

These events occurred at a sufficiently large scale to become *parapatric* (ie form suture zones). In other cases, the taxa either remained geographically isolated and constituted allopatric taxa or were too much differentiated to maintain the same ecological niches, being thus able to live syntopically in sympatry.

The scenario depicting the emergence of the suture zones mentioned above is based on the Darwinian hypothesis that evolution was gradual, i.e., the result of a series of little changes driven by natural selection towards one global direction; and, that speciation resulted in the differentiation between variations isolated by a barrier (a mountain, a river, etc). The glacial events, mainly invoked nowadays, were still unknown and could therefore not be mentioned at the time of Darwin [24]). One should keep in mind that in Darwin's time, the Mendelian segregation of characters was unknown even by Darwin, as it was discovered, very inconspicuously, by a contemporary monk named Mendel [81], and that the gene hypothesis, the origin of genetics, was to be proposed for the first time at the turn of the twentieth century. The source of variation was unknown and widely based on the hypothesis of blending hypothesis unquestioned even by Darwin himself, and nowadays completely overturned (see Part II). Still, largely modified because of the advent of genetics and of modern geology based on continental drift, the hypothesis of most evolutionary processes has many supporters, though debated.

12.8 Heterospecific Populations in the Sense of Components of a Community

12.8.1 A Difficult Problem Prone to Debate: Where Is the Frontier Between Conspecificity and Heterospecificity?

A number of biologists nicknamed *splitters*, perhaps the mainstream at present, already consider quasispecies or vicespecies as separate species though (1) they are fully interfertile, (2) intergrade in their sympatry zone, and (3) do not occupy distinct ecological niches. Most phylogeneticists with a cladist approach hold this point of view. On the opposite, other authors termed *lumpers* consider that two sets of populations are distinct species, not only when they show reduced or no interbreeding but also when their ecological niches are differentiated to such a point that they can be sympatric in a broad area without competing for food or space resources. Most synthetic evolutionists take a middle stand, considering that heterospecificity is already attained when the following situations are observed.

- The parapatric taxa are separated by a narrow sympatric belt where scarce gene exchange occurs via limited genetic introgression.
- No intergradation occurs, the parental taxa[37] prevail, and the hybrids form a small minority.
- The complete differentiation of the taxa's ecological niches is still not reached.

The authors have different philosophies and field or experimental experiences, inevitably leading to different taxonomic stands and, thus, to different ecological conclusions. A relatively recent article summarises this discussion about the number of genera of ferns worldwide [119]. We will call "fully heterospecific" two populations that have attained the species level in the sense of the lumpers. These heterospecific populations are, by definition, syntopic. This adjective defines populations that not only share the same area globally but also live in the same microhabitats, having the potential to meet without entering into competition for space, food or reproduction. In sum, taxa are syntopic when their ecological niches are fully differentiated.

From now on,

- heterospecific populations will be referred to populations of distinct syntopic species.
- These will build up the ecological communities.

Both will be treated in the next chapter.

References

1. Amundsen P et al (2004) Resource competition and interactive segregation between sympatric whitefish morphs.Ann Zool Fennici 41:301–307.
2. Bednarik RG (2011) The human condition. Springer.
3. Begon M et al (2006) Ecology: from individuals to ecosystems, 4th edn. Blackwell Publishing, Oxford
4. Bensch S et al (2002) Amplified fragment length polymorphism analysis identifies hybrids between two subspecies of warblers. Mol Ecol 11:473–481, https://doi.org/10.1046/j.0962-1083.2001.01455.x
5. Bernardi G (1980) Les catégories taxinomiques de la systématique évolutive. In: Bocquet C, Génermont J, Lamotte M (eds) Les problèmes de l'espèce, vol 40, Société Zoologique de France, Paris
6. Bernis F (1945) Datas sobre el *Phylloscopus collybita brehmii* (in Spanish) Bol Soc Esp Hist Nat 43:339–348.
7. Bertrand J-C et al (eds) (2018) Prokaryotes and evolution, Cham: Springer International Publishing.
8. Bravo-Cuevas VM, Ferrusquía-Villafraca I (2010) The oldest record of Equini (Mammalia: Equidae) from Mexico. Rev Mex Cienc Geol 27:593–602.

[37] That can bear a tiny percentage of the genes of the other taxon.

9. Buffon GL (1761) Animaux communs aux deux continents. In: Histoire naturelle générale et particulière avec la description du cabinet du Roy, tome IX, Imprimerie royale, Paris (In French)
10. Bush GL (1975) Modes of animal speciation. Ann Rev Ecol Syst 6:339–364.
11. Camacho J et al (2019) Peramorphosis, an evolutionary developmental mechanism in neotropical bat skull diversity. Dev Dynam 248:1129–1143, https://doi.org/10.1002/dvdy.90
12. Carnahan HL, Hill HD (1961) Cytology and genetics of forage grasses. Bot Rev 27:1–162.
13. Clements FE (1904) The development and structure of vegetation. In: Pound R, Woods AF, Marsland H (eds) Botanical survey of Nebraska. University of Nebraka, Lincoln
14. Clements FE (1916) Plant succession: an analysis of the development of vegetation. Carnegie Institution of Washington.
15. Cockell CS et al (2011) Microbiology of volcanic environments. In: Horikoshi K (ed) Extremophiles Handbook, pp 917–933, Springer Japan, Tokyo.
16. Cooper WS (1939) A fourth expedition to Glacier Bay, Alaska. Ecology 20:130–155, https://doi.org/10.2307/1930735
17. Costermans LF (1993) Native trees and shrubs of South-Eastern Australia. Reed New Holland, Sydney.
18. Cowles HC (1899) The ecological relations of the vegetation on the sand dunes of lake Michigan. part I.- Geographical relations of the dune floras. Bot Gaz 27:95–117, https://doi.org/10.1086/327796
19. Cox CB, Moore PD (2000) Biogeography: an ecological and evolutionary approach. Blackwell Science, Oxford.
20. Croizat L (1958) Panbiogeography. Wheldon & Wesley Ltd, London.
21. Cutler NA et al (2008) The spatiotemporal dynamics of a primary succession. J Ecol 96:231–246, https://doi.org/10.1111/j.1365-2745.2007.01344.x
22. Dacks JB, Doolittle WF (2001) Reconstructing/deconstructing the earliest eukaryotes: how comparative genomics can help. Cell 107:419–425, https://doi.org/10.1016/S0092-8674(01)00584-0
23. Darlington PJ (1957) Zoogeography: the geographical distribution of animals. J Wiley & Sons, New York
24. Darwin C (1859) On the origin of species by means of natural selection, or the preservation of favoured races in the struggle for life. Murray, London
25. David KT et al (2020) Patterns of gene evolution following duplications and speciations in vertebrates. PeerJ:e8813, https://doi.org/10.7717/peerj.8813
26. Deuve T (1994) Une classification du genre Carabus. Editions Sciences Nat, Venette (France)
27. DobzhanskyT (1937) Genetic nature of species differences. Am Nat 71:404–420, https://doi.org/10.1086/280726
28. Doolittle WF (1999) Phylogenetic classification and the universal tree. Science 284:2124–2128, https://doi.org/10.1126/science.284.5423.2124
29. Dorn RI, Oberlander M (1981) Microbial origin in desert varnish. Science 213:1245–1247, https://doi.org/10.1126/science.213.4513.1245
30. Dubois A (2008) Drôles d'espèces. Hybridation, perturbation de la méiose et spéciation. In: Prate D et al (eds) Peut-on classer le vivant? Linnée et la systématique (in French). Belin, Paris, pp 169–202
31. Dureau de la Malle A (1825) Mémoire sur l'alternance ou sur ce problème: la succession alternative dans la reproduction des espèces végétales vivant en société, est-elle une loi générale de la nature ? Ann Sci Nat 5:353–381
32. Ebach MC (2015) Origins of biogeography. Springer, Dordrecht
33. Ebach MC, Humphries CJ (2003) Ontology of biogeography. J Biogeogr 30:959–962, https://doi.org/10.1046/j.1365-2699.2003.00886.x
34. Ebach MC et al (2003) Phylogenetic biogeography deconstructed. J Biogeogr 30:1285–1296, https://doi.org/10.1046/j.1365-2699.2003.00928.x
35. Eggler WA (1941) Primary succession on volcanic deposits in southern Idaho. Ecol Monogr 11:277–298, https://doi.org/10.2307/1943205

36. Endler JA (1973) Gene flow and population divergence. Science 179:243–250, https://doi.org/10.1126/science.179.4070.243
37. Endler JA (1977) Geographic variation, speciation and clines. Princeton Univ Press.
38. Feer F (1989) Occupation de l'espace par deux Bovidés sympatriques de la forêt dense africaine (*Cephalophus callipygus* et *C. dorsalis*): Influence du rythme d'activité (in French). Rev Ecol (Terre Vie) 44:225–248, https://doi.org/10.3406/revec.1989.5516
39. Fernández-Palacios et al (2021) Scientists' warning – The outstanding biodiversity of islands is in peril. Global Ecol Conserv 31:e01847, https://doi.org/10.1016/j.gecco.2021.e01847
40. Frías DL (2010) Omissions in the synthetic theory of evolution. Biol Res 43:299–306, https://doi.org/10.4067/S0716-97602010000300006
41. Gan T et al (2021) Cryptic terrestrial fungus-like fossils of the early Ediacaran Period. Nat Commun 12:641, https://doi.org/10.1038/s41467-021-20975-1
42. Gardner SL, Campbell ML (1992) Parasites as probes for biodiversity. J Parasitol 78:596–600
43. Gould SJ (1977) Ontogeny and phylogeny. Belknap, Cambridge (MA).
44. Grant PR (1965) The adaptive significance of some size trends in island birds. Evolution 19:355–367.
45. Grant PR, Grant BR (2002) Unpredictable evolution in a 30-year study of Darwin's finches. Science 296:707–711, https://doi.org/10.1126/science.1070315
46. Grant PR, Grant BR (2009) The secondary contact phase of allopatric speciation in Darwin's finches. Proc Natl Acad Sci USA 106(48):20141–20148, https://doi.org/10.1073/pnas.0911761106
47. Grant V (1949) Pollination systems as isolating mechanisms in Angiosperms. Evolution 3:82–97.
48. Graur D, Li W (2000) Fundamentals of molecular evolution. 2nd edn, Sinauer.
49. Guénégou MC, Levasseur JE (1993) La nouvelle espèce amphidiploèdes *Spartina anglica* C. E. Hubbard: son origine, argumentation et implications (in French). Biogeographica 69:125–133.
50. Haeckel E (1866) Generelle Morphologie der Organismen. G.Reimer, Berlin
51. Haffer J (1986) Superspecies and species limits in vertebrates. Z zool Syst Evolutforsch 24:169–190
52. Haffer J, Prance GT (2001) Climatic forcing of evolution in Amazonia during the Cenozoic: on the refuge theory of biotic differentiation. Amazoniana 16:579–607
53. Helbig AJ et al (2001) Male-biased gene flow across an avian hybrid zone: evidence from mitochondrial and microsatellite DNA. J Evol Biol 14:277–287, https://doi.org/10.1046/j.1420-9101.2001.00273.x
54. Hull PM, Norris RD (2009) Evidence for abrupt speciation in a classic case of gradual evolution. Proc Natl Acad Sci USA 106(50):21224–21229, https://doi.org/10.1073/pnas.0902887106
55. Huxley JS (1938) Clines: an auxilliary taxonomic principle. Nature 142:219–220.
56. Huxley JS (1939) Clines: an auxilliary method in taxonomy. Bijdr Dierk 27:491–520.
57. Hwang HA, Choi DK (2005) Heterochrony of the Late Cambrian olenid trilobites from the Machari Formation, Yeongwol, Korea: implications for biostratigraphy and intercontinental correlation. Geosci J 9:215–222.
58. Isenmann P et al (2019) Nouvelles données sur les voies de migration et les quartiers d'hiver du pouillot ibérique *Phylloscopus ibericus* (in French). Alauda 87: 243–250.
59. Kelly L (2014) Psychrophile. In: Amils R, Gargaud M, Cernicharo Quintanilla J et al (eds) Encyclopedia of Astrobiology, pp 1–3. Springer Berlin Heidelberg, Berlin, Heidelberg
60. Key KHL (1968) The concept of stasipatric speciation. Syst Zool 17:14–22.
61. Kohanovski I et al.(2024) Aneuploidy can be evolutionary diversion on the path to adaptation. Mol Biol Evol 41(3):msae052, https://doi.org/10.1093/molbev/msae052
62. Kollmann J (1885) Das Ueberwintern von europäischen Frosch- und Tritonlarven und die Umwandlung des mexikansichen Axolotl. Verh Naturforsch Basel 7:387–398.
63. Koop BF et al (2008) A salmonid EST genomic study: genes, duplications, phylogeny and microarrays. BMC Genom 9:545, https://doi.org/10.1186/1471-2164-9-545

64. Kranstauber B et al. (2017) Similarity in spatial utilization distributions measured by the earth mover's distance. Methods Ecol Evol 8:155–160, https://doi.org/10.1111/2041-210X.12649
65. Kristensen DL at al (2014) Differential breeding investment in bridled and non-bridled common guillemots (*Uria aalge*): morph of the partner matters. Behav Ecol Sociobiol 68:1851–1858, https://doi.org/10.1007/s00265-014-1794-8
66. Lake JA, Rivera MC (2004) Deriving the genomic tree of life in the presence of horizontal gene transfer: conditioned reconstruction. Mol. Biol. Evol 21:681–690, https://doi.org/10.1093/molbev/msh061
67. Lamarck JBA (1809) Philosophie zoologique, ou exposition des considérations relatives à l'histoire naturelle des animaux. Dentu et L'Auteur, Paris
68. Landis MJ et al (2013) Bayesian analysis of biogeography when the number of areas is large. Sys Biol 62:789–804, https://doi.org/10.1093/sysbio/syt040
69. Lewontin RC (1974) The genetic basis of evolutionary change. Columbia University Press, New York
70. Lovelock J (1991) The ages of Gaia. Oxford University Press, Oxford
71. Mappes T et al (2008) Island selection on mammalian life-histories: genetic differentiation in offspring size BMC Evol Biol 8:296, https://doi.org/10.1186/1471-2148-8-296
72. Matsubayashi KW et al (2013) Micro-spatial and seasonal distributions of two sympatric host races of the phytophagous ladybird bettle *Henosepilachna diekei* (Coleoptera: coccinellidae) and their host plants in West Java, Indonesia. Treubia 40:9–24
73. Mayaud N (1943) L'avifaune des Landes. *Phylloscopus collybita.* L'Oiseau Rev Fr Ornithol 13:71–72 (in French).
74. Mayr E (1942) Systematics and the origin of species, from the viewpoint of a zoologist. Harvard University Press, Harvard
75. Mayr E (1970) Population, species, and evolution. Harvard Univ Press.
76. Mayr E, Diamond J(2001) The birds of northern Melanesia: speciation, ecology and biogeography. Oxford University Press.
77. McGreevy TJ et al (2024) Influence of geographic isolation and the environment on gene flow among phenotypically diverse lizards. Heredity 133:317–330, https://doi.org/10.1038/s41437-024-00716-y
78. McNamara KJ (1986) The role of heterochrony in the evolution of Cambrian trilobites. Biol Rev 61:121–156, https://doi.org/10.1111/j.1469-185X.1986.tb00464.x
79. McNamara KJ, Long JA (2012) The role of heterochrony in dinosaur evolution. In: Brett-Surman MK et al (eds) The complete dinosaur, p. 761–784, Indiana Univ Press.
80. Meneganzin A, Bernardi M (2023) Were Neanderthals and Homo sapiens 'good species'? Quart Sci Rev 303:107975, https://doi.org/10.1016/j.quascirev.2023.107975
81. Mendel G (1866) Versuche über Pflanzenhybriden. Verhandl Naturforsch Ver Brunn 4:3–44.
82. Meyer HW, Smith DM (2008) Paleontology of the Upper Eocene Florissant Formation, Colorado. Geological Society of America.
83. Miao L et al (2024) 1.63-billion-year-old multicellular eukaryotes from the Chuanlinggou Formation in North China. Sci Adv 10(4):eadk3208, https://doi.org/10.1126/sciadv.adk3208
84. Miller D et al (2022) Retrogene duplication and expression patterns shaped by the evolution of sex chromosomes in malaria mosquitoes. Genes 13:968, https://doi.org/10.3390/genes13060968
85. Miller SL (1953) A production of amino acids under possible primitive earth conditions. Science 117:528–529
86. Montagu A (1989) Growing young. McGraw-Hill.
87. Nadeau NJ, Jiggins CD (2010) A golden age for evolutionary genetics? Genomic studies of adaptation in natural populations. Trends Genet 26:484–492, https://doi.org/10.1016/j.tig.2010.08.004
88. Niculescu EV (1961) Quelques remarques sur *Pieris napi bryoniae* Hbn. (Lép. Pieridae). (In French) Bull Mens Soc Linn Lyon 30:232–234.
89. Nisbet EG, Sleep NH (2001) The habitat and nature of early life. Nature 409:1083–1091

90. Nojiri T et al (2021) On the sequence heterochrony of cranial ossification of bats in light of Haeckel's recapitulation theory, J Exp Zool B Mol Dev Evo 338:137–148. https://doi.org/10.1002/jez.b.23042.
91. Ottaviani G et al (2020) Linking Plant Functional Ecology to Island Biogeography. Trends Plant Sci 25:329–339, https://doi.org/10.1016/j.tplants.2019.12.022
92. Pan J et al (2024) Genome-wide population affinities and signatures of adaptation in hydruntines, sussemiones and Asian wild asses. Mol Ecol 33: e17527, https://doi.org/10.1111/mec.17527
93. Pavey SA et al (2010) The role of gene expression in ecological speciation. Ann N Y Acad Sci 1206:110–129, https://doi.org/10.1111/j.1749-6632.2010.05765.x. PMID:20860685;PMCID:PMC3066407
94. Pearce BKD et al (2018) Constraining the time interval for the origin of life on earth. Astrobiology 18:343–364, https://doi.org/10.1089/ast.2017.1674.
95. Petren K et al (1999) A phylogeny of Darwin's finches based on microsatellite DNA length variation. Proc R Soc Lond B 266:321–329.
96. Phillips J (1931) The biotic community. J Ecol 19:1–24
97. Piálek P et al (2001) Raciation and speciation in house mice from the Alps: the role of chromosomes. Mol Ecol 10(3): 613–625, https://doi.org/10.1046/j.1365-294x.2001.01209.x
98. Pierce SK et al (2009) Chlorophyll a synthesis by an animal using transferred algal nuclear genes. Symbiosis 49:121–131, https://doi.org/10.1007/s13199-009-0044-8
99. Poelstra J et al (2013) An extensive candidate gene approach to speciation: Diversity, divergence and linkage disequilibrium in candidate pigmentation genes across the European crow hybrid zone. Heredity 111:467–473, https://doi.org/10.1038/hdy.2013.68
100. Pufahl PK, Hiatt EE (2012) Oxygenation of the Earth's atmosphere-ocean system: A review of physical and chemical sedimentologic responses. Mar Petrel Geol 32:1–20, https://doi.org/10.1016/j.marpetgeo.2011.12.002
101. Ramade F (2009) Eléments d'écologie: Ecologie fondamentale-4e édn. Dunod, Paris
102. Ratcliffe LM, Grant PR (1983) Species recognition in Darwin's finches (Geospiza, Gould). I. Discrimination by morphological cues. Anim Behav 31:1139–1153, https://doi.org/10.1016/S0003-3472(83)80021-9
103. Ricklefs RE (1990) Ecology. Freeman & Co, New York
104. Riebel K et al (2015) Learning and cultural transmission in chaffinch song. Adv Stud Behav 47:181–227, https://doi.org/10.1016/bs.asb.2015.01.001
105. Rivera MC, Lake JA (2004) The ring of life provides evidence for a genome fusion origin of eukaryotes. Nature 431:152–155, https://doi.org/10.1038/nature02848
106. Sacchi R. et al (2023). Species-specific spatial patterns of variation in sexual dimorphism by two Lizards settled in the same geographic context. Animals 13(4):736, https://doi.org/10.3390/ani13040736
107. Safi R et al (2004) (1970) The Axolotl (Ambystoma mexicanum), a neotenic amphibian, expresses functional thyroid hormone receptors, Endocrinology 145:760–772, https://doi.org/10.1210/en.2003-0913
108. Sagan L (1967) On the origin of mitosing cells. J Theor Biol 14:225–274, https://doi.org/10.1016/0022-5193(67)90079-3.
109. Salomon M (1982) Spéciation allopatrique probable entre deux populations de pouillots véloces: le cas du Pouillot véloce ibérique *P. c. brehmii* (in French) (National Museum of Natural History of Paris, Paris 7 University). Unpublished DEA thesis (In French)
110. Salomon M (1987) Analyse d'une zone de contact entre deux formes parapatriques: le cas des pouillots véloces *Phylloscopus c. collybita* et *P. c. brehmii* (In French). Rev Ecol (Terre Vie) 42:377–420, https://doi.org/10.3406/revec.1987.5429
111. Salomon M (1989a) Recherches sur la spéciation allopatrique: le cas des Pouillots véloces (Aves, Muscicapidae) des formes européenne (*Phylloscopus c. collybita*) et ibérique (*P. c. brehmii*) dans leur zone de contact des Pyrénées occidentales (In French). Unpublished doctoral dissertation, Paris 7 University, Paris

112. Salomon M (1989b) Song as a possible reproductive isolating mechanism between two parapatric forms. The case of the chiffchaffs *Phylloscopus c. collybita* and *P. c. brehmii* in the western Pyrenees. Behaviour,111:270–290, https://doi.org/10.1163/156853989X00709
113. Salomon M (1991) Summary of my PhD dissertation (In French). Bull Soc Zool France 116:81–84
114. Salomon M (1997) Quel statut donner au Pouillot véloce ibérique ? Alauda 65:63–81 (In French)
115. Salomon M (2001) Evolutionary biogeography and speciation: essay on a synthesis. J Biogeogr 28:13–27, https://doi.org/10.1046/j.1365-2699.2001.00515.x
116. Salomon M (2004) Speciation centres and sustainable development. J Biogeogr 31:675–678, https://doi.org/10.1046/j.1365-2699.2003.01034.x
117. Salomon M, Dassy B (2016) Partitional classification: A complement to phylogeny. Evol Bioinform 12:149–156, https://doi.org/10.4137/EBO.S38288
118. Salomon M et al (2003) On the taxonomic status and denomination of the Iberian Chiffchaffs. Ibis 145:87–97, https://doi.org/10.1046/j.1474-919X.2003.00122.x
119. Schuettpulz E et al (2018) Are there too many fern genera? Taxon 67:473–480, https://doi.org/10.12705/673.1
120. Shantz HL (1906) A study of the vegetation of the mesa region east of pike's peak: The bouteloua formation. ii. development of the formation. Bot Gaz 42:179–207
121. Simmons L (2014) Sexual selection and genital evolution. Austr Entomol 53:1–17, https://doi.org/10.1111/aen.12053
122. Simpson GG (1953) The major features of evolution. Columbia University Press, New York
123. Sinervo B, Clobert J (2003) Morphs, dispersal behavior, genetic similarity, and the evolution of cooperation. Science 300:1941–1951, https://doi.org/10.1126/science.1083109
124. Solounias N, Semprebon G (2002) Advances in the reconstruction of ungulate ecomorphology with application to early fossil equids. Amer Mus Novit 3366:1–49, https://doi.org/10.1206/0003-0082(2002)366<0001:AITROU>2.0.CO;2
125. Svensson L (2001) The correct name of the Iberian Chiffchaff Phylloscopus ibericus Ticehurst 1937, its identification and new evidence of its winter grounds. Bull Brit Ornithol Club 121:281–296.
126. Tansley AG (1929) Succession: the concept and its values. In: Proc. 4th Int. Congr. Plant Sc, Ithaca, 1926, pp 677–686
127. Tansley AG (1935) The use and abuse of vegetational concepts and terms. Ecology 16:284–307
128. Thenius E (1974) Solipedes or Equidae. In: Grzimek B (ed) Animal Life Encyclopedia, vol. 12, Stauacher, Zürich (translated from German)
129. Theißen G (2009) Saltational evolution: hopeful monsters are here to stay. Theory Biosci 128:43–51, doi.:10.1007/s12064-009-0058-z
130. Thielcke G, Linsenmair KE (1963) Zur geographischen Variation des Gesanges des Zilpzalps, *Phylloscopus collybita*, in Mittel- und Süwesteuropamit einem Vergleich des Gesanges des Fitis, *Phylloscopus trochilus*
131. Thielcke G et al (1963) Reaktionen von Zilpzalp und Fitis (*Phylloscopus collybita, Ph. trochilus*) auf verschiedene Gesangsformen des Zilpzalps (In German). J. Ornithol 119:213–226.
132. Ticehurst CB (1937) *Phylloscopus collybita ibericus*, subsp. nov. Bull Brit Ornithol Club 57: 63–64.
133. Ticehurst CB, Whistler H (1928) On the avifauna of Galicia. Ibis 12:663–683.
134. Torsvik TH, Cocks LRM (2017) Earth history and palaeogeography. Cambridge University Press.
135. Trainer JM (1989) Cultural evolution in song dialects of yellow-rumped caciques in Panama. Ethology 80:190–204.
136. Turchin V (1977) The phenomenon of science. A cybernetic approach to human evolution. Columbia University Press, New York

137. Uroz S et al (2015) The min- eralosphere concept: mineralogical control of the distribution and function of mineral-associated bacterial communities. Trends Microbiol 23:751–762
138. Vagvolgyi J, Vagvolgyi M (1989) The taxonomic status of the small ground-Finch, Geospiza (Aves: Emberizidae) of Genovesa Island, Galápagos, and its relevance to interspecific competition. Auk 106:144–148.
139. Vaurie C (1954) The birds of Palearctic fauna. A systematic reference. Witherby, London.
140. Vernadsky VI (1997) The biosphere. English translation of the original in Russian published in 1926, Springer Verlag, New York.
141. Vershinina AO et al (2021) Ancient horse genomes reveal the timing and extent of dispersals across the Bering Land Bridge. Mol Ecol 30:6144–6161, https://doi.org/10.1111/mec.15977
142. Wagner DL, Liebherr, JK (1992) Flightlessness in insects. Trends Ecol Evol 7:216–220.
143. Walker LR, del Moral R (2003) Primary succession and ecosystem rehabilitation. Cambridge University Press
144. Wallace AR (1881) Island life. Macmillan, London.
145. White MJD (1968) Models of speciation. Science 159:1065–1070, https://doi.org/10.1126/science.159.3819.1065
146. Willdenow CL (1792) Grundriss der Kräuterkunde. Haude & Spener, Berlin
147. Williams W (2015). The Horse. Harper Collins, Toronto.
148. Williamson K (1976) Identification for ringers. The genus *Phylloscopus*. British Trust for Ornithology
149. Woese CR, Fox GE (1977) Phylogenetic structure of the prokaryotic domain: the primary kingdoms. Proc Natl Acad Sci USA 74:5088–5090
150. Woese CR et al (1990) Towards a natural system of organisms - proposal for the domains Archaea, Bacteria and Eucarya. Proc Natl Acad Sci USA 87:4576–4579
151. Wolfgramm H (2021) Asymmetric allelic introgression across a hybrid zone of the coal tit (*Periparus ater*) in the central Himalayas. Ecol Evol 11:17332–17351, https://doi.org/10.1002/ece3.8369
152. Wright N et al (2016) Predictable evolution toward flightlessness in volant island birds. Proc Natl Acad Sci USA 113 (17):4565–4770, https://doi.org/10.1073/pnas.1522931113
153. Zhang J (2003) Evolution by gene duplication: an update. Trends Ecol Evol 18:292–298, https://doi.org/10.1016/S0169-5347(03)00033-8

Chapter 13
The Biosociological Approach-II: Populations, Communities

13.1 Populations

13.1.1 Population: A Biological and a Statistical Concept

13.1.1.1 Populations and Individuals Are Biological Concepts

The living part of an ecosystem (or total biocoenosis) is composed of prokaryotes (bacteria and archaea), unicellular eukaryotes (namely, protozoans, green lineage unicells, and yeasts), and multicellular eukaryotes (multicellular fungi, animals, algae and plants). Each species of these groups of organisms is constituted of individuals organised in *populations* and sharing a given spatiotemporal unit where they are born, grow, have social relationships with their conspecifics, scatter, reproduce, feed, and sometimes coevolve with other species,[1] grow old and die.

An individual is (1) either the cell itself in unicellular organisms,[2] (2) the result of *fertilisation*, the zygote or egg, in diploid multicellular organisms or (3) the result of successive developmental mitoses in haploid or diploid multicellular organisms if they practise a form of pathenogenesis. A population is a system comprising interacting individuals of the same species living in a circumscribed area whose limit may be natural or conveniently defined by researchers for their study [48].[3]

Supplementary Information The online version contains supplementary material available at https://doi.org/10.1007/978-3-031-90656-5_13.

[1] In the acceptation I propose, cf. Sect. 4.1.

[2] Though the term may seem here paradoxical since its etymology refers to an organism that cannot divide, it is generally applied to any organisms if they are discontinuous units that are the seats or their biology, i.e. their physiology, metabolism and all their vital functions.

[3] p. 279.

M. Salomon, *Integrative Ecology*, https://doi.org/10.1007/978-3-031-90656-5_13

Its delineation is thus arbitrary and dependent on the spatial scale chosen for the study.

- What is meant by a population of the melliferous bee (*Apis mellifera*)? Is it the content of a beehive or the set of neighbouring beehives?
- Let us consider a vast forest like the Irati Forest in the French Basque Country, extending over about 17,200 ha. This place shelters hundreds of neighbouring Eurasian chaffinches (*Fringilla coelebs*). Do these birds form one population? Or a set of several contiguous ones?

One criterion to settle the question is to assess the strength of social interactions between the individuals and propose a threshold below which the group studied will be considered a single population or several others.

13.1.1.2 Populations and Individuals Are also Statistical Concepts

The concept of population and individual exposed in Sect. 13.1.1.1 refers to these notions in their literal sense. A metaphoric extension of these two concepts is used by the statisticians who use the term *population* for a mathematical set whose members are items identified with a set of descriptors. Statistical individuals are often numbers or vectors whose components are values of variables or attributes (the descriptors). This concept's approach involves describing a population by parameters (e.g., the mean or variance of descriptors, proportions, and rates) estimated with sampling techniques. In the sense of the statistician, we can speak, for example, of a population P of white and black balls whose individuals (the balls) are defined by an attribute value, "black" or "white". By analogy with the actual populations, the elements of a "population" in the statistical sense are called *individuals*. In order to remove the polysemy of the terms, we will qualify (1) the populations and individuals in their literal sense as *biological* and (2) the populations and individuals in the figurative sense of the statisticians as *statistical*.

The immense majority of the recent biological publications are based on statistical and computational methods. The populations they treated are *statistical populations* that represent the biological populations [1, 24, 26, 62]. The papers based on strictly biological populations are generally purely descriptive and generally ancient, with quantitative tools little involved.

Through its structure, a biological population is a statistical population whose primitives, the biological individuals at a given development level, can be characterised by a set of descriptors depicting aspects of their biology (e.g. their phenotype, ecotype, genotype). A biological population can also be processed as a statistical individual described by a set of variables, e.g. its sex ratio, age class distribution, degree of generation overlap, or many other descriptors. In this case, the statistical population will be a category subsuming the biological population, e.g. a taxon or a community. We can also have statistical individuals that are neither biological individuals nor groups of the latter, as in habitat surveys assessing the presence/absence or proportion of different species in examined habitats. Here, (1) the individuals are the habitats, and (2) the variables are the species or groups of biological individuals characterising the habitats.

13.1.1.3 Characterisation of Populations

Theoretically, a good representation of a population P described by a set of variable(s) V_1, V_2, $\cdots$ consists of a *simple random sample* of individuals drawn from it. Such a sample is said to be *representative* of P. In the case of the study of up to two variables, the tools of inferential statistics, whose purpose is to obtain the best estimation of the parameter(s) of interest of a population, are very commonly used. The parameters can be, for example, the central positions (means and median), different quantiles, correlations or dispersion of the variable(s). Randomisation is easily achieved when the reference population P is finite, containing thus N individuals, from which an experimenter can draw a simple random sample of n individuals. The technique to be applied is the following. Each individual of P is assigned a distinct number between 1 and N, then n different pseudorandom numbers[4] $s_1, s_2, \cdots, s_n$ $(n << N)$ are drawn from the set $\{1, 2, \cdots, N\}$ using a random function proposed by a suitable programme language. With R, for example, one such function is `sample`. Then, we select the individuals to which these numbers are assigned.[5]

Such techniques are widely used in agronomy [43] and clinical trials [16]. For example, a study on a population of 1000 laboratory mice wants to assess if a chemical significantly causes a disease inflating a given organ. The research team will first estimate the proportion of ill mice. To achieve this, they have to sacrifice animals. However, sacrificing the entire population is very costly, and the laboratory can only afford to sacrifice a simple random sample of 20 animals. The research team will proceed as follows. Each mouse will be assigned a distinct number between 1 and 1000. Then, we apply, for example, the R Random function `sample(1000, 20)` that generates a set of 20 pseudorandom numbers (sorted increasingly): 3, 20, 207, 252, 265, 309, 330, 335, 425, 515, 576, 616, 675, 699, 740, 824, 916, 936, 940 and 987. The mice to which these numbers are assigned are sampled and sacrificed, and the observed proportion of ill mice is estimated from the sample. This observed proportion under good application conditions is the estimated proportion of ill mice in the population.

Randomisation techniques like the one above are rarely used in *field ecology* since the reference populations are unknown, and also for obvious ethical reasons. We deal with finite populations that are theoretically considered samples of infinite

---

[4] A random number is theoretically drawn from a numeric sequence, a series of numbers that does not exhibit any pattern of regularities. However, as obtaining actual random numbers is impossible, we rely on algorithms known as pseudorandom number generators (PNGs) to approximate a sequence of random numbers convincingly.

[5] Readers proficient in mathematics can refer to [53] for the mathematical justification of this technique and other more sophisticated types of sampling from finite populations.

virtual populations. Researchers mostly rely on chance encounters of the individuals of study by trying to cover their research territories as thoroughly and systematically as possible; however, their prospectings are at risk of being biased by unconscious preferences. Therefore, this empirical method is not as efficient as randomisation. However, this bias can be mitigated by techniques used in polling institutes, such as panelling. The observer needs to interpret and express ecological data in understandable terms and logic. For this, two main approaches are modelling (see Chap. 6) and data mining (a technique that unravels the constant structures underlying a set of observed data without posing a priori assumptions).

13.1.2 Population Norm

Definition 13.1 A norm is a sensible hypothesis about a phenomenon, e.g. a parameter inductively inferred from the knowledge of the phenomenon studied, such as the 1:1 sex ratio in an animal species (meaning that there are as many females as males) or the outcome of an unbiased dice toss.

This concept can be applied to any physical character, like the colour of the balls, or to genetic or phenotypic character. The best way to check that $\pi = \frac{1}{2}$ is the norm of P for the character 'ball colour' is to randomly draw 20 balls from the urn containing P and calculate the observed proportion p of white balls in the sample. Let us do it; we observe nine white and 11 black balls in our sample. A classical test shows that from this result, we cannot reject the null hypothesis; we will thus accept that the estimated norm $\pi = \frac{1}{2}$ is correct. (cf. Problem 13.1 in Quizzes and Problems, p. 397).

This virtual population may be a good model for describing the sex composition of sexually reproducing *eukaryotes*. Let $N = 10,000$ be the size of P. We wish to verify whether there are as many females as males in P. This means that the probability of drawing a female or a male from P is $\frac{1}{2}$ (the null hypothesis). This probability is called the norm of population P. The biological objects or phenomena are processed as the balls just mentioned. The statistical concepts of "population" and "individual" are very useful and operational. However, it is an abstract and virtual simplification of biological populations and individuals.

Another key concept that finds practical application in biology is the *reaction norm* of an organism, which can be assessed at the level of a population or of a species.

Definition 13.2 (Reaction Norm) The reaction norm of a species is the pattern of phenotypes permitted by a coding genotype. This parameter can vary according to ecogeographical parameters.

Reaction norm is an old concept already proposed at the beginning of the twentieth century by the German zoologist Richard Woltereck (1877–1944) [60]. This scientist studied the daphna species *Daphnia longispina*, a planktonic crustacean.

Woltereck measured two characters: the length of the helmet and the sexual maturity onset. He conducted experiments to study the effect of different environmental factors on the two variables. These factors were light, temperature, salt content, gas content, quantity of available food and competition of conspecifics. He drew the phenotype curves of each of these factors and concluded the following:

- An increase in the quantity of available food or the number of generated individuals at the egg-hatching stage induces a proportional increase in the helmet's length (thus a rightward shift of the phenotype curve) and postpones the sexual maturity onset.
- The other factors (light, temperature, salt content, and gas content) did not directly impact the two characters studied.

The author posed the principle of reaction norm, as defined in Definition 13.2, as a hypothesis, which proved fruitful. This concept was later acknowledged as a fundamental characteristic of organismal populations [23, 39].

13.1.2.1 First Situation: The Norm of the Population Is Conserved

The norm of a given population P is said to be *conservative* when the degree of homogeneity of its phenotypes and ecotypes is high. This generally happens when the environment of P is homogeneous and stable, i.e. varies little and gradually or not significantly spatially and temporally. In this situation, speciation seems unlikely. Very frequently, simple biological features of an organism, generally measured on individuals but sometimes on groups of individuals, are Gaussian variables. This is the case, in particular, for weight, lengths of body parts, overall size, volume, and survival rate at a given age. It is often shown that when the norm of the population is conservative and the organisms under study have their phenotypes or ecotypes described by a set of relevant variables, the fittest individuals or demes are those showing values close to the average of these variables (the mean in the case of Gaussian variables). These variable values are hypothesised to constitute the population's optimal ecological conditions[6] of the population. The habitat corresponding to the individuals sharing the phenotypes and ecotypes described by these optimal values is called the optimal habitat.

Definition 13.3 The *ecological amplitude* of species S is, by definition, the range of the ecological habitats deviating from the optimal habitat to which S can adapt. The wider this range, the larger the ecological amplitude is.

Definition 13.4 Species with a broad ecological amplitude are called *eurytopic* and those with a narrow one are termed *stenotopic*.

---

[6] One ought to be cautious here. In biology, true optimality never exists since living beings are, by definition, thermodynamically unstable (see Sect. 9.2).

In Sect. 12.4, we already met the two odonate insects that will exemplify the concepts of Definitions 13.3 and 13.4. The common blue damselfly (*Enallagma cyathigerum*) is *eurytopic*, whereas the ruddy sympetrum (*Sympetrum sanguinem*) is *stenotopic*. The narrower the ecological amplitude of a species is, the less fit the individuals whose characteristics deviate from the average, thus living in deviating habitats, are. When the ecological amplitude of a species is broad as is the case in eurytopic species, the individuals will adapt to deviating habitats almost as well as to their optimal ones. On the contrary, stenotopic species are unfit for reproducing or even surviving at relatively small habitat deviations. This would explain and predict why eurytopic species tend to be polytypic (i.e. comprise several subspecies) and stenotopic species, on the contrary, tend to be monotypic (i.e. the species is not divided into subspecies).

"Optimal" phenotypes remain stable as long as the environmental factors do not vary significantly, with demes containing relatively large numbers of individuals and related by gene flow (even in relatively small amounts).

13.1.2.2 Second Situation: The Norm of the Population Is Novel

After geological, astrophysical or climatic upheaval, large populations can almost be annihilated, with only small groups of individuals stemming from these populations managing to survive in refuges. These groups, called refuge populations, are generally located in the periphery of the once central distribution zone of the species. Due to their small numbers, these relictual populations are subjected to genetic drift relative to the once-large population from which they originate. As a reminder, genetic drift is a statistical process due to the weak number of individuals in a small population, where the probability that an allele gets lost is high. This situation inevitably entails the loss of alleles within these small groups of individuals.

Furthermore, the probability of inbreeding increases with a decrease in population size. As time passes, if these populations do not disappear due to stochastic extinction,[7] mutations or non-vertical gene transfers (lateral gene transfer, endosymbiosis) accumulate within them. These relictual populations increase and form allopatric populations (i.e. their distribution zones are isolated) that mechanically will develop differentiated allele distributions for a significant number of their loci, resulting in divergent genetic norms between them. We say that these populations adopt *novel norms* or that the norm of the initial population has diverged.

[7] When a population is small and subject to genetic drift, inbreeding with homogamy for deleterious or lethal genes obtained by chance can cause its extinction. Such a process is called stochastic extinction.

13.1.3 Population Structure

13.1.3.1 Temporal Continuity and Spatial Continuity

Two situations concerning the direct descendants of a population can occur: (1) the offspring make their first reproduction in the territory of their parents, or (2) they perform it elsewhere, by joining other territories or founding new ones.

Case (1) is the general rule among poorly mobile species (e.g. animals that move little during their lifetime, plants whose seed dissemination is short-distanced), but also among *philopatric* mobile species, i.e. organisms whose offspring tend to make their first and subsequent reproductions in the territory of their parents (e.g. several long-distance migrating birds). Philopatric animals generally have vast home ranges or are migratory or erratic, periodically returning to their breeding areas. In this situation, the population presents a temporal continuity: the individuals of such a population are probably the descendants of individuals who have formed the same population during the past generations.

Case (2) is often encountered either in mobile non-philopatric species (e.g. migratory or resident birds like the window finches *Vidua chalybeata* [40], anemochorous plants, or poorly-mobile organisms when they are in an invasive phase (e.g. locusts). This phenomenon can be random or determined by a demographic equilibrium reached after long evolutionary processes.

In the case a population continuously remains in the same territory whatever the seasons and across a long period (e.g. thousands of generations), its individuals are related, thus sharing the same genetic pool: the individuals of such a population have more genetic traits than individuals from different populations of the species. In other words, their intrapopulation genetic variance is smaller than the interpopulation genetic variance of the species. If the area of the population is vast, the shorter the distance between individuals of such a population is, the more related the individuals will tend to be.

13.1.3.2 Ecological Interactions Within a Population

We will treat the relationships between the individuals of a population that are relevant for its sustainability. The intrapopulation ecological interactions are largely determined by the ecological sphere of influence of the individuals, i.e. the distances they cover in their home range. The aim is access to the available resources, protection against predators and reproduction. These features are intimately linked, and consequently, different strategies have been elaborated by the different species, namely,

- coloniality vs broad interindividual spacing or
- sociality vs territoriality.

According to the different lifestyles adopted by the species, different social behaviours organise the individuals differently.

- A strict hierarchy where dominance is involved in colonial species or subequalness of small social units practising territorial exclusion (the small social units are often pairs) can be observed.
- Modes of reproduction are diverse: monogamy, polygamy or promiscuity. In territorial species, spectacular nuptial parades can be observed among less conspicuous ones. In colonial species, where relationships of dominance play a major role, gatherings of the individuals in leks with clear dominance during the selection of sexual partners.

Antipredatory tactics can consist in

- collective defence, sometimes involving several species, or on the contrary, individual defence,
- the active mode (attack or flight) or passive (mimicry or camouflage).

The ecological interactions within a population significantly determine the mutual relationships of the individuals, namely their spatial position and movements, distribution area, and density.

13.1.4 Distribution Area

13.1.4.1 Overview of the Subject

The distribution area of a species is the complete area where all its populations are distributed. This definition is formal since only parts of this area are suitable for the species' individuals regarding their physiological needs (e.g., their shelters, foods, proper spots for breeding, or climatic conditions linked to altitude, local humidity, and other factors) [41].

In sedentary (or resident) species, the concept of distribution area is simple. Such an area corresponds to the set of sites, which is localised geographically to one specified portion of the Globe, where the individuals are born, live, reproduce, and die. The individuals remain in this area during their entire lifetime, or when they move from this area, will proceed to sites that are short-distanced from the birth sites. This case has been well documented in marine tropical invertebrate species (intertidal molluscs and crustaceans) with continuous reproductive cycles [5, 56]. Among sedentary species, many vertebrates of the temperate zone, namely bird species (e.g. house sparrow *Passer domesticus*, carrion crow *Corvus corone*, or grey partridge *Perdix perdix*) very rarely move from their birthplaces beyond 20 km in their entire lifetime [36].

With the migratory and erratic species, the concept of distribution area is more complex. Among the animals of the first category (migratory birds, salmonids, eels), the territories in which the individuals live during the breeding season are different from those in the other seasons when they do not reproduce. The organisms can carry out one or several back-and-forth trips between the breeding and non-breeding sites during their lifetime; the breeding sites are often where the individuals are born, but not always. The erratic species have home ranges that are moving across time; they are exemplified, for example, by species living in desert wetlands, i.e. temporary lakes or ponds existing only when short and violent downpours or provisory groundwaters fill valley bottoms and shallow hills for a brief period. The ambient and quasi-permanent drought rapidly transforms these areas into the barren territories that they usually are. These habitats shelter entire ecosystems that are very resilient to drought and desert conditions and able to wait for long periods before moving efficiently to the provisory wetlands when they appear. Several surveys of such species have been performed recently, notably a vegetation zonation in a desert of Saudi Arabia [10].

These considerations lead us to propose a more general definition of a distribution area.

Definition 13.5 The *distribution area* of a species is the set of all its home ranges during the entire life cycle of the individuals of that species.

- If the species is resident, the definition is straightforward. It corresponds to its single home range.
- The outline of the erratic species' distribution area is blurred. The set of home ranges corresponds to spots and trajectories endowed with fast-changing ecological conditions.
- The distribution area of migratory species comprises their breeding area, their non-breeding territories and all the stopovers in between. An important issue is whether the individuals gather in identical populations in the different subareas and stopovers of the distribution area, and the answer to this question depends on the species studied.

Different factors determine the distribution area of a species, e.g. (1) its geographical origin, (2) its life history and biogeographic evolution across time, (3) ecological conditions, namely selection pressure (e.g. from predators, abiotic hazards), (4) anthropogenic pressure (hunting and harvesting, direct or indirect habitat destruction, introductions) and (5) chance.

13.1.4.2 Geographical Origin of the Taxa

This factor is the first one to explain the natural distribution of a taxon. Each species probably has a single geographical origin due to a differentiation process within an ancestral species or a group of closely related ancestral species. The species has maintained its individuality thanks to geographical, behavioural, and genetic barriers [33]. In most species, the populations of a given area have been shaped by their life history, which explains that the same taxon cannot originate simultaneously from two different places. Thus, we can find many ecological equivalents belonging to relatively distant taxa.

Hence, the beech populations of the genus *Fagus* are met in their natural state only in the Northern Hemisphere, particularly in Eurasia. These species have ecological equivalents in the Southern hemisphere (neotropics), e.g. the southern beeches (genus *Nothofagus*) essentially found in Argentina and Chile. Sugar maple (*Acer saccharum*), the tree that produces maple sugar, is found only in North America; none of its populations grows naturally in Eurasia, even though the ecological and climatic conditions in some places can be similar. The Old World warblers belong to the genus *Sylvia* (family Sylviidae). As indicated in their names, they only live in the Old World; the Passerine birds that occupy the same niche in the New World are the New World warblers belonging to the family Parulidae. Many other examples like this can be enumerated.

13.1.4.3 Ecological Conditions

The distribution area of a taxon often has a discontinuous aspect since all the territories it comprises are not suitable for the taxon considered. We will mention several examples below.

- A passerine bird, the reed bunting *Emberiza schoeniclus*, has its distribution area encompassing Europe—except Iceland, the Italian Peninsula, the Dalmatian coast, and Greece—and an extensive part of northern Asia. However, the populations are never found in forests. In the spring and summer, this species lives and breeds only within the reed beds along ponds and marshes [54]. The rest of the year, the birds are found in open environments.
- A reptile, the viviparous lizard *Zootoca vivipara* has an extensive distribution across Eurasia, from Spain across the continent northwards to Southern Scandinavia and the British Isles and then eastwards continuously across Central and Eastern Europe to Eastern Siberia. Nevertheless, the only places piecemeal within this vast zone are peat bogs, wet meadows, heath, and dunes, generally in hygrophilous plant formations never distant from marshes and streams [18].
- Many migratory species (birds, eels) have limited and specific stopovers between the breeding sites and the non-breeding ones (e.g. wintering areas for birds, European and North American rivers for eels[8]) [2], which provide them in abundance the food they need. Several species of whales fast during the migration between the two areas.

[8] The spawning area of the European eel is located in the Sargasso Sea [61].

13.1.4.4 Selection Pressure from Predation and Anthropogenic Activities

Predation pressure can cause rarefaction of prey and even sometimes local extinctions, which can explain gaps in the distribution area of the latter at a given time t. However, since a too-great decrease of prey species entails a diminution of the number of their predators by starvation, a dynamic equilibrium ensues, and the preys are allowed to increase due to the drastic decrease of the predators; the gaps are generally filled by a new generation of prey (see Lotka-Volterra model, Sect. 13.2.3, p. 384).

On the other hand, anthropogenic pressure, i.e. agropastoral activities, sylviculture, hunting and diverse sorts of land use management, can eradicate the populations of a species in extensive areas and entail situations of artificial geographic isolation, fragmenting the genetic continuity of species by creating isolates comprising a small number of individuals. These groups are consequently isolated from the genetic pool, which enhances genetic drift and consanguinity. Thus, human activities simulate catastrophic events, like volcanic eruptions, earthquakes or typhoons, but on a more extended scale. Such a situation reduces the fitness of the impacted species, increasing the risk of extinction; in the less dramatic cases, it can strengthen local endemisms and insular phenotypic features.

Involuntary and Voluntary Alien Species Introduction

This factor confuses the understanding of the development of biomes. Therefore, it is of the utmost importance to know the history of the introductions to shed light upon the establishment of faunas and floras with the highest accuracy.

Since the neolithic, humans settling in a given habitat dragged many species with them, namely, (1) domestic species, among which a nonnegligible number of individuals became feral (horses, dogs, cats), (2) commensal species (rats, sparrows, nettles). These species can sometimes become invasive and compete with native species. In the modern age, man has practised voluntary introductions. The examples are numerous. We can mention several cases:

1. The barn owl *Tyto alba* was introduced in the Seychelles Islands in the 1940s to eradicate rats, themselves involuntarily human-introduced two centuries before.
2. The Indian grey mongoose *Urva edwarsii* was voluntarily introduced in Jamaica with the purpose of eradicating snakes.
3. Cervids and phasianids were introduced in various islands of the Indian and Pacific Oceans for hunting.
4. The release of the European starling *Sturnus vulgaris* as an ornamental bird in Central Park, New York, at the beginning of the 1900s resulted in the invasion of this species throughout the entire American continent.

13.1.4.5 Chance

This factor explains the settlement of pioneering organisms on islands like Hawaii, the granitic islands of the Indian Ocean, or the Galápagos archipelago. At a given time t, erratic populations invaded one or several neighbouring ecologically empty islands. Some of the invaders successfully colonised these territories and permanently settled on them. Adaptive radiation follows, during which their descendants, separated by ocean from the original, generally continental populations, undergo geographic isolation reinforced by genetic drift. In these circumstances, they can be subject to relatively rapid allopatric speciations.

In most cases, the implied taxa are volant organisms (birds, insects) that can establish themselves actively (via flapping or gliding flights where soaring on rising air is involved). Some terrestrial animals (micromammals, reptiles, spiders) can be transported to the islands on intertwined floating uprooted tree trunks and branches dragged by oceanic currents (masses forming natural rafts).

13.1.4.6 Intrapopulation Distribution of Individuals

Three types of intrapopulation distribution of individuals are observed, depending on the heterogeneity level of the environment of the population studied and the social interactions between its members [48, p. 283]: (1) the aggregative distribution, (2) equidistant distribution, and (3) the random distribution.

- Aggregative distribution is met with colonial or social species, or organisms that gather around relatively rare food sources concentrated in precise spots in the distribution areas.
- Conversely, the equidistant distribution between individuals reflects a territorial and individualistic lifestyle reinforced by strongly agonistic relationships; in this situation, the food resources are relatively evenly scattered in the distribution area.
- Random distribution describes a situation where the individuals have few mutual interactions (a case called neutralism). This tends to occur when the individuals are not numerous, the sexes are morphologically similar, and food is plentiful across the distribution area.

A continuum exists between the three states, but with proper statistical tests, the case studied boils down to one of the three situations. Let us base our hypothesis

$$\begin{cases} \mathbf{H_0}: & \text{Random distribution of the individuals} \\ & \text{vs} \\ \mathbf{H_1}: & \text{Individuals aggregated of equidistant} \end{cases} \tag{13.1}$$

We will assume that the distribution area is divided into subequal units that are the relevant statistical units. This situation is observed when the processed units are, for example, the equal squares in a grid, the leaves of the same tree, individual herbaceous plants or small ponds statistically of equal size. Hypothesis H_0 is tantamount to saying that each individual has the same probability of being on any unit. The law of probability of the number of individuals on such a unit is a Poisson distribution of parameter λ.

$$P(X = k) = e^{-\lambda}\frac{\lambda^k}{k!} \tag{13.2}$$

where $X = k$ is the number of individuals on a unit, and λ is the Poisson parameter.

Theorem 13.1 *If X follows a Poisson distribution with parameter* λ, *the theoretical mean* EX *is equal to the theoretical variance* $\mathrm{Var}(X)$; *that is,* $EX = \mathrm{Var}(X) = \lambda$.

Proof

Calculation of the mean:

$$EX = \sum_{k=0}^{\infty} k e^{-\lambda}\frac{\lambda^k}{k!} = \cancel{0 \times e^{-\lambda}\frac{\lambda^0}{0!}} + \sum_{k=1}^{\infty} k e^{-\lambda}\frac{\lambda^k}{k!} = \lambda e^{-\lambda}\sum_{k=1}^{\infty} \cancel{k}\frac{\lambda^{k-1}}{\cancel{k}(k-1)!}$$

$$= \lambda e^{-\lambda} e^{\lambda} = \lambda$$

Calculation of the variance: ($\hat{\lambda}$ is an unbiased estimator of the parameter λ very close to λ)

$$\mathrm{Var}(X) = E(X^2) - [E(X)]^2 = \sum_{k=0}^{\infty} k^2 e^{-\lambda}\frac{\lambda^k}{k!} - \lambda^2$$

$$= \cancel{0^2 \times e^{-\lambda}\frac{\lambda^0}{0!}} + \sum_{k=1}^{\infty} k^2 e^{-\lambda}\frac{\lambda^k}{k!} - \lambda^2$$

$$= \lambda e^{-\lambda}\sum_{k=1}^{\infty} k^{\cancel{2}}\frac{\lambda^{k-1}}{\cancel{k}(k-1)!} - \lambda^2 = \lambda e^{-\lambda}\sum_{k=1}^{\infty}\frac{(\hat{\lambda}^k)'_{\hat{\lambda}}}{(k-1)!} - \lambda^2$$

$$= \lambda e^{-\lambda}\left[\sum_{k=1}^{\infty}\frac{\hat{\lambda}^k}{(k-1)!}\right]'_{\hat{\lambda}} - \lambda^2 = \lambda e^{-\lambda}\left[\hat{\lambda}\sum_{k=1}^{\infty}\frac{\hat{\lambda}^{k-1}}{(k-1)!}\right]'_{\hat{\lambda}}$$

$$\lambda e^{-\lambda}\left(\hat{\lambda}e^{\hat{\lambda}}\right)' - \lambda^2$$

$$= \lambda e^{-\lambda}(e^{\hat{\lambda}} + \hat{\lambda}e^{\hat{\lambda}}) - \lambda^2 = \lambda \cancel{e^{-\lambda}}\cancel{e^{\hat{\lambda}}}(1 + \hat{\lambda}) - \lambda^2 \simeq \lambda + \cancel{\lambda^2} - \cancel{\lambda^2}$$

□

Let us consider a sample of size n, representative of the population studied where the random variable X, the number of individuals per surface unit [42, 48] has a Poissonian distribution, which is equivalent to saying that the scattering of the individuals is random (hypothesis H_0). In addition, let $\nu_k = P\{X = k\}$ be the relative frequency (estimated probability) of the individuals per surface unit. The efficient estimators of EX and Var(X) are constructed using the weighted arithmetic mean (WAM) of $X(\omega) = 0, 1, 2, \cdots, k \cdots$, the realisations of X (see hereunder).

- The efficient estimator of $E(X)$ is

$$\hat{\lambda} = \frac{1}{n}\sum_{k=0}^{\infty} \nu_k \times k \quad \text{[WAM of } X(\omega)\text{]} \tag{13.3}$$

- The efficient estimator of Var(X) is

$$\hat{\sigma}^2 = \frac{1}{n-1}\left[\sum_{k=0}^{\infty} \nu_k k^2 - \frac{\left(\sum_{k=0}^{\infty} \nu_k \times k\right)^2}{n}\right] \text{(Weighted empirical variance)} \tag{13.4}$$

Then, it is expected that if $X \rightsquigarrow \mathcal{P}(\lambda)$, $\hat{\lambda} \# \hat{\sigma}^2 \simeq \lambda \Rightarrow \frac{\hat{\sigma}^2}{\hat{\lambda}} \simeq 1$.

Theorem 13.2 *The ratio* $\frac{\hat{\sigma}^2}{\hat{\lambda}} \rightsquigarrow \chi^2_{n-1}$ *test.*

We will apply this analysis and testing to an example [48, p. 285, Table 15_1], the distribution of adult female red mites on apple trees. We wish to know whether we can reject the hypothesis of a random distribution of the mites (conclusion H_1) or on the contrary, not be conclusive, which means that we will have a parsimonious stance (acceptance of H_0) and admit that the distribution of the individuals is random.

If the test is conclusive, $\rho = \frac{\hat{\sigma}^2}{\hat{\lambda}} < 1$ means that the individuals are evenly spaced among the leaves. Conversely, $\rho > 1$ means that the individuals are clumped.

A sample of 150 leaves was prospected, where 172 mites were found: (1) In 70 leaves, the mites were absent. (2) In 38 leaves, only one mite was seen. (3) 17 leaves were occupied by two mites. (4) 10 leaves sheltered three mites. (5) 9 leaves sheltered four mites. (6) Five mites or more were observed in the remaining leaves of the sample.

We will write an R program to calculate the chi-square test and see whether the test is conclusive.

Program Code-1

```
rm(list=ls())
lambda <- 172/150 # Poissonian parameter
Nmites <- 0:5 # Number of mites per leaf: 0, 1, 2, 3, 4, 5 or more
Nth <- rep(0,5) # Initialisation of the theoretical numbers
Nobs <- c(70, 38, 17, 10, 9) # Definition of the observed numbers
Nobs <- c(Nobs, 150-sum(Nobs))
Nobs[5] <- Nobs[5]+Nobs[6]
Nobs <-Nobs[1:5]
for(i in 1:6)
Nth[i] <- exp(-lambda)*lambda ^ {i-1}/factorial(i-1)
# Construction of the theoretical numbers
Nth <- round(150*Nth,2)
Nth[5] <- Nth[5]+Nth[6]
Nth <- round(Nth[1:5],2); p=round(Nth/150,2)
chisq.test(x=Nobs, y=NULL, p, correct=TRUE) # Test carried out
```

The program is run and gives as result:

data: Nobs X-squared = 46.332, df = 4, p-value = 2.101e-09 (***)

Hypothesis H_0 is rejected, the test is conclusive. The program continues

Program Code-1 continued

```
...
for (k in 1:5) { S1 <- S1 + Nth[k] ∗ (k − 1)^2
S2 <- S2 + Nth[k]*(k-1)
}
S <- (S1 − (S2/150)^2)/149 # Calculation of σ̂^2
Res <- S/lambda
if(Res < 1){ # Value of σ̂^2/λ̂
print('even spacing')
}else {print('clumping')} # Decision on the type of the interindividual dispersion
```

The response of the program, about the type of mite scattering, is:

Res [1] 2.071509 [1] "clumping"

Res gives the value of $\rho = \frac{\hat{\sigma}^2}{\hat{\lambda}}$: $\rho > 1$, thus **the mites aggregate**.

13.1.5 Population Models

The individuals of a population are permanently interacting, and according to the circumstances:

- They compete for space, resources and sex when they belong to a sexually reproducing species.
- They cooperate in synergy.
- They develop strategies to reproduce.
- They contrive efficient strategies to obtain the necessary energy, ensuring their physical integrity (appropriate rest cycles and foraging, access to adequate food and liquid water, minimisation of stress or injury gravity between individuals involved in contests or fights for access to sex or a hierarchical position).

These points are schematic features of the dynamics of populations within a biocoenosis. Populations are characterised by the number of their member individuals, symbolised by the sequence N_t in discrete population models and $N(t)$ in continuous population models. The number of individuals in a population is called population size or *density*.

We will present hereunder models of population variation that are increasingly realistic (thus more and more complex). The simplest of these models, density-independent, is called the Malthusian model, named after Malthus (1766–1834) [31] . Such a model is frequency-independent, i.e. the number of offspring per reproductive unit is constant and does not depend on the density. We will see that in a Malthusian model, a consequence of this property is that the number of individuals in the population exponentially and indefinitely increases or decreases towards extinction. This model is unrealistic since any population P lives in a limited area where the resources are limited[9] and shared by predators and competitors, these factors having a given limitation effect on the density of the organisms. However, it is approximately valid at the beginning of any demographic process. The more sophisticated models, considering the variation of the number of offspring from one generation to another, are called frequency-dependent. Several models will follow, which will be increasingly realistic. We will discuss the simplest frequency-dependent models and the logistic models (owed to Verhust and Gompertz). A more detailed development of these models can be found in the pedagogic book *Mathematical Ecology* by the Canadian statistical ecologist Evelyn C. Pielou (1924–2016) [42].

Definition 13.6 Density-independent variation is a population variation (growth, constancy or decrease) whose development does not depend on the number of individuals in a population at any time t.

[9] One notable exception is human demography, which fits the Malthusian model well since the agricultural revolution [35].

13.1.5.1 A Density-independent Model: The Malthusian (or Exponential) Model

Let N_t be the number of individuals at generation t (running from 0 to ∞), and R be the increase rate of the population (mean number of offspring per individual). Sex ratio $= 1$.

Case 1: *Malthusian Discrete Model of a Semelparous Population, No Generation Overlap*

This growth model is based on the following axiom:

Axiom 13.1

(1) The population is semelparous (i.e. at a given generation, each individual reproduces and dies afterwards).
(2) Each adult produces R offspring.[10]
(3) There exists no immigration, emigration, or deaths within any generation before the reproduction.
(4) The generations do not overlap.

The equation of the Malthusian model, expressed in its recursive (or implicit) form [11] will thus be:

$$N_{t+1} = R\,N_t, \tag{13.5}$$

where N_t represents the number of individuals in a population $\mathcal{P}$ at time t (running from 0 to ∞) and R the instantaneous growth rate, i.e. the average number of products by parent (one of the sexes in sexually-reproducing organisms), which is constant in this model.

Equation (13.5) can also be expressed as

$$\frac{N_t}{N_{t+1}} = \frac{1}{R} \tag{13.6}$$

provided that $R \neq 0$. By the way, $R = 0$ is biologically uninteresting, so that we will discard this case.

Equation (13.6) allows us to study the average productivity of a population

$$\frac{N_t}{N_{t+1}} = \frac{\text{Number of parents}}{\text{Average number of their products}},$$

[10] R is a statistical number. The individuals of a population P can generate 0 to n offspring; the model will then attach the mean of the values to each individual, hence the possibility of having $R < 1$.

[11] For a detailed account, see [21].

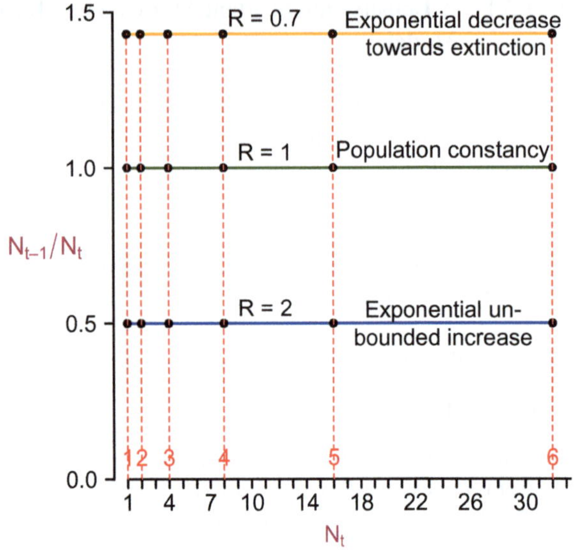

Fig. 13.1 Malthusian model where parent to offspring ratio is function of number of offspring

as a function of N_t, that is $f(N_t) = \dfrac{N_t}{N_{t+1}}$. The Malthusian model says that f is constant.

$$f(N_t) = \frac{N_t}{N_{t+1}} = \frac{1}{R} = \text{constant since } R \text{ is constant.}$$

The function f is thus illustrated in a graph by a horizontal line (cf. Fig. 13.1).

The process starts at time (or generation) $t = 0$ and proceeds indefinitely. Thus t runs through set $\mathbb{N}^* = \{0, 1, 2, 3, \cdots\}$.

Theorem 13.3 *Let a finite semelparous population P, whose number of individuals is N_t at generation t. Then*

$$N_t = R^t N_0 \tag{13.7}$$

Proof From the relationship (13.5), we have

$$\begin{cases} N_t = R\,N_{t-1} \\ N_{t-1} = R\,N_{t-2} \\ \cdots \\ N_1 = RN_0 \end{cases} \Rightarrow N_t = RN_{t-1} = R^2 N_{t-2} = R^3 N_{t-3} = \cdots = R^t N_0$$

□

According to the sign of R, we will have different behaviour patterns of the sequence of N_t.

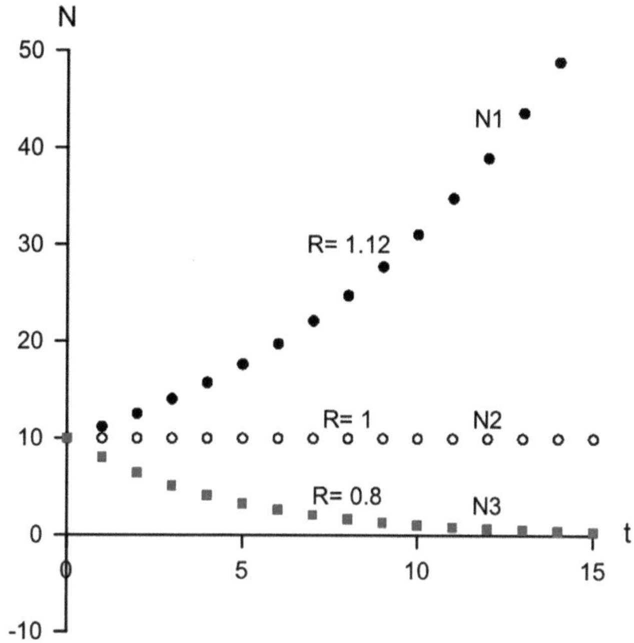

Fig. 13.2 Malthusian discrete model of a semelparous population: evolution of the number of individuals as a function of time. (1) $R > 1$ (*black bullets*), the population increases indefinitely. (2) $R = 1$ (*while bullets*), the population remains indefinitely stable at the level of the initial conditions. (3) $R < 1$ (*grey squares*), the population inexorably decreases towards its extinction

- If $R > 1$, N_t is indefinitely increasing from N_0: $N_0 < N_1 < N_2 < N_3 < \cdots$.
- If $R = 1$, N_t is constant and indefinitely equal to N_0.
- If $R < 1$, N_t is decreasing from N_0: $N_0 > N_1 > N_2 > N_3 > \cdots$: a process leading to the extinction of the population.

Figure 13.2 illustrates these behaviours.

Case 2: *Malthusian Discrete Model of an Iteroparous Population, No Generation Overlap, Pure Birth Process*

Axiom 13.2

(1) The population is iteroparous (i.e. at a given generation, the individual does not die after reproduction).
(2) At any generation, each individual reproduces.
(3) Each adult produces R offspring [same remark as Footnote 10].
(4) The evolution of the population is determined by internal causes: there exist no immigration and no emigration.
(5) This model is a pure birth process [42, p.8], the individuals do not die.
(6) The generations do not overlap.

An immediate consequence of this axiom is that $R > 0$. Such a population always increases.

Theorem 13.4 *If ΔN_t represents the increase of a population during one generation,*

$$\Delta N_t = RN_t \tag{13.8}$$

Proof At each generation t, the number of individuals is N_t. By point 3, the population at generation t produces RN_t offspring. Thus, at generation $t+1$, the number of offspring is $N_{t+1} - N_t = RN_t$, which is the increase of the population, i.e. ΔN_t. □

Theorem 13.5 *Let N_t represent the number of individuals of an iteroparous population at generation t, and R the number of offspring per adult. Hence*

$$N_{t+1} = rN_t, \ (\texttt{with}\ r = 1 + R) \tag{13.9}$$

Proof $\Delta N_t = N_{t+1} - N_t = RN_t \Rightarrow N_{t+1} = N_t + RN_t = \underbrace{(1+R)}_{r} N_t$ □

If we can fit the realised results of the process with an underlying theoretical curve, we will draw a tendency that can be described by variable *time*. Variable *generation* will just be a particular case of the latter. We can rewrite Eq. (13.8) as $\frac{\Delta N_t}{1} = RN_t$, and interpret that formula as the increase rate of the population during the lapse of time 1. Let Δt be an arbitrary lapse of time. By extension, we will state Axiom 13.3

Axiom 13.3 Let's consider Δt a lapse of time. The process at time (generation) t can be described, at the level of the underlying curve, by the *increase rate per unit of time*, that is:

$$\frac{\Delta N_t}{\Delta t} = \rho N_t \ (\text{where}\ \rho = \frac{R}{\Delta t}). \tag{13.10}$$

Definition 13.7 ρ is the variation rate per capita and unit of time, also called the *Malthusian constant.*

In this case, ρ is necessarily the increase rate per capita and unit time;thus $\rho > 0$.
Numerical application
Let $N_t = 50$ and $R = 1.4$, which means that at generation t, each individual generates 1.4 offspring, and $\Delta t = 5$. Let us consider the process during $\Delta t = 4.5$ generations, then the Malthusian constant will be $\rho = \frac{1.4}{4.5} = 0.333$, and the increase rate per time unit will be

$$\frac{\Delta N_t}{\Delta t} = 0.333 N_t$$

Case 3: *Malthusian Discrete Model of an Iteroparous Population, No Generation Overlap, Birth and Death Process*

Axiom 13.4

(1) The population is iteroparous.
(2) At any generation, each living individual reproduces.
(3) Each adult produces R offspring [same remark as Footnote 10].
(4) The evolution of the population is strictly internal. There exists no immigration, no emigration
(5) This model is a birth-death process [42, p.8]: the birth and death rate per capita and time unit is constant.
(6) The generations do not overlap.

Such a situation, where the parameter death is introduced, is hardly more complex than the birth-only model we treated above. A fundamental change is that, in this case, ρ can be positive, null or negative. Such a model can be written

$$\frac{\Delta N_t}{dt} = (b - d)N_t \tag{13.11}$$

where b is the number of births, and d is the number of deaths during a very short time-lapse δt. From Definition 13.7, we can see that ρ is the Malthusian constant, $\rho = b - d$. Equation (13.11) trivially shows that (1) when there are more births than deaths, the process is an exponential increase ($\rho > 0$), (2) when the numbers of births and deaths are equal, the population remains constant ($\rho = 0$) and (3) when there are fewer births than deaths, the process is an exponential decrease to extinction ($\rho < 0$).

Case 4: *Malthusian Continuous Model, Generation Overlap, Pure Birth Process*

Let N_t be the number of individuals at generation t (running from 0 to ∞) This growth model is based on the following axiom:

Axiom 13.5

(1) The population is iteroparous.
(2) At any generation, each individual reproduces.
(3) Each adult produces R offspring.
(4) This model is a pure birth process, i.e. there exists no immigration, no emigration, and no deaths.
(5) The generations continuously overlap. The process is described by a continuous function.

We should pose the following axiom as a natural extension of Axiom 13.3.

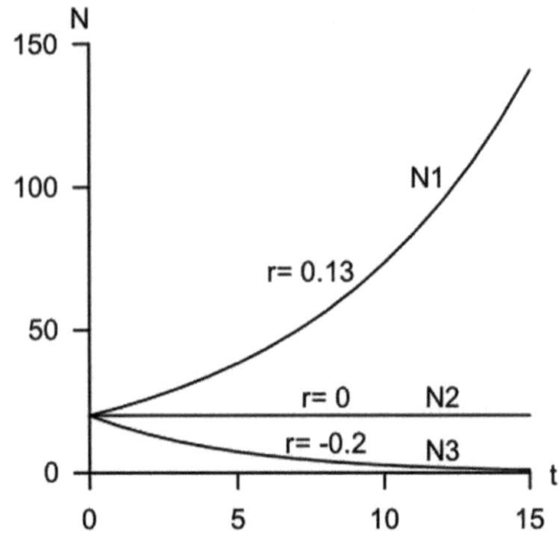

Fig. 13.3 Malthusian continuous model of a continuously iteroparous population, with continuously overlapping generations. Three populations: (1) P_1 with $N_1(t)$ exponentially increasing ($r > 0$), (2) P_2 with $N_2(t)$ constant ($r = 0$), (3) P_3 with $N_3(t)$ exponentially decreasing ($r < 0$) until extinction

Axiom 13.6

$$\frac{dN(t)}{dt} = r\,N(t) \tag{13.12}$$

Here, r is the variation rate per individual per time unit, exactly like the parameter ρ of the previous paragraph. $\frac{dN(t)}{dt}$ is the instantaneous variation rate.

Theorem 13.6 *Let P be a continuously iteroparous population following a Malthusian continuous model with continuous generation overlap, and $N(t)$ the number of individuals at time t. The initial condition of N is $N(0) = N_0$ Then,*

$$N(t) = N_0\,e^{rt} \tag{13.13}$$

Proof

$$\frac{dN}{dt} = rN \Rightarrow \frac{dN}{N} = rdt \Rightarrow \int_{N(0)}^{N(t)} \frac{dN}{N} = r\int_0^t dt = rt$$

$$\Rightarrow \ln\frac{N(t)}{N(0)} = rt \Rightarrow \frac{N(t)}{N(0)} = e^{rt} \Rightarrow N(t) = N_0 e^{rt}$$

□

Figure 13.3, hereunder, presents the graph of $N(t)$ with the three cases (different signs of the Malthusian constant): (1) $r > 0$: accelerating population increase without bounds, (2) $r = 0$: indefinite of the population, (3) $r < 0$: accelerating population decrease until extinction.

13.1.5.2 Verhulst (Logistic) Model

The Verhulst [57] model is the simplest density-dependent model. We will denote the density $N(t)$, i.e. the number of individuals and complexify the hypothesis of the Malthusian model presented in Axiom 13.5, preventing the catastrophic increase or decrease presented in the legend of Fig. 13.3, and pose Axiom 13.7.

Axiom 13.7

- The population is iteroparous, which means that it is ever-increasing..
- $N(t)$ is an increasing function of time t
- The birth and death rates, resp. $b(N)$ and $d(N)$ will not be constants anymore but increasing functions of time or density:
  - Variable N_t depends on the availability of resources, which is different in the initial stage than in later stages. Indeed, at the beginning of the process, the population has few individuals and plenty of available resources. As the process proceeds, the population increases, and the quantity of resources diminishes, which inevitably decreases births per time unit.
  - At the initial stage, when the population is limited, there is very little competition since there is food for each individual. Moreover, predators devote much time and energy to the search for prey among the population. In addition, epidemics have a reduced probability of affecting the population. As time passes, the population increases, and competition, predation/ parasitism, and epidemics have an increasing grip on it. Thus, $d(N)$ is expected to increase.
  - To let the population increase, the model must impose $b(N) > d(N)$.

  Verhulst proposes the simplest case; *b(N)* and *d(N)* are linear functions, thus also $g(N) = b(N) - d(N)$. This function is constructed so that it causes the continuous deceleration of $N(t)$, which tends to a supremum called carrier capacity, nearly reaching it and continuously approaching it without ever crossing it. Graphically, the function $N(t)$ is represented by an S curve (or logistic curve) and the carrier capacity by a horizontal asymptote representing the carrier capacity K.

Axiom 13.8 The Verhulst-Pearl logistic differential equation
It is a differential equation involving the population size N and time t:

$$\begin{aligned} f(N) &= \frac{dN(t)}{dt} = rN(t)\left(1 - \frac{N(t)}{K}\right) \\ &= N\underbrace{(r - sN)}_{g(N)} \text{ where } \begin{cases} r \text{ is the Malthusian constant} \\ K \text{ is the carrier capacity} \end{cases} \end{aligned} \tag{13.14}$$

The interpretation is as follows.

- When t is close to 0, $N(t)$ is negligible and $\frac{N}{K} \simeq 0$, and we see that $\frac{dN}{dt} = rN$ (hence the Malthusian equation). Let $\widehat{N} \in [0, K)$ the ordinate of an inflexion point where the concavity of $N(t)$ changes.
- When $N < \widehat{N}$, the function $f(N) = \frac{dN}{dt}$ is concave up, which corresponds to $f'(N) = \frac{d^2N}{dt^2} < 0$. On the contrary, when $N > \widehat{N}$, the function $f(N) = \frac{dN}{dt}$ is concave downwards, which corresponds to $f'(N) = \frac{d^2N}{dt^2} > 0$. Theorem 13.7 says that $\widehat{N} = \frac{K}{2}$.
- $\lim_{N \to K^-} f(N) = 0$, which means that function tends via lower values towards a horizontal asymptote (cf. Theorem 13.7).

Let us put it more formally.

Theorem 13.7 $\widehat{N} = \frac{K}{2}$ *and* $\lim_{N \to K^-} f(N) = 0$

Proof

- $\widehat{N} = \frac{K}{2}$

$$\frac{dN}{dt} = rN\left(1 - \frac{N}{K}\right) \Rightarrow \frac{d^2N}{dt^2} = r\Big[\underbrace{N(1 - \frac{N}{K})}_{U}\Big]'$$

$$U' = \frac{dN}{dt}\left(1 - \frac{N}{K}\right) + N\left(1 - \frac{N}{K}\right)' = \frac{dN}{dt}\left(1 - \frac{N}{K}\right) + N\left(-\frac{N}{K}\right)'$$
$$= \frac{dN}{dt}\left(1 - \frac{N}{K}\right) - \frac{N}{K}\frac{dN}{dt} = \frac{dN}{dt}\left(1 - \frac{2N}{K}\right) = rN\left(1 - \frac{N}{K}\right)\left(1 - \frac{2N}{K}\right).$$

Hence, $\frac{d^2N}{dt^2} = r^2N\left(1 - \frac{N}{K}\right)\left(1 - \frac{2N}{K}\right)$. This expression is equal to 0 at the inflexion point $(\widehat{t}, \widehat{N})$, which is theoretically possible for three values of $N(t)$:

$$\begin{cases} N = 0 & \text{A totally impossible a situation, no population} \\ & \quad \text{arises from no individuals} \\ 1 - \frac{N}{K} = 0 \Rightarrow N = K & \text{This case is impossible, K is never attained} \\ 1 - \frac{2N}{K} = 0 \Rightarrow N = \frac{K}{2} & \text{Thus, the only solution} \end{cases}$$

Hence,

$$\text{The ordinate of the inflexion point } \widehat{N} = \frac{K}{2} \tag{13.15}$$

- $\lim_{N \to K^-} f(N) = 0$

$$\lim_{N \to K^-} f(N) = \lim_{N \to K^-} r\left(N - \frac{N^2}{K}\right)$$

$$= r\left(\lim_{N \to K^-} N - \lim_{N \to K^-} \frac{N^2}{K}\right) \simeq r\left(K - \frac{K^{\not 2}}{\not K}\right)$$

$$= 0$$

□

We will now write the equation that delivers the process of the Verhulst model, the variable $N(t)$ as a function of t from the initial condition of the process, i.e. $N(0) = N_0$, the carrier capacity K and the Malthusian constant r.

Theorem 13.8 (Equation of the Verhulst Model) *Let $N(t)$ be the number of individuals of a population at time $t \in [0, \infty)$. The population follows a Verhulst logistic model, with as parameters: the carrier capacity K, and the increase rate r. The initial condition of the process is $N[0] = N_0$. Then, the variation of $N(t)$ is as follows:*

$$N(t) = \frac{K}{1 + \left(\frac{K}{N_0} - 1\right) e^{-rt}} \tag{13.16}$$

Proof The proof needs some mathematical proficiency and is left to the motivated reader [see Problem 13.3 in Quizzes and Problems (p. 398)]. □

We present hereunder the graph of the Verhulst model (cf. Fig. 13.4).

Theorem 13.9 *The inflexion point M has coordinates $(\widehat{t}, \widehat{N}) = (\frac{1}{r} \ln \frac{K - N_0}{N_0}, \frac{K}{2})$.*

Proof From Eqs. (13.8) and (13.15),

$$\widehat{N} = \frac{\not K}{1 + \left(\frac{K}{N_0} - 1\right) e^{-rt}} = \frac{\not K}{2}$$

$$\Rightarrow 1 + \left(\frac{K}{N_0} - 1\right) e^{-rt} = 2 \Rightarrow \left(\frac{K}{N_0} \quad 1\right) e^{-rt} = 1 \Rightarrow e^{-rt}$$

$$= \frac{1}{\frac{K}{N_0} - 1} = \frac{1}{\frac{K - N_0}{N_0}} (1) \Rightarrow e^{rt} = \frac{K - N_0}{N_0} \Rightarrow rt = \ln \frac{K - N_0}{N_0} \Rightarrow (2)$$

$$\widehat{t} = \frac{1}{r} \ln \frac{K - N_0}{N_0} (3) \tag{13.17}$$

□

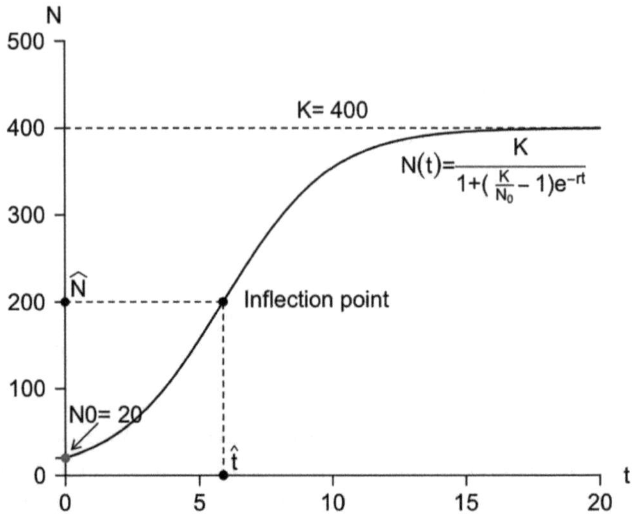

Fig. 13.4 Verhulst logistic model. The curve depicts a population increase where the death rate, negligible at the beginning of the process, increases as the population size $N(t)$ rises. Before the inflexion point $(\widehat{t}, \widehat{N})$, the curve rises almost like a Malthusian process and is concave up. As $N(t)$ increases with the passing time, and due to the rise in the death rate and decrease of birth rate, after the inflexion point $(\widehat{t}, \widehat{N})$, the curve is concave down and tends to the carrier capacity K

Numerical Application

Let P be a population with the following properties. (1) The initial population size N_0 is 30. (2) The carrier capacity K is 1000. (3) The time $\widehat{t}$, beyond which the rise per time unit decreases, is 200 in whatever unit, e.g. (day, month, year). Write the Verhulst model equation (to three significant figures), and give the population size at times 0, 10, 20, 50, 100, 250, 500, 750, 1000, 1500 and 2000 in the same unit (to 7 significant figures for $N(t)$s corresponding to $t \geq 750$, 3 significant figures for the other $N(t)$s).

Solution of the problem: from Theorem 13.9, $\widehat{t} = \frac{1}{r} \ln \frac{K - N_0}{N_0} = \frac{1}{r} \ln \frac{1000 - 30}{30}$ $= 200$. The unknown of this equation $r = 0.01694596$ (to 7 significant figures). Rounded to three significant figures, the Verhulst model equation is

$$N(t) = \frac{1000}{1 + \frac{1000}{30} e^{-0.016t}} = \frac{1000}{1 + 33.3e^{-0.016t}} \tag{13.18}$$

Table 13.1 shows the correspondence between time t and population size $N(t)$ of population P.

We have an illustration of the tendency of $N(t)$ towards $K = 1000$ via lower values when increasing t gets large.

Table 13.1 Correspondence between time $0 \leq t < \infty$ and number of individuals $N(t)$ of a population—Verhulst model

t	$N(t)$
0	30.0
10	35.5
20	41.9
50	68.7
100	150
250	705
500	995
750	999.9295
1000	999.9991
2000	1000^{-}

13.1.5.3 Gompertz (Logistic) Model

Another logistic model is the model of Gompertz [14]. It is based on the following hypothesis:

Axiom 13.9

(i) The population is iteroparous.
(ii) This model is a birth-death process.
(iii) The death rate increases exponentially as the population increases.
(iv) Generations overlap
(v) There is a carrier capacity: the population increases exponentially until a size threshold is reached; beyond this point, growth slows down and approaches a horizontal asymptote without crossing it..

Definition 13.8 (The Gompertz Differential Equation) Let $N(t)$ be the population size of a population P at time t, K the carrier capacity of the population, and r the increase rate of the population. Then

$$f(N) = \frac{dN}{dt} = rN \ln\left(\frac{K}{N}\right) \tag{13.19}$$

We see that when N represents a small population size (in the neighbourhood of 0 via positive values), then $N \ln\left(\frac{N}{K}\right) \simeq N$ since the polynomial growth wins over the logarithmic growth. Hence, at the beginning of the process, $N(t) \simeq r\,N(t)$ (approximation of Malthusian growth). On the contrary, if $N \to K$, then $N \ln\left(\frac{N}{K}\right) \simeq K$. In addition, f is an increasing function, and $N < K$. Indeed, by construction, $f(N$ is the product of positive terms.

Theorem 13.10

- $\widehat{N} = \dfrac{K}{e}$ *is the ordinate of the inflexion point.* $N(t)$ *is a concave up function when* $N < \dfrac{K}{e}$ *and concave down when* $N > \dfrac{K}{e}$.

Proof

- $$f'(N) = \left(rN\ln\frac{K}{N}\right)' = r\left(N\ln\frac{K}{N}\right)' = r\left[\ln\frac{K}{N} + N\frac{N}{K}\left(\frac{K}{N}\right)'\right]$$
$$= r\left[\ln\frac{K}{N} + \frac{N^2}{K}\left(-\frac{K}{N^2}\right)\right]$$
$$= r\left(\ln\frac{K}{N} - 1\right) = 0 \Rightarrow \ln\frac{K}{N} = 1 \Rightarrow \frac{K}{N} = e \Leftrightarrow$$

$$\widehat{N} = \frac{K}{e} \text{ Ordinate of the inflexion point.} \tag{13.20}$$

- $N < \dfrac{K}{e} \Leftrightarrow \dfrac{K}{N} < e \Leftrightarrow \ln\dfrac{K}{N} < 1 \Leftrightarrow \ln\dfrac{K}{N} - 1 = f'(N) = \dfrac{d^2N}{dt^2} < 0 \Leftrightarrow$ $N(t)$ is concave up.
- $N > \dfrac{K}{e} \Leftrightarrow \dfrac{K}{N} > e \Leftrightarrow \ln\dfrac{K}{N} > 1 \Leftrightarrow \ln\dfrac{K}{N} - 1 = f'(N) = \dfrac{d^2N}{dt^2} > 0 \Leftrightarrow$ $N(t)$ is concave down.

□

Theorem 13.11 (Gompertz Model of Population Increase) *Let* $N(t)$ *be the number of individuals of a population at time* $t \in [0, \infty)$. *The population follows a Gompertz logistic model, with as parameters: the carrier capacity* K, *and the increase rate* r. *The initial condition of the process is* $N[0] = N_0$. *Let us pose* $\gamma = \ln\left(\dfrac{N_0}{K}\right)$. *Then, the variation of* $N(t)$ *is as follows:*

$$N(t) = Ke^{\gamma e^{-rt}} \tag{13.21}$$

Proof We leave the proof to the reader [see Problem 13.4 in Quizzes and Problems p. (397)].

□

Theorem 13.12 *The inflexion point* M *has coordinates* $\left(\dfrac{1}{r}\ln\left(\ln\dfrac{K}{N_0}\right), \dfrac{K}{e}\right)$.

Proof From Theorems 13.10 and 13.11,

$$N(\widehat{t}) = \widehat{N} = Ke^{\gamma e^{-r\widehat{t}}} \Rightarrow \frac{K}{e} = Ke^{\gamma e^{-r\widehat{t}}}$$

$$\Rightarrow \ln\frac{1}{e} = -\ln e = -1 = \gamma e^{-r\widehat{t}} \Rightarrow e^{-r\widehat{t}} = -\frac{1}{\gamma} \Rightarrow -r\widehat{t}$$

Table 13.2 Correspondence between time $0 \leq t < \infty$ and number of individuals $N(t)$ of a population—Gompertz model

t	$N(t)$
0	30.0
10	154
20	368
50	858
100	993
250	999.9995
500	1000^-
750	1000^-
1000	1000^-
2000	1000^-

$$= \ln\left(-\frac{1}{\gamma}\right) = -\ln(-\gamma)$$

$$\Rightarrow \widehat{t} = \frac{1}{r}\ln(-\gamma) = \frac{1}{r}\ln(-\ln\frac{N_0}{K}) = \frac{1}{r}\ln\left(\ln\frac{K}{N_0}\right)$$

□

Numerical Application

Let P be a population with the following properties. (1) The initial population size N_0 is 30. (2) The carrier capacity K is 1000. (3) The time $\widehat{t}$, beyond which the rise per time unit decreases, is 50 in whatever unit (day, month, year). Write the Gompertz model equation (to three significant figures), and give the population size at times 0, 10, 20, 50, 100, 250, 500, 750, 1000, 1500, and 2000 in the same unit (to 7 significant figures for $N(t) \geq 999$, 3 significant figures for the other $N(t)$s).

Solution of the problem: from Theorems 13.11 and 13.10, $\widehat{t} = \frac{1}{r}\ln\left(\ln\frac{K}{N_0}\right) = \frac{1}{r}\ln\frac{100-30}{30} = 200$. The unknown of this equation $r = 0.06273175$ (to 7 significant figures). Rounded to three significant figures, the Gompertz model equation is

$$\begin{cases} \gamma = \ln\frac{30}{1000} = -3.51 \\ N(t) = 1000 * e^{-3.51\, e^{(-0.063\, t))}} \end{cases} \tag{13.22}$$

Hereunder is a table with the correspondence between time t and population size $N(t)$ of population P (Table 13.2).

We have an illustration of the tendency of $N(t)$ towards $K = 1000$ via lower values when t gets increasingly large. We see that with the Gompertz model, the

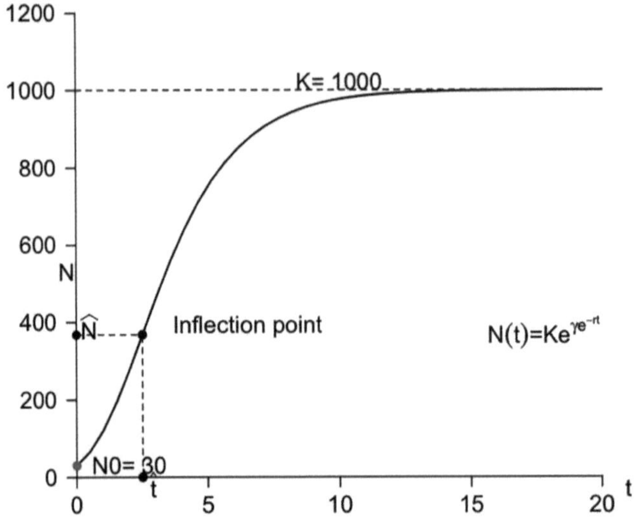

Fig. 13.5 Gompertz logistic model. Mathematically, function $N(t)$ is valid for $-\infty < t < \infty$, described by a sigmoid curve similar to Verhulst's. However, only the restriction of N to the domain $[0, \infty)$ has a biological sense. The curve is similar to the Verhulst's, but with a shorter initial concave up portion, a longer concave down portion tending to the carrier capacity K and an inflexion point $(\widehat{t}, \widehat{N})$ further from K (Theorem 13.10), thus closer to N_0

population reaches its carrier capacity faster. Hereunder, we present the graph of the Gompertz model for the data exemplified here (Fig. 13.5).

The Gompertz model and eight modernised variants have successfully been used to describe and predict the growth of tumour cells, bacteria counts, and fish and crustacean populations [55].

13.1.5.4 The Robustness of the Logistic Model Presented Above Despite Its Strong Assumptions: An Example of a Slightly More Realistic Model

The models mentioned above are robust and approximately verified in field ecology and experimental studies, even though they are based on strong assumptions. Pielou mentioned six of them [42, p. 35]:

1. The abiotic factors are evenly distributed within the population.
2. At a given time t, the population variation rate is unaffected by density and the spatial relationship between the individuals.
3. A given density instantly gives birth and death rates without lag or fluctuation
4. Population growth is density-dependent throughout the entire process.
5. The age distribution is stable.
6. In a sexually reproducing population, all females find mates even when the population size is low.

Each of these assumptions is approximately valid. Therefore, more elaborate models derived from the logistic functions have been described in a rich body of literature cited by Pielou [42] and Murray [35], with a body of increasingly realistic assumptions. Pielou cited the example of an article by Frederick E Smith [52] whose study object was an experimental population of a Daphnia species, *Daphnia magna*.[12] This author took into consideration the fact that the supply rate of food is higher in growing than in saturated populations, which is not accounted for in basic logistic models like those mentioned above.

This article derives his proposal from the Verhulst-Pearl differential equation (see Axiom 13.8), i.e.

$$\underbrace{\frac{1}{N}\frac{dN}{dt}}_{\text{Growth rate per individual}} = \nu = f(N) = r\frac{K-N}{K} = -\frac{N}{K} + r$$

This equation shows that the growth rate per individual decreases linearly with increasing N in the Verhulst model, which was not the case with the population of Daphnia that Smith studied. He discovered experimentally that ν was not a linear function but a hyperbolic one of N. He dropped the idea of calculating $N(t)$. He proposed instead a differential equation as follows:

$$\frac{dF}{dt} = r\left(1 - \frac{F}{T}\right) \tag{13.23}$$

where F and T are, respectively, the rates at which the population uses food (1) at instant t during growth on the one hand, and (2) at saturation (when it approaches its carrier capacity) on the other hand. From this equation, he derived a relationship where the unknown of the differential equation describing his Daphnia population was not density anymore but mass. Pielou [42, p. 39] made mathematical reasoning from Eq. (13.23), which looked more convincing. She equated this equation with the following:

$$\frac{1}{N}\frac{dN}{dt} = r\,\frac{T-F}{T} \tag{13.24}$$

Then, she posed that $\exists\, c_1, c_2 \in \mathbb{R}^+ \;\middle|\; F = c_1 N + c_2\frac{dN}{dt}$ from the argument that when a population grows, at the instant t, the availability of food to the population depends on the $N(t)$ adults present and the $\frac{dN}{dt}$ incoming offspring. When the population gets saturated, $\frac{dN}{dt} \simeq 0$. Thus,

[12] We already studied another species belonging to this genus of planktonic crustacean (see p. 346).

$$T = c_1 K \tag{13.25}$$

Theorem 13.13 *If we pose* $\alpha = \dfrac{c_1}{c_2}$, *then* $\dfrac{1}{N}\dfrac{dN}{dt} = r\dfrac{K-N}{K+\frac{r}{\alpha}N}$

Proof By replacing T and F in (13.23) with their values calculated in (13.24) and (13.25),

$$\frac{1}{N}\frac{dN}{dt} = r\frac{c_1 K - c_1 N - c_2\frac{dN}{dt}}{c_1 K} = r\frac{\alpha K - \alpha N - \frac{dN}{dt}}{\alpha K}$$

$$\Rightarrow \left(\frac{1}{N} + \frac{r}{\alpha K}\right)\frac{dN}{dt} = r\cancel{\alpha}\frac{K-N}{\cancel{\alpha}K} \Rightarrow \frac{\alpha K + rN}{N\alpha K}\frac{dN}{dt} = r\frac{K-N}{K}$$

$$\Rightarrow \frac{1}{N}\frac{dN}{dt} = \frac{\alpha\cancel{K}r(K-N)}{(\alpha K + rN)\cancel{K}} = \frac{\alpha r(K-N)}{\alpha K + rN} = r\frac{K-N}{K+\dfrac{r}{\alpha}N}$$

□

13.1.6 *Spatial Organisation of Populations*

We will focus on animal populations, which are by far the most varied and complex structures. Populations are organised as a set of social units that are structured in many ways. Let us consider the case of sexually reproducing animals.

- A number of populations are structured as a collection of colonies, e.g. (1) several alpha males with a harem (lions, ruffs), or (2) groups of agglutinated promiscuous individuals (gannets). Every degree of hierarchy is observed, from the strictest to the loosest, as far as the access to reproduction or feeding is concerned, i.e. from extreme dominance entailing strong inequality between the individuals to the quasi-equality between the group members.
  - If the social relationships are based on dominance and unequal status of the individuals, intraspecific competition between the members of the demonstrative sex (generally the males) for access to members of the other sex (generally the females) for reproduction will first tend to be strong, followed by the relative stability of the reproductive units once the rule of dominance established and only a few individuals of the demonstrative sex selected for breeding. In such a system, there is a tendency towards substantial sexual dimorphism, with a difference between male and female reproduction investment [3]. Different models show that when the ecological niche of a species is narrow or when the individuals are members of small groups, there is a trend of high competition between individuals of the demonstrative sex, generally the male, and correlatively of morphological difference between the sexes [51].

  - If, on the contrary, the species is promiscuous, the intraspecific competition for reproduction will be weak, the status of individuals will be quasi-equal, and the social units where reproduction takes place will be relatively unstable [25].

- In some insect species, generally among haplodiploids like the honey bee or most ant species, coloniality has taken an extreme form. Populations are colonies made up of (1) a large number of diploid females (up to 60,000)—generally only one or a very few of long-lived reproductive individuals (the queens), and many relatively short-lived and sexually undeveloped females composing the backbone of the colony and core of its social life—and (2) a few very short-lived males (counting up to several hundred) that are haploid (in the case of the Hymenoptera) and strictly limited to reproduction, playing thus no role in the social life of the colony. Females practice polyandry and multi-mating with males [6].
- Other populations are territorial: during the reproduction period, the social units consist of pairs, generally monogamous or small promiscuous groups with few animals of both sexes and their eventual progeny. These units exclude one another and constitute territories that are spatially contiguous and not or hardly overlapping. In some cases, the animals are solitary except during the reproduction period when they become territorial and monogamous (cf. robins [27]); in other cases, they are colonial outside the reproduction period and territorial within it (e.g. many migrating passerines).

  If the species is territorial, the individuals, generally monogamous, undergo moderate competition for sexual reproduction. Most individuals are paired, though several individuals can remain unpaired. Courtship displays, sometimes very conspicuous to the observer, lead to the pairing. The competition for food resources is limited once the spatial equilibrium of the breeding territories is achieved. This situation is attained after harsh territorial competition, with a level of aggressiveness depending on environmental factors. (In cichlids, for example, within a pair, females tend to be more aggressive than males against an intruder when the quality of the territory increases [28].) Cooperative behaviours are frequent within the pairs but infrequent outside since the social units are relatively independent [32].

The social structure and behaviours within populations in a species can vary according to “traditions”, as observed in bird songs that can geographically characterise populations. This fact has been observed in many publications on the Chaffinch (*Fringilla coelebs*) [49]. In this species, every population is characterised by a particular accent specific to a certain region. The list of the different spatial organisations of populations mentioned above is far from exhaustive and casts a summary picture of the social structures observed in the animal kingdom.

13.2 The Communities

In a given spatiotemporal unit, a population is not isolated but inserted in a group of populations that share the same area and with which they can potentially interact. This group of populations is called a community. A community is a part or the entirety of a biocoenosis. Let us consider several interacting populations belonging to the same community. They belong to the same species (i.e. conspecific populations), or to different species (i.e. heterospecific populations).

Here, we present mathematical and descriptive models of communities. The possible interactions (e.g. predation, parasitism, all types of coevolution, competition, cf. Sect. 4.2) will lead to a dynamic and relatively fragile equilibrium, tributary to extrinsic forces that sometimes can be destructive. For simplicity, we will adopt the hypothesis of parsimony: the description of a constant phenomenon with simplified assumptions. Thus, the following models are theoretical and must be considered critically, but if handled carefully, they can be useful. We will proceed with two mathematical models, the linear and the Lotka-Volterra models,[13] briefly evoke other models, and proceed afterwards with descriptive, verbose models.

13.2.1 The Linear Model

Let $N_1(t)$ be the size of population P_1 and $N_2(t)$ be the size of population P_2. The initial conditions on the number of individuals in P_1 and P_2 are, respectively, $N_1(0) = N_{10}$ and $N_2(0) = N_{20}$.

Axiom 13.10 The linear model involving these variables is described by the following differential equation system

$$\begin{cases} \dfrac{dN_1}{dt}(t) = a_{11}N_1(t) + a_{12}N_2(t) \\ \dfrac{dN_2}{dt}(t) = a_{21}N_1(t) + a_{22}N_2(t) \end{cases} \tag{13.26}$$

whose solutions will be $N_1^*(t)$ and $N_2^*(t)$.

Remark 13.1 Description of the coefficients of the system (13.26).

A coefficient $a_{ij} > 0$ is a positive contribution of P_i to the $\int$ size variation of population P_j. Conversely, a coefficient $a_{ij} < 0$ is negative contribution of P_i to the size variation of population P_j. Different cases occur:

Case 1 $a_{12} = 0$ and $a_{21} = 0$: We are in the simplest case. The system describes the absence of any interaction. Each population develops independently.

[13] These two types of mathematical model will describe communities of three agents, two animal populations and a dummy, the "vegetal" resources.

Case 2 $a_{12} > 0$ and $a_{21} < 0$: The system describes a predator-prey or parasite-host interaction in the situation where population P_1 is the prey (or host) and population P_2 is the predator (parasite).

Case 3 $a_{12} < 0$ and $a_{21} < 0$: The system describes a competition interaction between P_1 and P_2. Both species impair the other species' fitness (on the reproductive efficiency or life conditions).

Case 4 $a_{12} \geq 0$ and $a_{21} \geq 0$: The system describes a coevolutionary interaction between P_1 and P_2. Here, we have two subcases:

- If $a_{12} > 0$ and $a_{21} = 0$, the system describes an interaction called commensalism, where P_1 benefits P_2, but, in return, P_2 obtains neither benefit nor disadvantage from P_1.
- If $a_{12} > 0$ and $a_{21} > 0$, the coevolutionary interaction between P_1 and P_2 is beneficial mutualism.

This model is the simplest, but it is very reductionist. It assumes that the resources are unlimited and permanently equally distributed to each of the two interacting populations, which is a strong hypothesis. In addition, a very lengthy mathematical development shows that the theoretical population sizes N_is increase or decrease exponentially, reaching rapidly enormous numbers, in the order of the Avogadro's number with a positive or negative sign in front of it, or decline to extinction within few generations. Only at the very beginning of the process do we find solutions having a biological sense, the N_i^*s being positive and within the order of actual populations (see Appendix E in SI, via link p. 343).

To illustrate this, we present a concrete example of a linear model describing a prey-predator relationship:

$$\begin{cases} \dfrac{dN_1}{dt} = 3N_1 - 4N_2 \\ \dfrac{dN_2}{dt} = 2N_1 + 9N_2 \end{cases} \tag{13.27}$$

We are clearly in case 2 of Remark 13.1. The evolution of the theoretical population sizes as a function of t is depicted in Fig. 13.6.

In Fig. 13.7, the graph displays a curve representing the correspondence between the population sizes N_1 and N_2. Here, time does not act as a variable but as an implicit parameter associated with each point of the curve. Such a graph is called a *phase plane*. This curve is directional, its orientation determined by chevrons and successive t-bullets, i.e. bullets labelled with the phrases "t = 0.09", "t = 0.19", etc.

From the graph, we can see that such a community is extremely short-lived, it can exist only within less than a quarter of a generation ($0.22 \leq t \leq 0.23$); the prey's population is already extremely negative after one generation ($N_1(1) = -604, 388$), and we rapidly reach a population expanding exponentially while they have nothing to eat ($N_2(1) = 686, 016,\ N_2(2) = 829, 708, 443,\ N_2(3) = 921, 373, 054, 583$).

Thus, systems of linear differential equations are not appropriate for describing and predicting the mutual impact of two or more interacting populations. Models

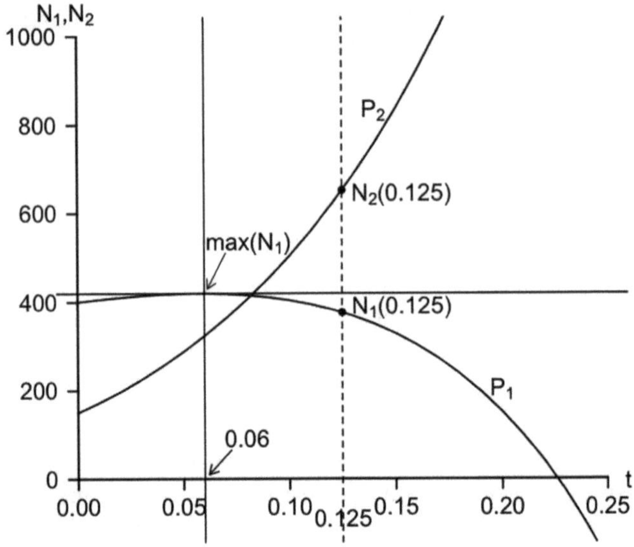

Fig. 13.6 Linear model: Evolution of the theoretical sizes of two populations involved in a trophic relationship. N_1 = population size of prey population (P_1) and N_2 = population size of predator population (P_2) feeding on P_1 at time t

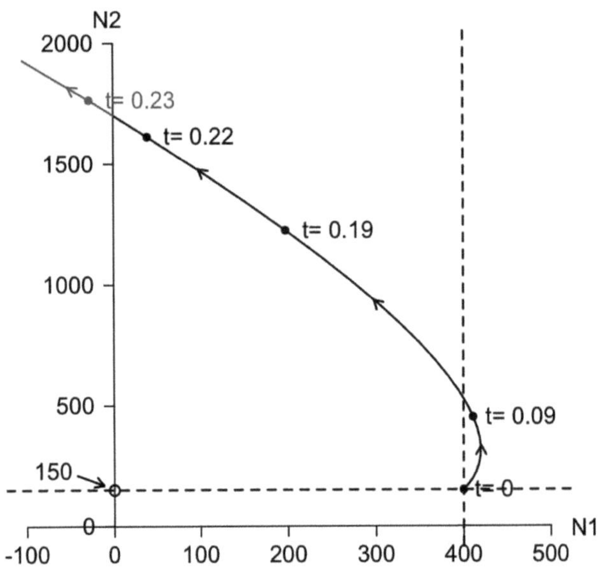

Fig. 13.7 Linear model—Phase plane of populations P_1 (prey) and P_2 (predators) N_1= population size of the prey population (P_1), N_2= population size of the predator population (P_2) that feeds exclusively on P_1. The implicit times correspond to t-bullets. The *black curve* corresponds to $N_1(t) > 0$, the only portion that could make biological sense. The rest of the curve appears in *grey*; the chevrons on the curve mark that the curve is directional

that are good for that are necessarily based on differential equations of order greater than or equal to 2 that can unravel limit cycles, i.e. closed lines strictly comprised in the quadrant one of a phase plane. The difficulty with such a system is the greatest difficulty to resolve them by analytical means, and we will have to apply computer-assisted numerical tools.

13.2.2 The Lotka-Volterra Model

The model of Lotka-Volterra [30, 58] is the oldest and simplest one acknowledged by the scientific community to describe a trophic [21, 42, 48] or competitive [26] relationship between two interactive heterospecific populations P_1 and P_2, which proposes limit cycles in the quadrant one of the phase plane. Indeed, this type of curve estimates the size of one population, knowing the size of the other. The model is based on the following hypothesis:

Axiom 13.11

(1) The vegetal resource on which the prey population P_1 is limited though perennial, in that it reproduces steadily. It can sustain P_1 if this population is "correctly" controlled by the predator population P_2.
(2) Population P_1 is the only food available to population P_2. The latter is consumed at a constant rate. There is no depletion or excess of food.
(3) The variation rate of each population is proportional to its size.
(4) The pressure of the environment on each population is identical.
(5) There is no migration in and out of the geographical unit where the interaction occurs. Birth is the only entry into the community, and death is the only exit from it.

This hypothesis is strong, but the model has proven robust. The following axiom settles the differential equation system.

Axiom 13.12 The two variables depicting the size of populations P_1 and P_2 are, respectively, $x(t)$ and $y(t)$. The equation system of the model, if we implicit time t, is:

$$\begin{cases} \dfrac{dx}{dt} = f(x, y) = \alpha x - \beta xy \\ \dfrac{dy}{dt} = g(x, y) = \gamma xy - \delta y \end{cases} \tag{13.28}$$

The differential equation system (13.28) of the 2nd order comprises two multiplied variables, x and y. It cannot be solved analytically like the linear model already studied but by an indirect numerical owed to the French mathematician Henri Poincarré (1854–1912) [44, 45]. Theorem 13.14 displays the equation of the closed curves representing the relation of the $x(t)$ and $y(t)$

Theorem 13.14 *We keep the variables and parameters of the relationship* (13.28).

$$\delta \ln x - \gamma x + \alpha \ln y - \beta y = C \text{ where constant } C \text{ determines each curve} \quad (13.29)$$

This equation is represented by a closed curve.

Proof

$$(13.28) \Rightarrow \frac{dx}{dy} = \frac{x(\alpha - \beta y)}{y(\gamma x - \delta)} \Rightarrow \frac{dx}{x} = \frac{\alpha - \beta y}{\gamma x - \delta}\frac{dy}{y}$$

$$\Rightarrow (\gamma x - \delta)\frac{dx}{x} = (\alpha - \beta y)\frac{dy}{y}$$

$$\Rightarrow -\delta\frac{dx}{x} + \gamma \not{x}\frac{dx}{\not{x}} = \alpha\frac{dy}{y} - \beta \not{y}\frac{dy}{\not{y}} \Rightarrow -\int\left(\frac{\delta}{x} - \gamma\right)dx = \int\left(\frac{\alpha}{y} - \beta\right)dy$$

$$\Rightarrow -\int \frac{\delta}{x}dx + \int \gamma dx = \int \frac{\alpha}{y}dy - \int \beta dy$$

$$\Rightarrow -\delta\int \frac{dx}{x} + \gamma\int dx = \alpha\int\frac{dy}{y} - \beta\int dy \Rightarrow -\delta \ln x + \gamma x = \alpha \ln y - \beta y + C$$

□

The resolution of the equation system (13.28) is graphical, i.e. its representation in the quadrant one of a phase plane. It consists of (1) estimating the two static equilibrium points (or fixed points) of the Lotka-Volterra equation systems, the null M_1 and non-null M_2 ones, defined as the points where the rates of population variation, i.e. $f(x, y)$ and $g(x, y)$, are equal to 0, and (2) constructing trajectories (oriented closed curves) around M_2.

Point M_2 cannot belong to any trajectory for the following reasons.

1. The predators (members of the P_2 population) eat individuals of the prey population, thus necessarily reducing the size of the P_1 population, a reduction compensated by the birth dynamic of the prey.
2. On the other hand, when too much prey is consumed, less food is available to the P_2 members, causing starvation or fewer births among the predators, thus a diminution of P_2, and, therefore, more chance for prey to escape from the teeth of the carnivores.
3. When the number of P_1 members rises, there is more food for P_2 members, favouring an increase in the number of predators.

A trajectory is thus a dynamical and oscillatory equilibrium resulting from a small perturbation from M_2. Close to this fixed point, a trajectory tends to be an ellipse. The more a trajectory deviates from M_2, the more oval it tends to be and the closer it will get to the x- or y-axis. Both populations will be extinct if a trajectory touches one of these axes. We will now proceed to the resolution itself. An important concept is that of isocline.

Definition 13.9 (Isoclines) Isoclines are straight lines in the phase plane that cancel the variation rates of each population. There are two isoclines per population size.

- x-isoclines are the two straight lines cancelling $f(x, y)$;
- y-isoclines are the two straight lines cancelling $g(x, y)$.

Theorem 13.15

- *The x-isoclines are* $x = 0$ *(the y-axis) and* $y = \frac{\alpha}{\beta}$*, which correspond to the red axes in Fig. 13.8.*
- *The y-isoclines are* $y = 0$ *(the y-axis) and* $x = \frac{\gamma}{\delta}$*, which correspond to the blue axes in Fig. 13.8.*

Proof From the equation system (13.28), we infer:

- $f(x, y) = x(\alpha - \beta y) = 0 \Rightarrow \begin{cases} x = 0 & \text{(the y-axis)} \\ y = \frac{\alpha}{\beta} \end{cases}$
- $g(x, y) = y(\gamma x - \delta = 0 \Rightarrow \begin{cases} y = 0 & \text{(the x-axis)} \\ y = \frac{\gamma}{\delta} \end{cases}$

□

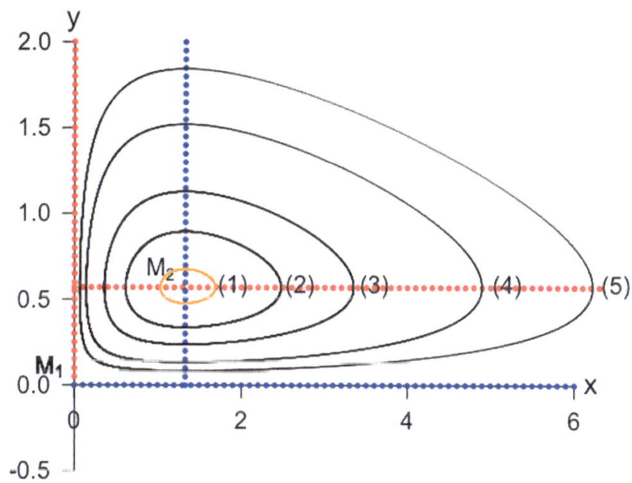

Fig. 13.8 Lotka-Volterra model represented in the phase plane of $(x, y.)$ The *orange* trajectory labelled (1), closest to M_2, is an ellipse. The further the closed curves (2) towards (5), the less elliptical and the more *egg-shaped* they are. The x-isocline passing through M_2 intersects the trajectories where $\frac{dx}{dt} = 0$, and the y-isocline passing through M_2 intersects the trajectories where $\frac{dy}{dt} = 0$. The *curves* are directional. By convention, the sense of the movement is anticlockwise

Theorem 13.16 *The Lotka-Volterra system has two fixed points corresponding to* $f(x, y) = g(x, y) = 0$, $\mathrm{M}_1 = (0, 0)$ *and* $\mathrm{M}_2 = (\frac{\gamma}{\delta}, \frac{\alpha}{\beta})$.

Proof There are four ways of associating an x-isocline with a y-isocline, among which only two meet in an intersection point corresponding to a fixed point; hence, we have two fixed points: $\mathrm{M}_1 = (0, 0)$ and $\mathrm{M}_2 = (\frac{\gamma}{\delta}, \frac{\alpha}{\beta})$. □

Remark 13.2 Only M_2 can be circumscribed by closed curves.

Now, we will devise an algorithm constructing several closed curves. Let point M be at a given distance from the fixed point M_2. This point can dynamically be considered a perturbation of the couple (x, y) from the static equilibrium, triggering the oscillation of x and y following the Lotka-Volterra model. For simplicity, let us deviate M from M_2 with the same deviation p. We will have thus obtain

$$\mathrm{M} = \Bigg(\underbrace{\frac{\gamma}{\delta} - p}_{x}, \underbrace{\frac{\alpha}{\beta} - p}_{y}\Bigg). \tag{13.30}$$

We will join to M, $\mathrm{M}^{(1)} = \mathrm{M} + (\mathrm{f(x, y)}\Delta \mathrm{x}, \mathrm{g(x, y)}\Delta \mathrm{y})$, a point adjacent to M in the trajectory about to be traced. For simplicity, we pose that $\Delta x = \Delta y = 0.001$,[14] and

$$\mathrm{M}^{(1)} = \mathrm{M} + (\mathrm{f(x, y)} \times 0.001, \mathrm{g(x, y)} \times 0.001) = (\mathrm{x}^{(1)}, \mathrm{y}^{(1)}).$$

We will then action two nested loops: Loop 2 is included within Loop 1.

- Loop 1 constructs the closed curves at different distances, from the closest to the furthest relatively to M_2. It is controled by the variable p of the equation system (13.30).
- Loop 2 constructs the trajectory for a given p. At a given value of t, from 0 to N (a large number):
  - to (x, y), we join $(x^{(1)}, y^{(1)})$,
  - we affect the content of $(x^{(1)}, y^{(1)})$ to (x, y),
  - we proceed to the next t until N is reached.

13.2.2.1 Numerical Application

$$\begin{cases} \dfrac{dx}{dt} = f(x, y) = 4x - 7 \\ \dfrac{dy}{dt} = g(x, y) = \dfrac{3}{2}xy + 2y \end{cases} \tag{13.31}$$

[14] We approach an infinitesimal dx and dy.

I used the following R-code to construct several trajectories (closed curves oriented anticlockwise)

Program Code-2

```
rm(list=ls())
alpha «- 4; betaa <- 7; gammaa «- 1.5; delta «-2 # Global variable assignment
f <- function(x,y) {val <- alpha*x-beta*y; val } # 1st Lotka-Volterra (LV) equation
g <- function(x,y) {val <- gammaa*x*y-beta*y; val }# 2nd LV equation
N <- 44; N1 <- 128 # lengths of the horizontal and vertical isoclines
# Construction of the isoclines
isox1 <- rep(0, N); isoy1 <- rep(0, N1) # x-isoclines
isox2 <- rep(alpha/betaa, N1); isoy2 <- rep(delta/gammaa, N) # y-isoclines
# Construction of an empty plot plane (type='n'), default axes removed (axes=F)
# absissae extending from -0.5 to 7 (xlim), and ordinates from -0.5 to 2 (ylim)
plot(x, y, type='n', axes=F, xlab="", ylab='', xlim=c(-0.5, 7), ylim=c(-0.5,2))
# Position and characteristics of the axes constructed by the user, two next lines
axis(side=1, pos=0) # side = 1 meaning x-axis and pos= 0, passing through abscissa
0
axis(side=2, pos=0) # side = 2 meaning y-axis and pos= 0, passing through ordinate
0
# Plotting of the x-isoclines, red and pearly appearance (pch=20), two next lines
points(isox1[1:N-3], seq(0,2,0.05), col='red', pch=20) # x-isocline confused with
the y-axis
points(seq(0,2,0.05),isox2[1:N2/2],col='red',pch=20) # x-isocline \\ to x-axis pass-
ing through M2
# Plotting of the y-isoclines, blue and pearly appearance (pch=20), two next lines
points(isoy2[1:N-3],seq(0,2,0.05),col='blue',pch=20) # y-isocline \\ to y-axis pass-
ing through M2
points(seq(0,2,0.05),isox2[1:N2/2-3],col='red',pch=20) # y-isocline confused with
the x-axis
X1 <- c(isox1[1], isoy1[1]); X2 <- c(isoy2[1], isox1[2]) # Construction of the two
fixed points
text(X1[1]-0.3, X1[2]+0.1, expression(M[1]) # Naming and labelling of the first
fixed point, i.e. M1
text(X2[1]-0.3,X2[2]+0.1,expression(M[2]) # Naming, labelling of the second fixed
point, i.e. M2
# Graphing of trajectories around M1 at increasing distances from M2, two nested
loops 1 and 2.
for(step in c(0.1,0.3,0.5,0.8,1)) {#Loop 1
Step=distance of initial point of trajectory from M2
x <- delta/gammaa-step; y <-alpha/betaa +step # Coordinates of the initial point of
trajectory
z <- ' ' # Colour of the trajectory
```

```
NN <- 7274 # NN large; the larger NN, the closer the adjacent points forming the trajectory is
for(t in 0:NN) { #Loop 2
#t = rank of the processed point constructing the trajectory
if(step==0.1) {z <- 'orange2'} else {z <- black} # Curve closest to M1 orange, the other ones black. points(x,y,pch=20,cex=0.05,col=z) # Point of trajectory tiny (cex) and round (pch) for a linear aspect
x <- x + f(x,y)*0.001; y <- y + g(x,y)*0.001 # For turning loop 2
} # End of loop 2
} # End of loop 1
mtext('x', side=1, line=-9.8, at=6.2, cex=1.4) # Labelling of x-axis
mtext('y', side=2, line=-3, at=2.15, cex=1.4, las=2) # Labelling of y-axis
```

One of the results of this program is Fig. 13.8. Though based on strong hypotheses, the Lotka-Volterra model revealed robustness and could explain many situations, such as the trophic relationship between sardine and shark populations in the Adriatic Sea in the early 1920s. After the four years of World War I, during which the Italian fisheries were at a standstill, the Italian zoologist Umberto d'Ancona (1896–1964) unexpectedly observed a significant decline in the sardine stocks and, conversely, a surprising rise in the biomass of sharks, their predators [7, 8]. These observations led Vito Volterra (1860–1940) to propose his model, later called the Lotka-Volterra model, which was proposed after these paradoxical observations [58, 59].

Other examples of frequently cited trophic interactions, which are well-explained by the Lotka-Volterra, are the relationship (1) between the Snowshoe hare *Lepus americanus* and the Canada lynx *Lynx canadensis* already mentioned in Chap. 2 or (2) between the grey wolf *Canis lupus* and the moose *Alces alces* [20].

Though the Lotka-Volterra model has often been successfully applied, some trophic relationships do not fit it. The relationship between a host (e.g., an aphid) and a parasitoid (e.g., a wasp) is such a case. The Nichoson-Bailey model has been proposed to describe this.

13.2.3 The Nicholson-Bailey Model

The Nicholson-Bailey pair of difference equations (see Axiom 13.14) is the classical model successfully describing a trophic phenomenon intermediate between parasitism and predation, *parasitoidism*. In such a trophic interaction, the parasitoid, like, for example, a female wasp, lays her eggs into the body of its host, e.g. a small insect like an aphid. The larvae hatching out of these eggs grow within the tissue of the unfortunate host, feed on it, and eventually kill it, bore the corpse, and free themselves from it. This type of relationship is thus lethal to the host after a long agony. Nearly all the parasitoids and most of their victims are insects, though other arthropods and molluscs can also be victims of parasitoids. Over half of them are hymenopterans (sawflies, ants, bees and wasps) [13].

Axiom 13.13 Hypothesis of the Nicholson-Bailey model:

(1) Parasitoids search hosts randomly.
(2) Parasitoids and hosts are initially independently distributed.

Axiom 13.14 Difference equation describing Nicholson-Bailey model
Since the model is based on difference equations, the population sizes are discrete magnitudes. Let H_t be the host population size at time t and P_t be the parasitoid population size.

$$\begin{cases} H_{t+1} = RH_t e^{-aP_t} \\ P_{t+1} = cH_t(1 - e^{-aP_t}) \end{cases} \tag{13.32}$$

where

- R is the reproduction rate of hosts,
- a is the area of discovery, i.e. a parameter assessing the capacity of a female parasitoid to find a host[15] and
- c is the average number of viable eggs that a parasitoid oviposits on one host.

The parameters k, a and c are nonnegative.

Remark 13.3 The expression e^{-aP_t} in the first equation is the probability that the host will survive P_t predators; thus, the expression $1 - e^{-aP_t}$ is the probability that the host will not survive P_t predators [29].

A more recent version of Nicholson and Bailey's model [37], dating back to 1935, was proposed at the end of the 1960s by Hassell and Varley [17], which appeared to explain a great number of cases of parasitoidism. This model, more predictive, has been successfully applied in biological controls of pests, essentially phytophagous insects. The Hassel-Varley modification of Nicholson and Bailey's model is based on the hypothesis of Axiom 13.13 and two added assumptions:

- $R \geq 1$: At each generation, the host generates at least one offspring,
- The search area per parasitoid is inversely proportional to $\sqrt{P_t}$

[15] This parameter is a measure describing the searching efficiency of the parasitoid.

Axiom 13.15 Modified Nicholson-Bailey model

$$\begin{cases} f(H_t, P_t) = H_{t+1} = RH_t e^{-a\sqrt{P_t}} \\ g(H_t, P_t) = P_{t+1} = H_t(1 - e^{-a\sqrt{P_t}}) \end{cases} \tag{13.33}$$

The ecologists are only interested in the case where the system (13.15) is stable, with a unique equilibrium point $(\overline{H}, \overline{P})$. Asymptotical stability means that when t increases indefinitely, the pair (H_t, P_t) converges to $(\overline{H}, \overline{P})$. This means that for large values of t,

$$H_{t+1} \simeq H_t = H \ \# \ \overline{H} \text{ and } P_{t+1} \simeq P_t = P \ \# \ \overline{P}$$

Definition 13.10 (The Jacobian Matrix) If f and g are twice differentiable, then the Jacobi matrix of system (13.33) is

$$J = \begin{pmatrix} \dfrac{\partial f}{\partial H}(H, P) & \dfrac{\partial f}{\partial P}(H, P) \\ \dfrac{\partial g}{\partial H}(H, P) & \dfrac{\partial g}{\partial P}(H, P) \end{pmatrix} \tag{13.34}$$

Theorem 13.17 (Calculation of the Jacobian Matrix with the Same Data) *P and H are variables respectively in the neighbourhood of $\overline{P}$ and $\overline{H}$. They are linearised forms of the variables.*

$$J = \begin{pmatrix} Re^{-a\sqrt{P}} & -\dfrac{aRHe^{-a\sqrt{P}}}{2\sqrt{P}} \\ 1 - e^{-a\sqrt{P}} & \dfrac{aHe^{-a\sqrt{P}}}{2\sqrt{P}} \end{pmatrix} \tag{13.35}$$

Proof

- $\dfrac{\partial f}{\partial H}(H, P) = \left(RHe^{-a\sqrt{P}}\right)'_H = Re^{-a\sqrt{P}}$
- $\dfrac{\partial f}{\partial P}(H, P) = \left(RHe^{-a\sqrt{P}}\right)'_P = RHe^{-a\sqrt{P}}\left(-a\sqrt{P}\right)'_P$

  $= -aRHe^{-a\sqrt{P}}\left(p^{1/2}\right)'_P$

  $= -aRHe^{-a\sqrt{P}} \times \dfrac{1}{2}p^{-1/2} = -\dfrac{aRHe^{-a\sqrt{P}}}{2\sqrt{P}}$
- $\dfrac{\partial g}{\partial H}(H, P) = \left[H\left(1 - e^{-a\sqrt{P}}\right)\right]'_H = \left(H - He^{-a\sqrt{P}}\right)'_H = H' - (He^{-a\sqrt{P}})'_H$

  $= 1 - (He^{-a\sqrt{P}})'_H = 1 - e^{-a\sqrt{P}}$

- $$\frac{\partial g}{\partial P}(H, P) = \left[H\left(1 - e^{-a\sqrt{P}}\right)\right]'_P = \left(H - He^{-a\sqrt{P}}\right)'_P = -\left(He^{-a\sqrt{P}}\right)'_P$$
$$= -H\left(e^{-a\sqrt{P}}\right)'_P = -He^{-a\sqrt{P}}(-a\sqrt{P})' = aHe^{-a\sqrt{P}}(P^{1/2})'$$
$$= aHe^{-a\sqrt{P}} \times \frac{1}{2}P^{-1/2} = \frac{aHe^{-a\sqrt{P}}}{2\sqrt{P}}$$

□

Lemma 13.1 *Let λ_1 and λ_2 be the eigenvalues of J, then system* (13.33) *is stable with a simple equilibrium point* $(\overline{H}, \overline{P})$ *if and only if* $|\lambda_1| < 1$ *and* $|\lambda_2| < 1$.[16]

The value of the unique fixed point is given by Theorem 13.18.

Theorem 13.18 *If* $(\overline{H}, \overline{P})$ *is the unique equilibrium point of system* (13.37), *with the conditions mentioned above,*

$$(\overline{H}, \overline{P}) = \begin{bmatrix} \frac{R}{R-1}\left(\frac{1}{a}\ln(R)\right)^2 \\ \left(\frac{1}{a}ln(R)\right)^2 \end{bmatrix} \tag{13.36}$$

Readers interested in the rigorous developments and analyses of Lemma 13.1 and Theorem 13.18 can refer to the study carried out by a Pakistani scientific team [47].

13.2.3.1 Numerical Application

The initial conditions of a system, which follows the Nicholson-Bailey with the correction of Hassell and Varley (see (13.37)), are $H_0 = 3.2$ and $P_0 = 2.8$. The growth constant of the host is $R = 3.99$, and the searching efficiency of the parasitoid is $a = 1.5$.

1. Write the system
2. Is the system stable with a unique equilibrium point? If it is, find it.
3. Write the algorithm calculating the joint evolutions of the host and the parasitoid population across time and graph it.
4. Write the algorithm calculating the phase plane of the host and the parasitoid population across time and graph it.

[16] The | | symbol representing the absolute value if the eigenvalue is real or the modulus if the eigenvalue is complex.

Solution of the problem:

1. The system

$$\begin{cases} H_{t+1} = 3.99\, H_t\, e^{-1.5\sqrt{P_t}} \\ P_{t+1} = H_t\,(1 - e^{-1.5\sqrt{P_t}}) \end{cases} \tag{13.37}$$

2. From Theorem 13.17,

$$J = \begin{pmatrix} Re^{-a\sqrt{P}} & -\dfrac{aRHe^{-a\sqrt{P}}}{2\sqrt{P}} \\ 1 - e^{-a\sqrt{P}} & \dfrac{aHe^{-a\sqrt{P}}}{2\sqrt{P}} \end{pmatrix}$$

$$\simeq \begin{pmatrix} 3.99e^{-1.5\sqrt{\overline{P}}} & -\dfrac{1.5 \times 3.99\overline{H}e^{-1.5\sqrt{\overline{P}}}}{2\sqrt{\overline{P}}} \\ 1 - e^{-1.5\sqrt{\overline{P}}} & \dfrac{1.5\overline{H}e^{-1.5\sqrt{\overline{P}}}}{2\sqrt{\overline{P}}} \end{pmatrix}$$

From Theorem 13.18,

- $\overline{H} = \dfrac{R}{R-1}\left(\dfrac{1}{a}\ln(R)\right)^2 = \dfrac{3.99}{3.99-1} \times \left(\dfrac{1}{1.5} \times \ln 3.99\right)^2 = 1.135$
- $\overline{P} = \left(\dfrac{1}{a}\ln(R)\right)^2 = \left(\dfrac{1}{1.5} \times \ln 3.99\right)^2 = 0.851$

Thus, $J = \begin{pmatrix} 3.99e^{-1.5\sqrt{0.851}} & -\dfrac{5.99 \times 1.135e^{-1.5\sqrt{0.851}}}{2\sqrt{0.851}} \\ 1 - e^{-1.5\sqrt{0.851}} & \dfrac{1.5 \times 1.135e^{-1.5\sqrt{0.851}}}{2\sqrt{0.851}} \end{pmatrix} = \begin{pmatrix} 0.251 & -0.923 \\ 0.749 & -0.231 \end{pmatrix}$

The characteristic polynomial is $P(\lambda) = \lambda^2 - 0.019\lambda + 0.634$, whose roots are the eigenvalues λ_1 and λ_2. The determinant, $\Delta = (-0.019)^2 - 4 \times 1 \times 0.634 = -2.535$, is negative. Hence, λ_1 and λ_2 are two conjugate complex numbers:

$\lambda_{1,2} = \dfrac{0.019 \pm \sqrt{2.535}i}{2} = 0.01 \pm 0.796i$, thus, of the same modulus $0.796 < 1$.

Thus, the Nicholson-Bailey equation system is convergent, tending to a unique fixed point, the one which has been calculated: $(\overline{H}, \overline{P}) = (1.135, 0.851)$.

3. Joint evolutions of H_t and P_t across time.

The essentials are processed in the following code in R, where

- H0 and P0 are the variables respectively affecting the initial conditions of the sequences H_t and P_t;
- R and a are the variables respectively affecting the parameters R and a as defined in the formula (13.33);
- limH is the variable affecting $\overline{H}$; limP is the variable affecting $\overline{P}$;
- H, P and tt are respectively the vectors comprising the terms H_t, P_t and the succeeding times t.

Program Code-3

```
H0 «-3.2; P0 «-2.8; R «-3.99; a «- 1.5 #parameters of the system
f <- function(x,y) { val <- R*x*exp(-a*sqrt(y)); val } 1st equation of the system
g <- function(x,y) { val1 <- x*(1-exp(-a*sqrt(y))); val1 } 2nd equation of the system
N <- 200 Number of terms of the sequence followed
H <- rep(0, N); P <- rep(0, N) Definition of the sequences Ht and Pt as R-vectors, initialisation
H[1 <- H0; P[1] <-P0First terms of the sequences Ht and Pt
for(t in 1:(N-1)) {H[t+1]<-f(H[t],P[t]); P[t+1]<-g(H[t],P[t])} Construction of the sequences
tt <- 1:N time sequence
limH <- (R/(R-1))*((1/a)*log(R))^2; limP <- ((1/a)*log(R))^2 Fixed point (limH,limP)
Plotting of the joint evolutions
plot(tt, H, type='n', axes=F, xlab="", ylab="", xlim=c(-10,100), ylim = c(-1,4))
axis(side=1, pos=0); axis(side=2, pos=0, las=2); lines(tt, H, col='red'); lines(tt, P, col='blue')
```

We obtain the graph of the joint evolutions (Fig. 13.9).

4. Display of the phase plane.
   I constructed the following R-program to obtain it, with the following features.

   - The same variables with the same contents were applied.
   - The same functions were carried out.

   Thus, we constructed and graphed the phase plane using the same data and equations. We represented the common feature of Program Code 2 and Program Code 3 (hereunder mentioned), i.e., the first nine lines of Program Code 2, with an ellipsis.

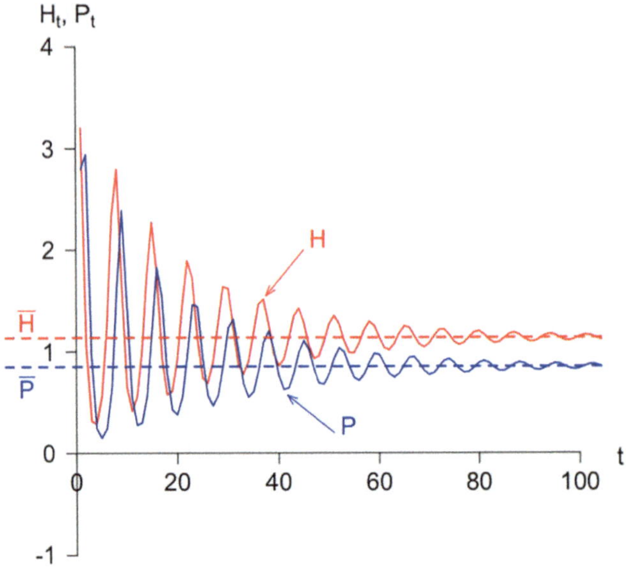

Fig. 13.9 Nicholson-Bailey model—Joint evolutions of the host (H_t) and parasitoid (P_t) populations across time (t). $\overline{H}$ is the fixed point of the host population, toward which its size converges; $\overline{P}$ is the fixed point of the parasitoid population, toward which its size converges

Program Code-3 continued

...

Plotting of the phase plane. We represent 100 terms of the pair (H_t, P_t) (as a reminder, $N = 200$ (see Program Code 2)

```
plot(H[1:N/2],P[1:N/2],type='n', axes= F, xlab='', ylab='', xlim=c(-0.5,4), ylim = c(-0.5,3.5))
axis(side=1, pos=0); axis(side=2, pos=0, las=2); lines(H[1:N/1.8],P[1:N/1.8])
```

Evolution of the first eleven phases $(H_t, P_t)_{0 \le t \le 10}$

```
for(t in 1:11) points(H[t], P[t], cex=1.1)
text(H[1], P[1]+0.15, expression(M[0])): text(H[2], P[2]+0.15, expression(M[1]))
text(H[3]-0.15, P[3]+0.15, expression(M[2])); text(H[11], P[11]+0.15, expression(M[10]))
```

Convergence towards the fixed point $(\overline{H}, \overline{P})$

```
segments(0, limP, limH, limP, lty=2, lwd=2, col='blue')
segments(limH, 0, limH, limP, lty=2, lwd=2, col='blue')
points(limH, limP, pch=3, col='red', cex=2, lwd=2); text(limH, -0.2, expression(bar(H)), col='red')
text(-0.15, limP, expression(bar(P)), col='red')
```

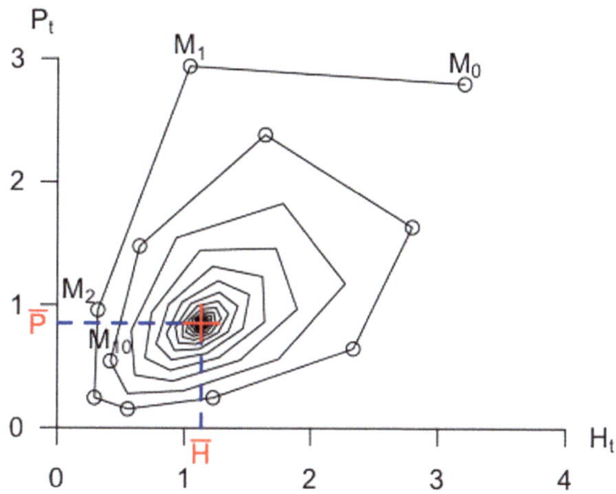

Fig. 13.10 Example of a phase plane (H_t, P_t). The *circles* M_t represents the phases (H_t, P_t) that converge to the fixed point $(\overline{H}, \overline{P})$ when $t \to \infty$

Fig. 13.10, just above, is the phase plane obtained.

13.2.4 Limitation of the Mathematical Models in Ecology

A broad literature has proposed a wide range of models to explain the interactions that construct an ecological community, or even a limited biocoenosis, taking into account the age structure, the fitness differences between the individuals, behavioural parameters like dominance, and many other factors. We will not study them here since this book is a general textbook with limited space that aims to browse important aspects of the discipline. Readers who want to delve into more mathematical ecological models may consult the books mentioned earlier [21, 42]. However, when the community gets relatively complex (e.g. comprising more than four interacting populations), the mathematical treatment becomes very difficult, and most ecologists resort to descriptive analysis supported by tools of inferential statistics or more or less sophisticated descriptive statistics, e.g. multivariate statistics like principal component analysis or factorial analyses of different sorts, for example.

We will now treat two old models discussed in Chap. 3 within a historical overview. These models, which are still applied nowadays, oppose a holistic vision against an individualistic viewpoint on how a community forms. They are the so-called (1) closed model of Clements, and (2) open model of Gleason.

13.2.5 Clement's and Gleason's Models of Communities

13.2.5.1 Closed Clementsian Model

Frederic E. Clements [4] conceives a community as an ecological unit isolated from adjacent similar units that are structurally analogous: all these units are forged by their trophic, competitive and/or coevolutionary relationships constrained in a definite ecogeographical area. Clements views a community as an organism and the interactions between its constitutive species as a phenomenon comparable to the physiological communications occurring within a live plant. The constitutive populations are the analogues of the organs or cells of an organism. Two neighbouring communities are separated from one another by a clear boundary, the *ecotone*.

Climatic, geological, edaphic[17] and biochemical factors are among the essential determinants of a community's uniqueness. Every community is characterised by dominant plant types that determine a unique structure pattern of the plant and animal populations. These features entail a strict relationship between community species, notably via food chains. The entire space occupied by a community self-organises into a functional unit. Figure 3.1 (p. 32) gives a good feel of the concept of the Clementsian closed model.

13.2.5.2 Open Gleason Model

Henry A. Gleason [11, 12] proposed a concept of open community, where the spatial association of sympatric species is fortuitous. A community is not viewed as the analogue of an organism but as a number of species contained in an area whose boundary is arbitrary. The encounter between different populations, whatever the relationship that ensues, is an adaptive process which is local and fixed by definite biological and ecological conditions. The geographical distribution of one species is independent of the geographical distribution of another species, which is sympatric to it. The sympatry between different species is partial and the result of conjuncture.

The geographical distribution of each species is carried out as a function of its genetic heritage and adaption capacity. The result is a clinal metric distribution pattern where determining peaks are observed. When the shape of two of these curves pertaining to two different species is observed within the same section of a gradient, this is a clue that these species interact. Thus, a community is the sectioning of a continuum. Figure 13.11 gives a feel of the Gleason model.

13.2.5.3 Validity of Both Models

The two types of communities are found.

---

[17] Edaphic means "related to the physical properties of the soil".

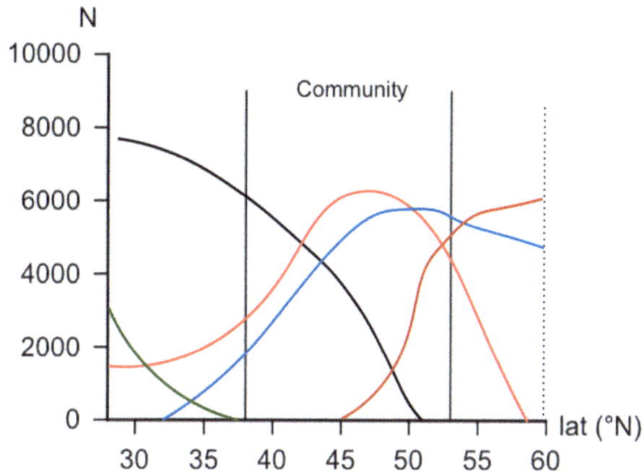

Fig. 13.11 Open Gleason model. Each *coloured curve* represents a different population. A community is a section of a continuum here between 28° N and 60° N latitude. In this example, we arbitrarily defined a community along a cline between 38° N and 53°30'N latitude comprising the *black, orange, blue* and *brown* species. It does not include the *green* one

- The closed model is suitable for describing rainforest communities relative to neighbouring savannah or uprooted area communities in the intertropical zone. It also describes *ubacs* relative to *adrets*[18] in mountainous areas.
- In contrast, the open model adequately describes communities in the temperate deciduous forest zone. Thorough studies have shown it for the deciduous forests of North America [38].

13.3 The Ecological Niche

13.3.1 Ecological Characteristics of a Species

Whether closed or open, communities consist of species that use resources differently. We have already seen that the community forming its biocoenosis in a given ecosystem is associated with a physicochemical substrate. In addition, it is penetrated by a flux of energy and a flow of inert matter. The latter is transformed into organic matter, the intrinsic physicochemical structure of living beings. This organic matter is transferred between organisms via food chains, allowing them to construct their live matter.

- Each species carries out this operation in a characteristic way, utilising specific strategies. It possesses a morphology adapted to its feeding mode (e.g. the strategy and tactic to obtain food, the way to take and absorb food, and the food itself). For example, the different prey species have ways to escape their predators by specific means adapted to their organism: (1) fleeing if the species has developed running abilities, (2) camouflage if the species is slightly mobile

[18] Between two faces of a mountain, the slope facing the sun is called *adret*; and the shadowy one, which is not exposed to the sun, is termed *ubac*.

but endowed with a body that can modify its appearance and colour to blend with its environment or (3) attack if the species has offensive means securing its defence (e.g. teeth, tusk, venom, sting).

- Two *syntopic*[19] species at identical trophic levels eat the same type of food and are hunted by the same predators. Non-syntopic species can avoid mutually destructive competition by spatial or temporal segregation.
  - They can exploit different tree strata, which offers them the same type of food.
  - They can choose different food items within the entire range of food suited to them (thus using ethological means to restrict their food potential). One strategy is that each species chooses a habitat containing slightly different resources remaining in the range of the nutrients adapted to them.
  - They practice temporal segregation: one species is diurnal, whereas the other is nocturnal.

Species have a limited habitat choice. The temperature, salinity, atmospheric, or pressure intervals suited for their biology depend on their homeostasis, which is highly species-specific. The optimality of the physical parameters is also species-specific. The abovementioned intervals can be broad for some species and narrow for others. The parameters chosen here are far from exhaustive; they are countless. Some authors, belonging to the sociobiological school of thought and applying evolutionary game theory, interpret the habitat choice by the species as a selection that offers them the highest fitness payoff, where factors would come into play, like density, habitat area and habitat shape [22].

13.3.2 The Hutchinsonian Ecological Niche

Thus, the parameters determining the habitat choice of a species, in fact, all its ecological needs and requirements, are numberless. Defined as proportions or measurable quantities, they theoretically constitute a species-specific n-dimensional hypervolume, thus a set of vectors comprising n components, each representing an individual. Such a hypervolume represents a specific species unit if it gathers vectors whose values are close to each other beyond a threshold. This object is the modern *ecological niche* concept conceived by the British ecologist George E. Hutchinson (1903–1991). In a seminal article [19, p. 416], this author defined this concept as a species' *fundamental niche*. In the same article [19, p. 418], Hutchinson acknowledged that this definition is too complex to be practical and builds up a method defining a *realised niche*, a subset of vector extracted from the fundamental niche. A realised niche is constructed with no more than two or three variables among the n components of the fundamental niche that an ecologist would consider the most relevant for his study. Multivariate statistics is presently the best tool

[19] Syntopic means sympatric and sharing totally or partially the same resources.

for studying about ten dimensions of the fundamental niches. The realised niches that current ecologists reveal are the factorial planes of this type of analysis (e.g. Principal Component Analysis, Factor Analysis).

The fundamental idea is that each species occupies its ecological niche as a general rule. This feature entails that the differentiation of an ecological niche reveals current speciation, which brings us to an important point. Usually, the ecological niches of closely related species are slightly divergent, except in the case of relatively uncommon phenomena called *adaptive radiation*, which constituted real milestones in the past evolutionary processes.

13.3.2.1 Adaptive Radiation

Adaptive radiation is a series of events that occur either (1) after a catastrophic event that wipes out quasi-entire ecosystems like a volcanic eruption or the fall of a gigantic meteorite or (2) when an oceanic island or an archipelago emerges. After such a phenomenon, the impacted area becomes approximately void of its previous fauna and flora, thus of ecological niches. Very quickly, a series of alien species invasions follows, and an ecological succession ensues, rapidly filling the empty niches. If the new habitat is heterogeneous, accelerating speciation processes take place, and from a single invading species, in a few generations, distinct species evolve, occupying different niches. We witness a type of ecological differentiation within a genus that, in continental speciation, is observed within a family. A dramatic example of this phenomenon is the case of the Darwin's finches (or Geospizines) forming 18 species of very closely related passerine birds belonging to the same genus or very close genera,[20] related to the Tanagers family of South America and living in the Galápagos archipelago. The ancestor of these species is closely related to the dull-coloured grassquit (*Asemospiza obscura*), a tanager species widespread in tropical and subtropical rainforests of western South America (from Argentina to Venezuela through Bolivia, Peru, Ecuador and Columbia) [50]. In ca one million years, one species developed thus 18 very different niches with a dramatic evolution of their beaks, which revealed different types of feeding behaviours. This case was one of Darwin's arguments to elaborate his evolution theory. Spectacular adaptive radiations occurred after the mass extinction, e.g., following the end of the Cretaceous period with the disappearance of the dinosaurs and their replacement by their descendants, the birds, and the mammals.

Now, let us return to the most frequent manifestation of ecological niches. Among territorial species, two types of organisms compete and exclude each other. One type has already been mentioned: the conspecific populations and the closely

[20] As usual, a debate has occurred here between poolers, who gather birds based on their genetics, songs, and general morphology [34], and splitters, who base their analysis more on ecological and behavioural considerations [15]. An interesting article takes an intermediate stand, showing that ground Darwin's finches show overlapping niches [9].

related species that did not evolve sufficiently to differentiate their ecological needs (and correlatively did not differentiate the morphologies adapted to these needs). The other type is the *ecological equivalents*, i.e. unrelated species superficially resembling each other and sharing the same ecological needs.

13.3.2.2 Examples of Ecological Equivalent

We will give a few examples of ecological equivalents, first in Australia. This continent had long been isolated from the other continents, where most mammals that had successfully and permanently settled are placentals. Due to its isolation, Australia has been populated by few placental mammals. The native mammals are overwhelmingly marsupials. We find, among the Australian marsupials, ecological equivalents to placentals. Some marsupials sometimes have a look curiously similar to their placental ecological equivalent. Still, this resemblance is superficial, not due to their relatedness since these species are very distant taxonomically, but rather due to environmental pressure on the bodies that impose similar bodily architectonics (e.g. similar legs, jaws, teeth). We can mention two spectacular examples, e.g. (1) the extinct marsupial thylacine (*Thylacinus cynocephalus*) vs the placental grey wolf (*Canis lupus*) and (2) the marsupial sugar glider (*Petaurus breviceps*) vs the placental flying squirrels (50 worldwide distributed species of rodents).

Other examples of ecological equivalents can be found among placental mammals, namely a pair of taxa where the first is African or Asian, and the second is South American:

- the pangolins (order Pholidota) vs the armadillos (Order Cingulata),
- the royal antelope *Neotragus pygmaeus* (order Artiodactyla) vs the agoutis (Order Rodentia),
- the mouse-deer or chevrotain (genus *Tragulus*, order Artiodactyla) vs the paca (*Cuniculus paca*) (Order Rodentia),
- the duikers (22 African species, subfamily Cephalophini, family Bovidae) vs the neotropical brockets (10 species of the genus *Mazama*, family Cervidae).

Just to underline the fundamental difference between actual resemblance due to kinship on the one hand, and mock and superficial resemblance between ecological equivalents on the other hand. Closely related species are really similar because they share identical features in great numbers, obtained through common heritage. Conversely, the apparent resemblance between ecological equivalents is due to similar corporal responses to identical ecological pressure (e.g., acquisition of the same type of food, similar survival strategies against predation, or any other sorts of adverse events) in different types of unconnected habitats. Only a restricted number of salient and functional characters look similar and can give a superficial but spurious impression of identity. Nevertheless, in these two cases, the ecological niches are equivalent.

13.4 The Food Web

Definition 13.11 (Food Web) The *food web*, or *trophic web*, links biotic agents (populations of different organismal species) of a given biocoenosis, by transfering through them matter and energy.

Axiom 13.16 A food web is directional. It passes from trophic level (TL) 1 to TL n (the trophic level or an apex predator, n being a finite integer, generally ≤ 4).

Definition 13.12 (Food Link) If a population a_i pertains to trophic level (TL) i and population a_{i+1}, to TL $i + 1$, then the pair (a_i, a_{i+1}) forms a *food link* if a_i is predated by a_{i+1}.

Definition 13.13 (Food Chain) A food web is composed of food chains, which are a sequence of food links relating primary producers to apex predators. Hereunder, I propose a formal definition of a food chain.
Let a_1, a_2, ..., a_n be populations (distinct species) of a biocoenosis $\mathcal{B}$ with the following properties:

- a_1 pertains to trophic level 1 and is thus a primary producer; and a_n belongs to trophic level n and is thus an apex predator.
- $\forall i \in [n]$, $\ell_i \stackrel{\text{def}}{=} (a_i, a_{i+1})$ are food links, which means that a_2 predates a_1, a_3 predates $a_2, \ldots, a_n$ predates a_{n-1}.

A formal definition of food chain C is as follows:

$$C = (\ell_1, \ell_2, \cdots, \ell_n) \text{ is a food chain relating } a_1 \text{ to } a_n. \tag{13.38}$$

Naively, we can conclude that n is the food chain C length. That is the classical stand of most ecology textbooks, and ours adopted in Chaps. 2 and 4).

However, an interesting analysis points out that the different paths leading from a given primary producer to a given apex predator do not necessarily comprise the same number of relays, so that n is not unique. Besides, the same publication stresses that the intensity of the nutrients and energy transit from one TL to the succeeding one is not equal. Hence, researchers face difficulty defining and evaluating the actual food length, an issue for conservation, economics and health. A good evaluation of this parameter can help assess the impact of the concentration of pollutants on the tissues of apex predators, like fishes captured for human consumption) [46].

Quizzes and Problems

13.1 In a population P of 10 000 balls, a statistician hypothesises that there are as many white balls as black balls. The norm of P is $\frac{1}{2}$. He randomly draws a sample of $n = 20$ balls and observes nine black and 11 white balls.

1. Will this result lead him to accept the null hypothesis? Answer his question via
   (a) a direct statistical reasoning,
   (b) an algorithm that simulates the event described in the exercise statement, coded in R or Python.
2. What type of population do we have here?

13.2 We will draw the graph of a Malthusian discrete model of an iteroparous population. With an algorithm and coding language of your choice (R or Python), draw the curve of N, the density of population P, knowing that the generation t runs from 0 to 15, with a path of 1. Let Δt be a lapse of time, calculated in fractions of generation. Let $\Delta t = 2.5$ generations. Calculate the increase rate of the population during Δt and the Mathus constant ρ; report these three quantities in the graph.

13.3 Prove Theorem 13.8: In a given population following the Verhulst model, if *(1)* $N(t)$ is the population size at time t with, as initial condition $N(t = 0) = N_0$, *(2)* K the carrier capacity and *(3)* r the Malthusian constant, then

$$N(t) = \frac{K}{1 + \left(\dfrac{K}{N_0} - 1\right) e^{-rt}}$$

13.4 Prove Theorem 13.11: In a given population following the Gompertz model, if *(1)* $N(t)$ is the population size at time t with, as initial condition $N(t = 0) = N_0$, *(2)* K the carrier capacity and *(3)* r the population's growth rate, then, if $\gamma = \ln \frac{N_0}{K}$,

$$N(t) = K e^{\gamma e^{-rt}}$$

References

1. Andrewartha HG (1970) Introduction to the study of animal populations, 2nd edn. Chapman & Hall, London.
2. Briedis M et al (2020) Integrating light-level geolocation with activity tracking reveals unexpected nocturnal migration patterns of the tawny pipit. J Avian Biol 51(9), https://doi.org/10.1111/jav.02546
3. Chu CYC, Lee RD (2012) Sexual dimorphism and sexual selection: a unified economic analysis. Theor Popul Biol 82:355–363, https://doi.org/10.1016/j.tpb.2012.06.002
4. Clements FE (1916) Plant succession: an analysis of the development of vegetation. Carnegie Institution of Washington.
5. Collin et al (2017) Reproductive cycles in tropical intertidal gastropods are timed around tidal amplitude cycles. Ecol Evol 7:5977–5991, https://doi.org/10.1002/ece3.3166
6. Crozier RH, Fjerdingstad EL (2001) Polyandry in social Hymenoptera - disunity in diversity? Ann Zool Fennici 38:277–285.

7. D'Ancona U (1926) Dell' influenza dell stasi peschereccia del periodo 1914–18 sul patrimonio ittico dell'Alto Adriatico. Memoria, R Comitato Talassographica Italiano, 126. Officine Grafichen, C Ferrari, Venice (In Italian).
8. D'Ancona U (1954) The struggle for existence. Brill, Leiden (English version).
9. De León LF et al (2014) Darwin's finches and their diet niches: the sympatric coexistence of imperfect generalists. J Evol Biol 27:1093–1104, https://doi.org/10.1111/jeb.12383
10. Galal TM et al (2021) Vegetation zonation along the desert-wetland ecosystem of Taif Highland, Saudi Arabia. Saudi J Biol Sci 28:3374–3383, https://doi.org/10.1016/j.sjbs.2021.02.086
11. Gleason HA (1926) The individualistic concept of the plant association. Bull Torrey Bot Club 53:7–26.
12. Gleason HA (1939) The individualistic concept of the plant association. Am Midl Nat 21:92–110
13. Godfray HCJ (2007) Parasitoids. In: Encyclopedia of Biodiversity, p. 674–682, https://doi.org/10.1016/B978-0-12-384719-5.00152-0, Elsevier.
14. Gompertz B (1825) On the nature of the function expressive of the law of human mortality, and on the new mode of determining the value of life contingencies. Phil Trans R Soc 115:513–583, https://doi.org/10.1098/rstl.1825.0026
15. Grant PR, Grant BR (2008) How and why species multiply: The radiation of Darwin's finches. https://doi.org/10.2307/j.ctvx5wc23, Princeton University Press, Princeton, NJ.
16. Hannan E (2008) Randomized clinical trials and observational studies. JACC Cardiovasc Interv 1:211–217, https://doi.org/10.1016/j.jcin.2008.01.008
17. Hassell MP, Varley GC (1969) New inductive population model for insect parasites and its bearing on biological control. Nature 223:1133–1137, https://doi.org/10.1038/2231133a0
18. Heulin B, Guillaume C (1989) Extension géographique des populations ovipares de *Lacerta vivipara*. [Geographical extension of the oviparous populations of *Lacerta vivipara*] (In French) Rev Ecol (Terre Vie) 44:283–289.
19. Hutchinson GE (1957) Concluding remarks. Cold Spring Harb. Symp. Quant. Biol vol. 22(2):415–427.
20. Jost C et al (2005) The wolves of Isle Royale display scale-invariant satiation and ratio-dependent predation on moose. J Anim Ecol 74:809–816, https://doi.org/10.1111/j.1365-2656.2005.00977.x
21. Kaplan D, Glass L (1995) Understanding nonlinear dynamics. Springer Science & Business Media, https://doi.org/10.1093/beheco/ary046
22. Katz N, Scharf I (2018) Habitat geometry and limited perceptual range affect habitat choice of a trap-building predator. Behav Ecol 29:958–964, https://doi.org/10.1093/beheco/ary046
23. Kingsolver J et al (2015) Curve-thinking: understanding reaction norms and developmental trajectories as traits (ch 3) In: Martin LB et al (eds) Integrative organismal biology, John Wiley & Sons, Inc.
24. Kok PJR et al (2024) Lost in space and time: robust demography and enhanced resilience buffer adverse environmental effects in a highly isolated and sedentary pre-pleistocene landscape vertebrate. Ecol Evol 24:125, https://doi.org/10.1186/s12862-024-02314-2
25. Krasncc MO et al (2012) Mating systems in sexual animals. Nat Educ Knowl 3(10):72.
26. Krebs CJ (1985) Ecology: the experimental analysis of distribution and abundance. 3rd edn, Harper & Collins.
27. Lack DL (1943) The life of the robin. HF & G Witherby, London
28. Leese JM, Blatt T (2021) Sex differences in how territory quality affects aggression in convict cichlids. Integr Org Biol 3(1):obabo28, https://doi.org/10.1093/iob/obab028
29. Logan JD, Wolesensky WR (2009). Mathematical methods in biology, John Wiley & Sons.
30. Lotka AJ (1925) Elements of physical biology. Williams & Wilkins Co, Baltimore.
31. Malthus TR (1798) An essay on the principle of population as it affects the future improvement of society with remarks on the speculations of Mr. Godwin, M. Condorcet, and other writers. J. Johnson, London

32. Mathews LM (2002) Territorial cooperation and social monogamy: factors affecting intersexual behaviours in pair-living snapping shrimp. Anim Behav 63:767–777, https://doi.org/10.1006/anbe.2001.1976
33. Mayr E (1963) Animal species and evolution. Harvard University Press, Cambridge, MA
34. McKay BD, Zink RM (2014) Sisyphean evolution in Darwin's finches. Biol Rev 90:689–698, https://doi.org/10.1111/brv.12127
35. Murray JD (2001) Mathematical biology. I. An introduction, 3rd edn. Springer, New York
36. Newton I (2010) Introduction. In: The migration ecology of birds. Academic Press
37. Nicholson AJ, Bailey VA (1935) The balance of animal populations. Part I. Prc Zool Soc Lond 3:551–598.
38. Nicolson M, McIntosh RP (2002) H.A. Gleason and the individualistic hypothesis revisited. Bull Ecol Soc Amer 83:133–142, https://doi.org/10.2307/20168699
39. Oomen RA, Hutchings JA (2022) Evolution of reaction norms. Oxford bibliographies, website article: https://doi.org/10.1093/OBO/9780199941728-0130
40. Payne RB (1985) Song populations and dispersal in steelblue and purple windowfinches. Ostrich 56:135–146, https://doi.org/10.1080/00306525.1985.9639582
41. Pfenninger M et al (2011) Factors and processes shaping the population structure and distribution of genetic variation across the species range of the freshwater snail *Radix balthica* (Pulmonata, Basommatophora). Evol Biol 11:135, https://doi.org/10.1186/1471-2148-11-135
42. Pielou EC (1977) Mathematical ecology. 2n edn, John Wiley, New York
43. Piepho HP et al (2013) Why randomize agricultural experiments? J Agron Crop Sci 199(5):374–383, https://doi.org/10.1111/jac.12026
44. Poincarré H (1881) Mémoire sur les courbes définies par une équation différentielle (In French). J Math Pures Appl 7:375–422.
45. Poincarré H (1882) Mémoire sur les courbes définies par une équation différentielle (suite) (In French). J Math Pures Appl 8:251–296.
46. Post DM (2002) The long and short of food-chain length. Trends Ecol Evol 17:269–277, https://doi.org/10.1016/S0169-5347(02)02455-2
47. Qureshi MN et al (2014) Asymptotic behavior of a Nicholson-Bailey model. Adv Differ Equ 2014:62, https://doi.org/10.1186/1687-1847-2014-62
48. Ricklefs RE (1990) Ecology. Freeman & co, New York
49. Riebel K (2015) Learning and cultural transmission in chaffinch song. Adv Stud Behav 47:181–227, https://doi.org/10.1016/bs.asb.2015.01.001
50. Sato A et al (2001) On the origin of Darwin's finches. Mol Biol Evol 18:299–311, https://doi.org/10.1093/oxfordjournals.molbev.a003806
51. Slatkin M (1984). Ecological causes of sexual dimorphism. Evolution 38:622–630, https://doi.org/10.2307/2408711
52. Smith FE (1963) Population dynamics in *Daphnia magna* and a new model for population growth. Ecology 44:651–663, https://doi.org/10.2307/1933011
53. Solomon H, Zacks S (1970) Optimal design of sampling from finite populations: a critical review and indication of new research areas. J Amer Stat Assoc 65:653–677, https://doi.org/10.2307/2284578
54. Surmacki A (2004) Habitat use by reed bunting *Emberiza schoeniclus* in an intensively used farmland in Western Poland. Ornis fenn 81:137–143.
55. Tjørve KMC, Tjørve (2017) The use of Gompertz models in growth analyses, and new Gompertz-model approach: An addition to the Unified-Richards family. Plos One 12(6):e0178691, https://doi.org/10.1371/journal.pone.0178691
56. Varadajan S, Subramoniam T (1982) Reproduction of the continuously breeding tropical hermit crab *Clibanarius clibanarius*. Mar Ecol Prog Ser 8:197–201, https://doi.org/10.3354/meps008197
57. Verhulst PF (1845) Recherches mathématiques sur la loi d'accroissement de la population (In French). Nouv Mém Acad R Sci B-lett Brux 18:14–54, https://doi.org/10.3406/marb.1845.3438

58. Volterra V (1928) Variations and fluctuations of the number of individuals in animal species living together. Translated from Italian, original published in 1926. J Cons Int Explor Mer 3:3–51, do:10.1093/icesjms/3.1.3
59. Volterra V, d'Ancona U (1935) Les associations biologiques au point de vue mathématique (In French). Hermann, Paris.
60. Woltereck R (1909) Weitere experimentelle Untersuchungen über Artveränderung, speziell über das Wesen quantitativer Artunterschiede bei Daphnien (In German). Verh Dtsch Zool Ges 19:110–172, https://doi.org/10.1007/BF01876686
61. Wright RM et al (2022) First direct evidence of adult European eels migrating to their breeding place in the Sargasso Sea. Sci Rep 12(1):15362, https://doi.org/10.1038/s41598-022-19248-8
62. Wu L et al (2023) Limited song mixing without genomic gene flow in a contact zone between two songbird species. Mol Biol Evol 40(3):msad053, https://doi.org/10.1093/molbev/msad053

Chapter 14
Epilogue

This book had the ambition to be a contribution to improve the pedagogy of ecology, and I hope my essay will be convincing and foster reflections, constructive criticisms and comments from ecology teachers and researchers. It aimed at giving a more rigorous basis to the discipline. But it is just a proposition, certainly neither a panacea replacing the former ecology essays, nor a comprehensive work. I am aware of the many loopholes in it.

The book treats only a tiny part of the biome-likes. If I had treated the 108 subbiomes (ecological functional groups) acknowledged by the IUCN, it would have doubled the volume of the book, which was planned to contain ca 400 pages. I chose to select fourteen subbiomes, belonging to terrestrial biomes, to give a feel of a type of analysis to be applied to all the subbiomes in future essays (by me or another author).

The history of ecology, which certainly would foster great interest in the public, is far from complete. Many aspects highlighting the evolution of the science, may be added. This chapter can be the object of another book.

My proposition to insert, in the chapter dedicated to energy, a part on thermodynamics applied to ecology was risky, as I am not a physiscist. But I think it is a fundamental and crucial aspect of ecology. It still needs to be refined and ameliorated by authorised researchers in the domain of physics; I am open to any constructive criticism and comments that can come from them and ask for indulgence for the potential weaknesses in this chapter. This topic may be the beginning of reflections on this domain that in my view had not sufficiently been developed. I am aware that many aspects on the subject of food web, ecological niche, and r/K selection strategies had to be developed, since they are fundamental aspect in the functionality of the ecosystems. But only a summary development of these topics would need at least the space of this book.

Supplementary Information The online version contains supplementary material available at https://doi.org/10.1007/978-3-031-90656-5_14.

M. Salomon, *Integrative Ecology*, https://doi.org/10.1007/978-3-031-90656-5_14

This book should be followed by popularisation essays in different languages. I am convinced that the topics addressed in it can raise interest in the public and be useful to politics and local policy makers.

Besides it can foster ideas to other disciplines, namely the humanities, philosophy or other sciences.

Glossary

This glossary gives very summary definitions aimed at helping the more thorough descriptions found in the text.

Abiotic Factors These parameters are (1) the *chemical agents* and (2) the *physical agents* of an ecosystem, enabling a biocoenosis (live part of this ecosystem) to exist and last:

- In a perpetual cycle, *chemical agents*, originally in a mineral state as ions or mineral salts, are transformed into the organic chemical compounds necessary to the lives of the biotic agents. These molecules are eventually released in the environment, and their constituents return to their initial states. And the cycle continues likewise.
- The *physical agents* are physical variables, such as temperature, pressure, and pH, which, in a given interval, constitute the physical conditions of life of the organisms underlying a biocoenosis.

The abiotic factors are interrelated and form a system that was partly favoured and organised by the biocoenoses during evolution. These considerations explain why one can say that the abiotic factors fuel the ecosystems.

Aphotic This adjective describes the absence of light in a water body, at the deepest layers where no sun rays penetrate. In the aphotic zones of oceans, the trophic chains start with primary producers only depending on redox chemical energy produced in the geochemical emanations of the sea floor.

Biocoenosis A biocoenosis is a system comprising interacting populations, each of different species, that form a community and are constructed to last for a long period of time (relative to the average generation of the member species of the biocoenosis).

M. Salomon, *Integrative Ecology*, https://doi.org/10.1007/978-3-031-90656-5

Biogeochemical Cycle The biogeochemical cycle of a given chemical element (e.g. O, H, N, C, S, Na) or simple chemical compound (e.g. H_2O, CO_2) is properly described in the entry *Abiotic factors*. Any of these chemicals starts its cycle in its mineral form, is incorporated in live organism in an organic molecule (e.g. nucleic acid, protein, carbohydrate, lipid) through a vast *trophic* chain, and eventually returns to its initial mineral state. The cycle proceeds as follows:

- The primary producers (e.g.plants) carry out the transformation from the mineral form to the organic form, a process called biosynthesis.
- The consumers 1 (e.g. herbivores) consume the primary producers through consumption, i.e. they digest the organic compounds and recompose their proper organic molecules from the products of their digestion (simple chemicals).
- Recursively, a chain of consumers proceed likewise, consumers 2 (e.g. predators) feeding on consumers 1, consumers 3 (e.g. superpredators) feeding on consumers 2, etc, up to the apex consumers (i.e. populations that are not preyed upon).
- From the plant litter, carrion and excreta, a chain of organisms (the decomposers/mineralisers) return the components of the organic molecules into their initial states; and the cycle resumes.

Biome This term is polysemic and varies between authors. The biome concept I have adopted throughout this book is the biome in the sense if the IUCN. It consists in one of the 25 entities defined with a series of drivers and physiognomy traits by a panel of 103 scientists and experts internationally recognised in the field of ecology and conservation biology.

Biosphere The term has classically designated the portion of the Earth comprising all the live organisms. Throughout this book, I proposed to use it with another acceptation: for me, the biosphere is the resultant of all the ecosystems of our planet.

Climax In the vision of the holistic ecologists (Clements and followers), an ecological succession functions like a true organism, e.g. a plant, with a number of stages, which can start from a portion of the Earth devoid of life because of a catastrophe or mass extinction or when an oceanic island emerges (bare rock, no plant or animal). The first stage is the invasion of the territory by pioneering species (lichens, mosses) that construct a soil. The following stages correspond to a continuous development of the ecosystem, i.e. soil deepening and enrichment allowing the growth of an increasingly dense, complex and high vegetation, increasing plant and animal species richness (all these species would be originally invasive). These successive stages would to correspond to juvenile stages. When the soil is fertile, the adult stage called the *climax* would correspond to a forest containing the maximal biodiversity. When the soil is less fertile, the climax could correspond to one of the "juvenile" stages depicted above, e.g. a meadow or a moor. This holistic vision has been largely debated and challenged by different authors. Nevertheless, the notion of *climax* is still frequently used but with caution and a critical eye, and without the holistic vision of the pioneering ecologists mentioned above.

Competition Inhibition In such a situation, a given species has a negative effect on another one within a community; it slows or inhibits a successional change by preventing the impacted species to establish in the next stage. This results is slowing down and in some times arresting the succession.

Complex Organism This phrase is a synonym of biotic community, what I call biocoenosis throughout this book.

Cryptogams A category now completely abandoned as a taxon, gathering allegedly inferior plants, due to the fact that they reproduce by spores (e.g. ferns, mosses and the green algae, algae and fungi), and not from seeds obtained by pollination (from cones or flowers). The group "cryptogam" (meaning "with hidden sexual organs"), is now proven to be artificial, containing organisms that are plants and others that are not. For example, ferns, mosses and the green algae are plants; but red and brown algae and fungi are not. Worse, red and brown algae are related to plants and belong to the green lineage. But fungi are phylogenetically close to animals, the latter taxa being members of the opistokonts.

Deciduous Adjective referring to trees or other plants getting leafless during the season which is unfavourable to photosynthesis (winter in the temperate, and dry season in the tropical climatic zone).

Demersal This adjective refers to aquatic species that live just above the benthic zone of a water body (i.e. the latter being the deepest layer of the water column) and that possesses, as a characteristic, to regularly move to more superficial layers of the water body in question, e.g. fish species like the Atlantic cod (*Gasus morhua*) or several zooplanktonic species that can adjust their buoyancy and can thus drift upwards and downwards.

Diazotrophs Nitrogen-fixing prokaryotes involved in the cycle of nitrogen across an ecosystem. Nitrogen is an atmospheric element that cannot directly be fixed by the phototrophic organisms (plants, algae, phytoplankton), but need to be processed by these microorganisms to be incorporated. This is due to the fact that nitrogen is an inert gas.

Ecological Niche

- Joseph Grinnel (1877–1939) [3] proposes the oldest concept of "niche" as the only portion of the habitat that fits a given species, where he can find its food and find protection against predators. He applied this concept to a species, the California Thrasher *Toxostoma redividum*, to explain why this birds lives in specific bushes of the Chaparral.
- In Elton's view, "the niche of an organism is its mode of life in the sense that we speak of trades or jobs or professions in a human community" [1, p. 31]. Elton explains that, I quote [2, p. 64],

the " niche " of an animal means its place in the biotic environment, its relations to food and enemies.

- The modern concept of ecological niche is owed to G. Hutchinson [4]. The fundamental niche is an ecomathematical concept, a vector containing all the ecological needs of a species, described by n variables that cannot be represented graphically. The realised niche is a subset of these vectors, constructed from two or three variables among the n components, for example pH $\times$ temperature to evaluate the niche difference between two aquatic taxa.

Ecophenotype This concept defines the correspondence between the phenotype of an organism (all his physical and chemical features resulting from its genome) and its ecological needs associated to it.

Energy When we refer to energy in biology, and in particular in ecology, it is the potential and kinetic energy of the electron and proton occurring in the metabolic pathways; and the phenomena related to the specificity of the thermodynamics of the living being.

Ericoid Adjective referring to the aspect of a branch of heather; It applies to the aspect of the leaves of a number of shrubs. E.g. the South-African shrub *Struthiola myrsinites.*

Euphotic Adjective describing the presence of sufficient light in a water body, generally at the upper layers where the sun can penetrate, to sustain photosynthesis. The euphotic zone concentrates the primary producers (phytoplankton and all

Evergreens Adjective referring to trees or shrubs having their leaves during the whole year.

Giga-annum (Ga) Measure of geological (or paleontological time)—billion years.
Example: 3 Ga = 3×10^9 years.

Graminoid This group comprises grass-like plants like the families Poaceae (domesticated cereals like maize, wheat, barley, etc and wild relatives), Cyperaceae (sedges) and Juncaceaea (rushes). These plants are angiosperm monocots.

Great Oxidation Event Abbreviated GOE. This phenomenon occurred 2.4 billion years ago. It was the release of gaseous oxygen in the primitive Earth's atmosphere devoid of this gas and essentially composed of carbon dioxide (with a concentration 10 to 200 times higher than the present atmosphere), methane, and hydrogen sulfide. This event was very probably caused by cyanobacteria, which were the first oxic photosynthetic organisms that appeared on Earth.

Homeostasis Homeostasis is a self-regulating mechanism that allows an organism to survive to an environmental change. The internal environment of the body must remain constant, also when the physical abiotic factors deviate from their optimal value for the species. An example of homeostasis in the human being is the impact of altitude on the number and size of the red blood cells.

Hydrophilicity Capacity of an object to absorb water.

K- and R-demographic Strategies

- In K-strategy, individuals live in low densities and have on average long life-spans and slow growth rates, their motabolism rate is slow; they give birth to a limited number of offspring (one or two) per life cycle. K-strategy would be often adopted in stable environments.
- In R-strategy, individuals live in high densities and have on average short life-spans and rapid growth rates, their motabolism rate is high. They give birth to a large number of offspring (ten and more, in extreme case hundreds) per life cycle. R-strategy would be often predicted in unstable environments.

Generally, large animals and plants tend to adopt K-stategy, and small-sized organisms to R-strategy.

Leaf Area Index Abbreviation: LAI. Projected area of the leaves on trees, i.e. A', within a definite habitat area A more extensive than A', i.e. the ratio $\frac{A'}{A}$.

Mega-annum (Ma) Measure of geological (or paleontological time)—million years.
Example: 7 Ma = 7×10^6 years.

Mesophotic Adjective describing the state of light concentration found just below the euphotic zone. Solar energy is less concentrated than in the euphotic zone, making photosynthesis activities far more reduced; but this zone is rich in nutrients, i.e. many autochthonous species of fish, coral and zooplankton, and allochthonous resource subsidies.

Pelagic Adjective referring to the open sea also called pelagic zone. It is the part of the ocean that is beyond the continental shelf. The water column is stratified, with (1) an epipelagic zone which concentrates most of the solar light and energy (> 200 m), (2) a mesopelagic zone between −200 and −1000 m (concentrating most of the nutrients needed to the marine species), (3) a bathypelagic zone (between −1000 and −4000 m) and (4) the abyssopelagic zone (< −4000 m). With depth, pressure increases, temperature and light decreases, life is typically organised according to these abiotic conditions.

Periphyton The periphyton is a community of microorganisms (protists, microalgae, cyanobacteria, plant detritus and minute invertebrates and plants) that form a compact organic complex covering rocks, branches and all kinds of things and plants soaked in rivers, lakes and all kinds of water bodies.

Phanerogams A category now completely abandoned as a taxon, gathering allegedly superior plants, due to the fact that they practice a true sexual reproduction from seeds obtained by pollination (from cones or flowers). This group includes two main clades, the Gymnosperms and the Angiosperms.

Podzol Before defining the podzol, one needs to sketch briefly some elements of pedology. The soil system comprises three layers, i.e. the *horizons* identified by their biological composition, colour and texture.

The surface soil: This layer is just below the humus (or 0 horizon), which does not belong to the soil system but nurtures it. The surface soil contains (1) withered rocks, (2) elements like iron, aluminium, simple organic compounds, clay minerals and (3) most of the organic matter in a phase of decomposition, amidst most of the soil organisms, i.e. burrowing animals, scavengers, decomposers and mineralisers. This layer, if the soil is acid, leeches, resulting in the oxidisation of iron and other elements and the loss of some chemical elements, mostly iron oxides which move downward to the layer just below, the subsoil.

The subsoil: This layer accumulates iron oxides coming from above, which explains the reddish aspect of the subsoil. It contains less organic life and tests the acidity level of the soil, largely explaining the type of vegetation cover and ecosystem above.

The substratum: This layer forms the border with the bedrock, mainly composed of unweathered or poorly weathered rocks, and little chemical and biological activities.

Podzol is a very acidic and highly leached soil. This situation results in a strong accumulation of iron oxides in the surface subsoil, under the influence of humid and cold climatic conditions as found in taigas. The soil comprises a thick 0 horizon containing a large quantity of undecomposed plant litter, which forms a very low pH, very dark-coloured substrate, poor in mineral salts. Combined with water, this very poor humus entails the decay of the clay comprised in the surface soil, which leads to the migration of oxidized soil into the substratum. The surface soil gets a typical ashy texture and a greyish colour.

Sclerophyll Type of vegetation characterized by hard, leathery, evergreen foliage specially adapted to prevent moisture loss" (Encyclopaedia Britannica, 2024)

Seedbank Concentration of seeds gathered in a hard fire-resistant fruit lying either in the canopy of a tree, or in the soil.

Semi-deciduous Adjective referring to trees or other plants getting partially leafless during the season which is unfavourable to photosynthesis (winter in the temperate, and dry season in the tropical climatic zone).

Serotinous Adjective referring to the ability of *seedbanks* (see the proper entry in this glossary), kept in a fire-resistant woody fruit within the canopy of a tree, to resist fire or the death of the tree killed by fire.

Species Richness Biodiversity is a complex concept that has multiple facets. It means the diversity of living beings in a given portion of the Globe. But diversity can mean many things, for example, (1) the number of species, (2) the taxonomic or genetic diversity within species or (3) the diversity of the taxa notwithstanding their ranks (e.g. number of genera, of classes, of phyla). Within these categories, the species richness of a given geographic unit is the total number of species comprised in it. The ecogeographical unit can be a geographical region, a community, a biocoenosis, a biome-like, etc.

References

1. Begon M et al (2006) Ecology: from individuals to ecosystems, 4th edn. Blackwell Publishing, Oxford.
2. Elton CS (1927) Animal ecology. McMillan, New York.
3. Grinnell J (1917) The niche-relationships of the California thrasher. Auk 34:427–433.
4. Hutchinson GE (1957) Concluding remarks. Cold Spring Harb. Symp. Quant. Biol vol. 22(2):415–427.

Author Index

M. Salomon, *Integrative Ecology*, https://doi.org/10.1007/978-3-031-90656-5

Taxonomic Index

M. Salomon, *Integrative Ecology*, https://doi.org/10.1007/978-3-031-90656-5

R

S

T

Subject Index

M. Salomon, *Integrative Ecology*, https://doi.org/10.1007/978-3-031-90656-5

MIX
Papier aus verantwortungsvollen Quellen
Paper from responsible sources
FSC® C105338

If you have any concerns about our products,
you can contact us on
ProductSafety@springernature.com

In case Publisher is established outside the EU,
the EU authorized representative is:
Springer Nature Customer Service Center GmbH
Europaplatz 3, 69115 Heidelberg, Germany

Printed by Libri Plureos GmbH
in Hamburg, Germany